AUTOMOTIVE CONTROL SYSTEMS

This engineering textbook is designed to introduce advanced control systems for vehicles, including advanced automotive concepts and the next generation of vehicles for Intelligent Transportation Systems (ITS). For each automotive-control problem considered, the authors emphasize the physics and underlying principles behind the control-system concept and design. This is an exciting and rapidly developing field for which many articles and reports exist but no modern unifying text. An extensive list of references is provided at the end of each chapter for all topics covered. This is currently the only textbook, including problems and examples, that covers and integrates the topics of automotive powertrain control, vehicle control, and ITS. The emphasis is on fundamental concepts and methods for automotive control systems rather than the rapidly changing specific technologies. Many of the text examples, as well as the end-of-chapter problems, require the use of MATLAB and/or Simulink.

A. Galip Ulsoy is the C. D. Mote Jr. Distinguished University Professor and the William Clay Ford Professor of Manufacturing at the University of Michigan. He served as director of the Ground Robotics Reliability Center and deputy director of the Engineering Research Center for Reconfigurable Manufacturing Systems. He has been on the faculty of the Department of Mechanical Engineering at Michigan since 1980 and was the founding director of the Program in Manufacturing. He served as technical editor of the American Society of Mechanical Engineers' (ASME) *Journal of Dynamic Systems, Measurement, and Control* and is the founding technical editor of the *ASME Dynamic Systems and Control Magazine*. Professor Ulsoy is a member of the National Academy of Engineering and a Fellow of the ASME, the International Federation of Automatic Control, and the Society of Manufacturing Engineers; a Senior Member of IEEE; and a member of several other professional and honorary organizations. He is the past president of the American Automatic Control Council. He co-authored, with Warren R. DeVries, *Microcomputer Applications in Manufacturing*, and he is a co-author, with Sun Yi and Patrick W. Nelson, of *Time Delay Systems*. He has published more than 300 refereed technical articles in journals, conferences, and books.

Huei Peng is a Professor in the Department of Mechanical Engineering at the University of Michigan. He served as the executive director of interdisciplinary and professional engineering programs. His research interests include vehicle dynamics and control, electromechanical systems, optimal control, human-driver modeling, vehicle active-safety systems, control of hybrid and fuel-cell vehicles, energy-system design, and control for mobile robots. He has received numerous awards and honors, including the Chang-Jiang Scholar Award, Tsinghua University; a 2008 Fellow of the ASME; the Outstanding Achievement Award, Mechanical Engineering Department, University of Michigan (2005); the Best Paper Award, 7th International Symposium on Advanced Vehicle Control (2004); and the CAREER Award, National Science Foundation (July 1998–June 2002). He has published more than 200 refereed technical articles in journals, conferences, and books. Professor Peng is co-editor of *Advanced Automotive Technologies* with J. S. Freeman and co-author of *Control of Fuel Cell Power Systems – Principles, Modeling, Analysis and Feedback Design*, with Jay T. Pukrushpan and Anna G. Stefanopoulou.

Melih Çakmakcı is a professor of Mechanical Engineering at Bilkent University in Ankara, Turkey. His research areas include modeling, analysis and control of dynamic systems, control systems, smart mechatronics, modeling of manufacturing systems and their control, automotive control systems, optimal energy-management algorithms, and design and analysis of network control systems. Prior to joining Bilkent University, he was a senior engineer at the Ford Scientific Research Center.

Automotive Control Systems

A. Galip Ulsoy
University of Michigan

Huei Peng
University of Michigan

Melih Çakmakcı
Bilkent University

CAMBRIDGE
UNIVERSITY PRESS

CAMBRIDGE
UNIVERSITY PRESS

32 Avenue of the Americas, New York, NY 10013-2473, USA

Cambridge University Press is part of the University of Cambridge.

It furthers the University's mission by disseminating knowledge in the pursuit of
education, learning, and research at the highest international levels of excellence.

www.cambridge.org
Information on this title: www.cambridge.org/9781107686045

© A. Galip Ulsoy, Huei Peng, and Melih Çakmakcı 2012

First published 2012
First paperback edition 2014

A catalog record for this publication is available from the British Library.

Library of Congress Cataloging in Publication data
Ulsoy, Ali Galip.
Automotive control systems / A. Galip Ulsoy, University of Michigan, Huei Peng,
University of Michigan, Melih Çakmakcı, Bilkent University.
    p.   cm.
Includes index.
ISBN 978-1-107-01011-6 (hardback)
1. Automobiles – Automatic control.   2. Adaptive control systems.
3. Automobiles – Motors – Control systems.   I. Peng, Huei.   II. Çakmakcı,
Melih.   III. Title.
TL152.8.U47   2012
629.25'8–dc23   2011052559

ISBN 978-1-107-01011-6 Hardback
ISBN 978-1-107-68604-5 Paperback

Contents

Preface

This textbook is organized in four major parts as follows:

I. *Introduction and Background* is an introduction to the topic of automotive control systems and a review of background material on engine modeling, vehicle dynamics, and human factors.

II. *Powertrain Control Systems* includes topics such as air–fuel ratio control, idle-speed control, spark-timing control, control of transmissions, control of hybrid-electric vehicles, and fuel-cell vehicle control.

III. *Vehicle Control Systems* covers cruise control and headway-control systems, traction-control systems (including antilock brakes), active suspensions, vehicle-stability control, and four-wheel steering.

IV. *Intelligent Transportation Systems (ITS)* includes an overview of ITS technologies, collision detection and avoidance systems, automated highways, platooning, and automated steering.

With multiple chapters in each part, this textbook contains sufficient material for a one-semester course on automotive control systems. The coverage of the material is at the first-year graduate or advanced undergraduate level in engineering. It is assumed that students have a basic undergraduate-level background in dynamics, automatic control, and automotive engineering.

This textbook is written for engineering students who are interested in participating in the development of advanced control systems for vehicles, including advanced automotive concepts and the next generation of vehicles for ITS. This is an exciting and rapidly developing field for which numerous articles and reports exist. An extensive list of references, therefore, is provided at the end of each chapter for all topics covered. Due to the breadth of topics treated, the reference lists are by no means comprehensive, and new studies are always appearing. However, the lists cover many major contributions and the basic concepts in each sub-area. This textbook is intended to provide a framework for unifying the vast literature represented by the references listed at the end of each chapter. It is currently the only textbook, including problems and examples, that covers and integrates the topics of automotive powertrain control, vehicle control, and ITS.

The emphasis is on fundamental concepts and methods for automotive control systems rather than the rapidly changing specific technologies. For each automotive-control problem considered, we emphasize the physics and underlying principles behind the control-system concept and design. Any one of the many topics covered (e.g., engine control, vehicle-stability control, or platooning) could be discussed in more detail. However, rather than treating a specific control problem in its full complexity, we use each automotive control application as an opportunity to focus on a key engineering aspect of the control-design problem. For example, we discuss the importance of regulating the air–fuel ratio in engine control, the benefits for vehicle dynamics of reducing the vehicle side-slip angle in four-wheel steering and vehicle-stability control, the importance of predictive/preview action in the material on driver modeling, the concept of string stability for platoons and autonomous cruise-control systems, and the role of risk homeostasis in active-safety-systems design.

We also use various automotive-control applications to focus on specific control methodologies. For example, the Smith predictor for control of time-delay systems is introduced in air–fuel ratio control; linear quadratic optimal estimation and control is introduced for active suspensions; adaptive control using recursive least squares estimation is introduced in the chapter on cruise-control systems; and sliding-mode control is introduced in the discussion of traction-control systems. However, all of these methods can be applied to many other automotive-control problems.

End-of-chapter problems are included and many are used in our courses as homework and/or examination problems. Throughout the text, we include examples to illustrate key points. Many of these examples, as well as the end-of-chapter problems, require the use of MATLAB and/or Simulink. It is assumed that students are familiar with these computational engineering tools; for those who are not, we highly recommend the *Control Tutorials for MATLAB and Simulink* Web site (www.engin.umich.edu/class/ctms) for self-study.

This textbook is based on course notes originally developed by A. Galip Ulsoy during the mid-1990s, then refined and added to by both Ulsoy and Huei Peng during a period of fifteen years of teaching this material to beginning graduate students at the University of Michigan, Ann Arbor. The students are primarily from mechanical engineering disciplines, but students with a suitable background from other engineering disciplines also are included, as well as practicing engineers in the automotive industry who take the course through distance-learning programs and short courses.

We sincerely thank all of our former students for their useful feedback, which led to many improvements in and additions to this material. We also welcome your comments so that we can continue to improve future versions. The current textbook was rewritten extensively from those course notes in collaboration with Melih Çakmakcı, who was not only a former student who took the course but also has worked in the automotive industry as a control engineer for a decade. He brings an additional perspective to the material from his extensive industrial experience.

A. Galip Ulsoy
Huei Peng
Melih Çakmakcı

PART I

INTRODUCTION AND BACKGROUND

1   Introduction

The century-old automobile – the preferred mode for personal mobility throughout the developed world – is rapidly becoming a complex electromechanical system. Various new electromechanical technologies are being added to automobiles to improve operational safety, reduce congestion and energy consumption, and minimize environmental impact. This chapter introduces these trends and provides a brief overview of the major automobile subsystems and the automotive control systems described in detail in subsequent chapters.

1.1 Motivation, Background, and Overview

The main trends in automotive technology, and major automotive subsystems, are briefly reviewed.

Trends in Automotive Control Systems

The most noteworthy trend in the development of modern automobiles in recent decades is their rapid transformation into complex electromechanical systems. Current vehicles often include many new features that were not widely available a few decades ago. Examples include hybrid powertrains, electronic engine and transmission controls, cruise control, antilock brakes, differential braking, and active/semi-active suspensions. Many of these functions have been achieved using only mechanical devices. The major advantages of electromechanical (or mechatronic) devices, as opposed to their purely mechanical counterparts, include (1) the ability to embed knowledge about the system behavior into the system design, (2) the flexibility inherent in those systems to trade off among different goals, and (3) the potential to coordinate the functioning of subsystems. Knowledge about system behavior – in terms of vehicle, engine, or even driver dynamic models or constraints on physical variables – is included in the design of electromechanical systems. Flexibility enables adaptation to the environment, thereby providing more reliable performance in a wide variety of conditions. In addition, reprogrammability implies lower cost through exchanged and reused parts. Sharing of information makes it possible to integrate subsystems and obtain superior performance and functionality, which are not possible with uncoordinated systems.

Today's electrical and electronic devices have evolved into systems with good reliability and relatively low cost. They feature many new benefits including increased safety, reduced congestion and emissions, improved gas mileage, better drivability, and greater driver satisfaction and passenger comfort. Safety is perhaps the most important motivation for the increased use of electronics in automobiles. On average, one person dies every minute somewhere in the world due to a car crash. The cost of crashes totals 3 percent of the world's gross domestic product (GDP) and was nearly $1 trillion in 2000. Clearly, the emotional toll of accidents and fatalities is immeasurable (Jones 2002). Data from the National Highway Transportation Safety Association (NHTSA) show that 6,335,000 accidents (with 37,081 fatalities) occurred on U.S. highways in 1998 (NHTSA 1999). In 2008, the same statistic improved by about 10 percent to 5,811,000 accidents (with 34,017 fatalities) (NHTSA 2009). Data also indicate that although various factors contribute to accidents, human error accounts for 90 percent of all accidents (Hedrick et al. 1994).

Delays due to congestion are a major problem in metropolitan areas, providing strong motivation for an increase in automotive electronics. Traffic-information systems can reduce delays significantly by alerting drivers to accidents, congested areas, and alternate routes. Automated highway systems (AHS) at on ramps and tollbooths also can improve traffic flows. Significantly higher traffic flows can be achieved by closely packing automatically controlled vehicles in "platoons" on special highway lanes. These AHS concepts, developed and demonstrated in California, require automatic longitudinal and lateral control of vehicles (Rajamani et al. 2000).

In 1970, only 30 million vehicles were produced and 246 million vehicles were registered worldwide; by 1997, these numbers had increased to 56 million and 709 million, respectively. By 2005, 65 million vehicles were produced and more than 800 million were registered (Powers and Nicastri 2000). Consequently, another major factor that contributes to the increased use of electronics is the expanding government regulation of automotive emissions. For example, the 2005 standard for hydrocarbon (HC) emissions was less than 2 percent of the 1970 allowance; for carbon monoxide (CO), it was 10 percent of the 1970 level; and for oxides of nitrogen (NO_x), it was 7 percent of the 1970 level. The California requirements for ultra-low emission vehicles (ULEV) reduced the levels approximately by half again. Spilling 5.7 liters of gasoline on a driveway produces as many HC emissions as a ULEV vehicle driven more than 160,000 kilometers. At the same time, government regulations also require improved fuel economy. Advanced control technologies (e.g., fuel injection, air–fuel ratio control, spark-timing control, exhaust-gas recirculation [EGR], and idle-speed control) are and will continue to be instrumental in reducing emissions and improving fuel economy (e.g., hybrid-electric, all-electric, and fuel-cell vehicles).

This evolution (some might say "revolution") of automotive electronics also is enabled by recent advances in relevant technologies, including solid-state electronics, computer technology, and control theory. Table 1.1 summarizes developments in automotive electronics from 1965 through 2010. The already-evident trend toward increased automotive electronics can be expected to continue in the foreseeable future (Cook et al. 2007; Ford 1986; Powers and Nicastri 2000). In the next decade, significant advances are expected in the use of power electronics, advanced control systems, and alternative powertrain concepts. Among others, new technologies are

Table 1.1. *Historical development of automotive electronics*

Year	Examples of automotive electronics available
1965	Solid-state radio, alternator rectifier
1970	Speed control
1975	Electronic ignition, digital clock
1980	Electronic voltage regulator, electronic engine controller, electronic instrument cluster, electronic fuel injection
1985	Clock integrated with radio, audio graphic equalizer, electronic air suspension
1990	Antilock brakes, integrated engine and speed control, cellular phones, power doors and windows
1995	Navigation systems, advanced entertainment/information systems, active suspensions
2000	Collision avoidance, autonomous cruise control, vehicle stability enhancement, CVT
2005	Hybrid electric vehicles, driver monitoring, drive-by-wire, integrated vehicle controls
2010	Driver-assist systems (e.g., automated parallel parking), integrated telematics (i.e., location-aware vehicles via mobile devices), plug-in hybrid electric vehicles

being developed for fuel-efficiency management, integrated chassis control, power management of hybrid vehicles, electrical power steering, collision warning and prevention, automatic lane following, rollover and lane-departure warnings, and fuel-cell vehicles.

In the near future, it is anticipated that these advancements may reach beyond individual vehicles and eventually lead to the development of Intelligent Transportation Systems (ITS) (Jurgen 1995). Due to rapidly increasing highway congestion, it is necessary for automotive and transportation engineers to devise ways to increase safety and throughput on existing highways. The term *ITS* (previously referred to as Intelligent Vehicle/Highway Systems [IVHS]) defines a collection of concepts, devices, and services to combine control, sensing, and communication technologies to improve the safety, mobility, efficiency, and environmental impacts of vehicle and highway systems. The importance of ITS is in its potential to produce a paradigm shift in transportation – that is, away from individual vehicles and roadways and toward the development of those that can cooperate effectively, efficiently, and intelligently.

Major Automobile Subsystems

To provide background for subsequent chapters, this section is an introductory overview of an automobile and its major subsystems. Refer to other sources, including Bastow et al. (2004), Bosch (2009), Dixon (1992), Ellis (1969), Gillespie (1992), Mizutani (1992), Ribbens (2003), Segel (1986), Washine (1989) and Wong (2008), for more in-depth discussions. The functional systems of an automobile are shown in Figure 1.1 and are classified as follows:

Chassis or Body. This basic structure of an automobile supports many other systems described herein, as well as passengers and loads. It is supported by the suspension, which connects it to the axles and the wheels. The design of the chassis also affects vehicle dynamics, aerodynamic drag, fuel efficiency, and passenger comfort. The current trend is toward lighter body structures, including

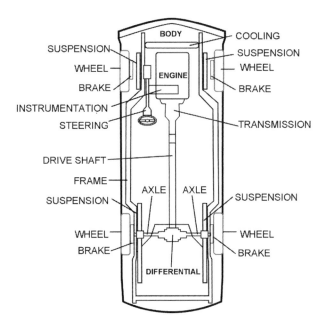

Figure 1.1. Vehicle subsystems.

the more efficient use of lighter-weight materials that nonetheless are durable and crashworthy.

Engine. This component provides the power for moving a vehicle as well as operating various subsystems. The most prevalent of the many engine designs is the piston-type, spark-ignited, liquid-cooled, internal combustion engine with four strokes per cycle and gasoline fuel. Engine controls, which improve engine performance in various ways, are used widely. Many new engine technologies (e.g., homogeneous-charge compression-ignition, electric, hybrid, and fuel-cell) also are being developed (Ashley 2001).

Drive Train or Powertrain. This system consists of the engine, transmission, driveshaft, differential, and driven wheels. The transmission is a gear system that adjusts the ratio of wheel speed to engine speed to achieve near-optimum engine performance. Automatic transmissions already are commonplace, and electronic transmission-control systems and continuously variable transmissions (CVT) are being introduced. A driveshaft is used in front-engine, rear-wheel-drive systems to transmit the engine power to the drive wheels. The differential provides not only the right-angle transfer of the driveshaft rotary motion to the wheels but also a torque increase through the gear ratio, thereby allowing the driven wheels to turn at different speeds (e.g., when turning a corner). The wheels and pneumatic tires provide traction between the vehicle and the road surface. Traction-control systems have been developed to provide good traction under a variety of road-surface conditions.

Steering. Steering allows a driver to change the orientation of a vehicle's front wheels to control the direction of forward motion. A rack-and-pinion steering-system design is typical in many modern automobiles. Power-assisted steering is now commonplace and four-wheel-steering (4WS) vehicles are emerging.

Suspension. The two major functions of the suspension system are to (1) provide a smooth ride inside the automobile, and (2) maintain contact between the wheels and the road surface. An independent-strut–type suspension design is common. A semirigid axle suspension system also is typical on the rear wheels of front-wheel-drive (FWD) vehicles. The suspension design also influences vehicle dynamics. Active and semi-active suspensions, which use electronic controls, are currently available on some vehicles.

Brakes. The brakes are the means for bringing a vehicle to a stop. Two common designs are drum and disk brakes. Antilock brakes, which use electronic controls to limit wheel slip, are now common on many commercial vehicles.

Instrumentation. A modern vehicle includes many electronic sensors, actuators, and other instrumentation. In today's cars, there are more than two dozen sensors in the powertrain alone. Most vehicles also now include dozens of microprocessors (e.g., for electronic engine control and diagnostics). Technologies such as the global positioning system (GPS) are starting to be used in automobiles. The average value of automotive electronics per vehicle, which was less than $100 in the 1960s, reached approximately $1,000 in 1990 and more than $2,000 by 2000 (Ford 1986). Due to increasing power needs, today's 14-volt (V) electrical systems (with a 12-V battery) eventually may be replaced with a 42-V system (with a 36-V battery).

1.2 Overview of Automotive Control Systems

The automobile is rapidly becoming a complex electromechanical system due in part to advances in computing and sensing technologies as well as advances in estimation and control theory. Vehicles now include hierarchically distributed, onboard computing systems, which coordinate several distinct control functions. Among these are control functions associated with the engine and transmission, cruise control, traction control, and active suspensions, which are discussed in subsequent chapters of this book.

The control functions in an automobile can be grouped as follows: (1) powertrain control, (2) vehicle control, and (3) body control (Mizutani 1992). Before discussing each topic in detail, we briefly introduce these control systems, the basic concepts, and the terminology associated with control-system design.

Powertrain control consists of engine- and transmission-control systems and is discussed in Part II. The engine-control systems may include fuel-injection control, carburetor control, ignition or spark-timing control, idle-speed control, antiknock-control systems, and exhaust-gas recirculation (EGR) control. The goal of engine-control systems is to ensure that an engine operates at near-optimal conditions at all times. Electronic transmission control is used primarily in automatic transmissions. Transmission-control systems determine the optimal shift point for the torque converter and the lockup operation point based on throttle-angle and vehicle-speed measurements. Often, a single electronic control unit (ECU) handles both engine and transmission control functions. Four-wheel drive (4WD) systems are used (1) to obtain the optimal torque–transmission ratio, (2) in braking and acceleration, and (3) between the front and rear wheels. This optimal ratio depends on the vehicle forward velocity.

Table 1.2. *Automotive control functions and variables*

	Controlled variable	Control input	Control algorithm	Sensors and actuators
Fuel Control	Air–fuel ratio	Injected fuel	Smith Predictor	Airflow, EGO, fuel injector
EGR Control	EGR rate	EGR valve opening	Optimal control	Valve position, EGR valve
Spark-Timing Control	Spark timing	Primary current	Rule-based, optimal control	Crank angle, vibration
Idle-Speed Control	Idle speed	Airflow rate	PI, linear quadratic regulator	Engine speed, idle speed control valve, throttle
Cruise Control	Vehicle speed	Airflow rate	PI, adaptive PI	Vehicle speed, throttle
Transmission	Gear ratio	Pressure, current	Rule-based	Vehicle speed, MAP
All-wheel drive, four-wheel drive	Torque distribution	Pressure, current	Rule-based, P, PI, PID	Engine speed, steering angle, control valve
Four-wheel steering	Wheel angle	Stepper motor	Feed forward, PI	Vehicle speed, wheel angle, stepper motor
ABS	Slip ratio	Pressure, current	Rule-based, sliding mode	Vehicle speed, wheel speed, control valve

Typically, 4WD systems have been achieved using mechanical rather than electromechanical components and are not discussed in this book.

Vehicle control systems, discussed in Part III, include suspension control, steering control (e.g., 4WS), cruise control, braking control (e.g., antilock brake systems [ABS]), and traction control. These systems improve various vehicle functions including response, steering stability, ride, and handling; many were introduced in recent decades or are currently being developed.

Body control refers to systems such as automatic air conditioning, electronic meters, multi-instrument displays, energy control systems, security systems, communication systems, door-lock systems, power windows, and rear-obstacle detection. The intent of these systems is to increase driving comfort and convenience and to improve the value of the automobile. These features often are perceived immediately by drivers as a benefit and typically are introduced first in luxury vehicles. Body-control systems are not discussed in detail in this book.

Several of the systems listed here are shown in Table 1.2, including the controlled variable, the manipulated variable (i.e., control input), the control logic (or control algorithm) used, the measured variables, and the actuators used to generate the control input. These vehicle control systems can be compared to the "generic" control-system block diagram shown in Figure 1.2. A typical feedback-control system consists of four basic elements: (1) controller, (2) actuator, (3) controlled system, and (4) sensor. The controller receives a reference (or set-point) input, which defines the desired value of the controlled variable, and a feedback signal from the sensor, which is a measurement of the controlled variable. The controller then applies a particular control logic (or law or algorithm) to compute a control signal. The control

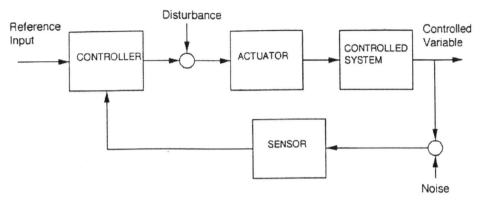

Figure 1.2. Control-system block diagrams.

signal is sent to the actuator, which supplies energy to the system by converting this information-type input signal to a power-type input to the controlled system. The controlled system responds to the actuator input as well as any other uncontrolled inputs (i.e., disturbances) that act on it. The sensor provides a measurement of the controlled variable for the purpose of feedback to the controller.

The control system illustrated in Figure 1.2 is a simple feedback loop with a single-input/single-output (SISO) system, which attempts to control only one variable. In reality, many automotive control systems consist of several such loops that interact in a complex manner. For example, an electronic engine-control system includes many controlled variables, actuators, and sensors; in fact, they are multi-input/multi-output (MIMO) control systems. The detailed analysis of these control problems can be carried out but is a complex process. Instead, the first stage of control-system analysis or design can be performed by neglecting the interactions among the various control tasks and treating each as an independent SISO control system. In a typical electronic engine-control system, the following SISO control systems can be identified (see Table 1.2):

Air–Fuel Ratio Control. The air–fuel ratio is the controlled variable and it is controlled by fuel injection at each cylinder. A mass airflow sensor is used and the fuel injector is the actuator. An optimal control is used to maintain the air–fuel mixture at *stoichiometry* (i.e., air–fuel ratio = 14.7); this reference is selected because in conjunction with a catalytic converter, it provides near-optimal performance in an engine. Thus, accurate air–fuel ratio control is important from the perspective of reducing emissions as well as other performance measures. Current systems use an EGO sensor for air–fuel ratio control.

EGR Control. The controlled variable is the EGR rate, the EGR control valve is the actuator, and the engine temperature and speed measurements are used to compute the proper EGR rate. EGR effectively reduces peak combustion temperature, thereby reducing emissions. The drawbacks of EGR include increased HC emission, deteriorated fuel economy, and combustion instability at idle or low engine speed and/or when an engine is cold. Typically, the EGR function may be turned off when an engine is cold or when a vehicle is accelerating or idling.

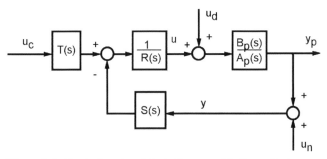

Figure 1.3. Control-system block diagram: Mathematical representation.

Spark-Timing Control. Spark timing is adjusted to affect engine-torque output. Moreover, the response is usually much faster than throttle-angle manipulations. It also is used to affect emission and fuel economy and to minimize engine knock. Because the "timing" is with respect to top dead center (TDC), the crank angle must be measured.

Electronic-Transmission Control. The hydraulic pressure and solenoid status can be controlled for fuel economy (i.e., shift-point control) and comfort (i.e., torque control during shifting). The shift point typically is regulated based on two measurements: vehicle speed and manifold absolute pressure (MAP), or throttle angle. The latter measurement is an indicator of engine load.

Idle-Speed Control. The purpose of the idle-speed control function is to maintain idle speed in the presence of load disturbance as well as to minimize speed for reduced fuel consumption and emission. An idle-speed controller typically measures the idle speed and adjusts the airflow rate using either the throttle or an idle-speed control valve.

Subsequent chapters describe in more detail not only these powertrain (i.e., engine and transmission) control functions but also vehicle-control functions such as cruise control, traction control, active suspensions, and 4WS. For design purposes, each function is treated as a stand-alone SISO control system. In fact, they are interacting MIMO systems, which must be accounted for when they are integrated in an automobile. In addition, they become the building blocks for even higher-level control functions, such as those described in Part IV, the ITS chapters.

Control Structures and Algorithms

Figure 1.3 is a typical mathematical representation of the physical elements shown in Figure 1.2. Note that the actuator, controlled system, and sensor blocks are combined into a process (or plant) transfer function:

$$G_p(s) = \frac{B_p(s)}{A_p(s)} \tag{1.1}$$

where s is the Laplace transform variable.

In addition, the controller is represented by:

$$u = \left(\frac{T(s)}{R(s)}\right)u_c - \left(\frac{S(s)}{R(s)}\right)y \tag{1.2}$$

where u_c is the reference (or command) input and y is the controlled variable. Depending on the particular form of the polynomials R, S, and T, particular control laws are obtained. Simple control laws include Proportional (P) control:

$$u(t) = K_P(u_c(t) - y(t)) = K_P\, e(t) \tag{1.3}$$

which corresponds to $R = 1$, $T = S = K_P$ in Eq. (1.2). Integral (I) control is given by:

$$u(t) = K_I \int_0^t e(t)dt \tag{1.4}$$

which corresponds to $R = s$, $T = S = K_I$ in Eq. (1.2). Similarly, a Derivative (D) controller is:

$$u(t) = K_D \frac{de(t)}{dt} \tag{1.5}$$

which corresponds to $R = 1$, $T = S = K_D s$ in Eq. (1.2). Common combinations of these basic control laws are as follows:

Proportional plus Derivative (PD) control:

$$u(t) = K_P\, e(t) + K_D \frac{de(t)}{dt} \tag{1.6}$$

Proportional plus Integral (PI) control:

$$u(t) = K_P\, e(t) + K_I \int_0^t e(t)dt \tag{1.7}$$

and Proportional plus Integral plus Derivative (PID) control:

$$u(t) = K_P\, e(t) + K_I \int_0^t e(t)dt + K_D \frac{de(t)}{dt} \tag{1.8}$$

A more complete description of feedback-control concepts is provided in basic textbooks (see also Appendix A). Of course, several other control algorithms not mentioned herein are used in vehicle-control applications; for example, optimal control methods are widely used. Current trends include the use of robust control and adaptive control methods to handle process-parameter variations, which are common in automotive control applications. In addition, many automotive systems have nonlinear characteristics (e.g., the engine) and require the use of nonlinear methods for effective control.

Examples of Automotive Control Systems

Air–Fuel Ratio Control. The air–fuel ratio is the controlled variable and it is controlled by fuel injection at each cylinder. An airflow sensor is used and the fuel injector is the actuator. An optimal control is used to maintain the air–fuel

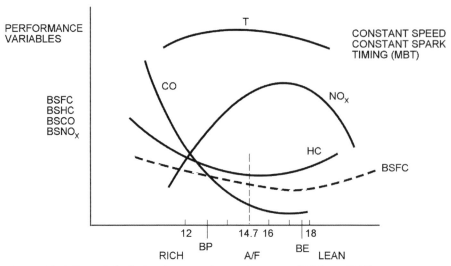

Figure 1.4. Effect of air–fuel ratio on engine performance (Ribbens 2003).

mixture at stoichiometry (i.e., air–fuel ratio = 14.7). As shown in Figure 1.4, this value provides near-optimal performance in an engine; that is, engine torque (T) is near maximum, brake-specific fuel consumption (BSFC) is near minimum, and hydrocarbon (HC) and carbon monoxide (CO) emissions are near minimum. However, nitrous oxide (NO_x) emissions are maximized. The importance of the catalytic converter found on most modern vehicles is illustrated in Figure 1.5, which shows that the conversion efficiency is also very high for HC, CO, and NO_x emissions in a very narrow band near stoichiometry. Thus, accurate air–fuel ratio control is important from the perspective of reducing emissions as well as other performance measures. Current systems use an EGO sensor for air–fuel ratio control (Grizzle et al. 1991). The problem of air–fuel ratio control is discussed in detail in Chapter 6.

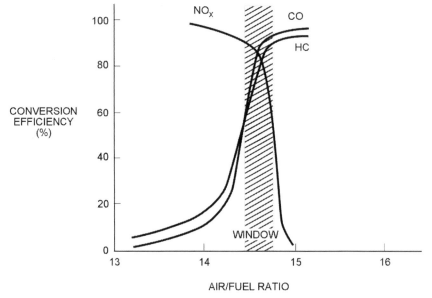

Figure 1.5. Conversion efficiency of two-way catalytic converter (Ribbens 2003).

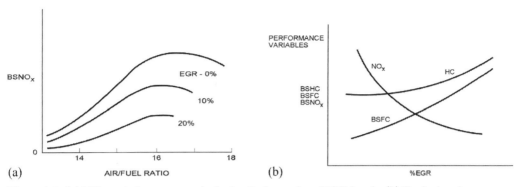

Figure 1.6. (a) NO$_x$ emissions versus air–fuel ratio for various EGR levels; (b) Typical variation in engine performance with EGR (Ribbens 2003).

EGR Control. The controlled variable is the EGR rate, the EGR control valve is the actuator, and the engine temperature and speed measurements are used to compute the proper EGR rate (Figure 1.6). EGR effectively reduces peak combustion temperature, thereby reducing the NO$_x$ emission. The drawbacks of EGR include increased HC emission, deteriorated fuel economy, and combustion instability at idle or low engine speed and/or when an engine is cold. Open-loop (i.e., offline optimization) control typically is used, and the EGR function may be turned off when an engine is cold or when a vehicle is accelerating or idling.

Spark-Timing Control. This function is a mechanism or, rather, an objective. It has been observed that up to a certain point, engine torque increases with spark advance. In Figure 1.7, the minimum spark advance for best torque (MBT) denotes the spark advance for optimal torque. Therefore, spark timing can be adjusted to affect engine torque output. Moreover, the response is usually much faster than throttle-angle manipulations. Spark advance also can be used to affect emissions and fuel economy and to minimize engine knock. Because the "timing" is with respect to TDC, the crank angle must be measured.

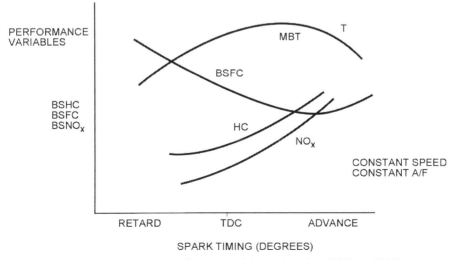

Figure 1.7. Typical variation of performance with spark timing (Ribbens 2003).

When engine torque or fuel economy is a concern, the *peak-pressure concept* is a simple but accurate rule to follow. It was found that under varying humidity, engine-speed, load, air–fuel ratio, and other conditions that the *angle-to-peak pressure* generated by optimal spark timing is roughly constant (which does make sense!). Therefore, if we can measure engine-cylinder pressure and adjust spark timing to achieve constant angle-to-peak pressure (i.e., 10 to 12 degrees), all of these perturbations will be rejected. Spark-timing control is discussed in Chapter 7.

Automatic Transmission Control. Hydraulic pressure and solenoid status can be controlled for fuel economy (i.e., shift-point control) and comfort (i.e., torque control during shifting). The shift point is usually regulated based on two measurements: vehicle speed and MAP or throttle angle. The latter measurement is an indicator of engine load. Again, the control strategy is the open-loop table lookup type. Torque control also is concerned with ride quality during gear shifting. The basic idea is to control the hydraulic pressures at both oncoming and offgoing clutches to transfer smoothly the load while minimizing the torque disturbance at the output shaft. By directly controlling clutch pressures, it is possible to enable the use of transmissions with low mechanical and hydraulic complexity. Transmission-control problems are discussed in Chapter 9.

Idle-Speed Control. The purpose of the idle-speed control function is to maintain idle speed in the presence of load disturbance, as well as minimize possible speed for fuel consumption and emission. An idle-speed controller typically uses a feed-forward lookup table to handle engine loads (i.e., air conditioning compressor, automatic transmission, power steering charging system, and vehicle speed). A PI control algorithm in the feedback loop measures the idle speed and adjusts the airflow rate using either the throttle or an idle-speed control valve.

The idle-speed control is important because variations can cause stalling and affect fuel economy and are easily perceived by human drivers. Idle-speed control is discussed in detail in Chapter 8.

Cruise Control. A cruise-control system (see Chapter 12) adjusts the throttle angle, using a throttle actuator, to maintain a desired vehicle forward velocity. The velocity sensor typically is driven by a flexible cable from the driveshaft, and the process model (i.e., $G_p(s)$ in Eq. [1.1]) must represent how the vehicle forward velocity changes as the throttle actuator input is varied (i.e., y = vehicle forward velocity and u = throttle actuator input). A PI control algorithm typically is used in many cruise-control applications.

Sensors, Actuators, and Controller Modules

It is clear that many sensors and actuators are required to implement the various control systems discussed in this chapter. For engine and vehicle control, for example, measurements of the following variables are required.

Manifold Pressure. A MAP sensor produces a voltage signal that is proportional to the average manifold absolute pressure. A variety of sensor designs is used that typically are based on measuring the deflection of a diaphragm, which is deflected by the manifold pressure.

Crankshaft Angular Position and Engine Speed. This can be measured at the crankshaft or at the camshaft. A typical sensor is of the noncontacting magnetic reluctance or hall-effect type. It produces a voltage proportional to the distance between a fixed magnetic sensor and protruding metal tabs on a rotating disk attached to the crankshaft. For example, four tabs might be attached with each corresponding to the TDC position for a specific cylinder on a four-cylinder engine. The number of pulses then indicates angular position and their frequency is proportional to the engine speed. Optical techniques also can be used to measure crankshaft angular position and speed.

Airflow Rate. The mass flow rate of air into the engine is measured using a mass airflow sensor (MAS), which is based on the same principles as the classical hot-wire anemometer.

Throttle Angle. A potentiometer (i.e., rotary variable resistor) can be used to measure the throttle angle.

Exhaust gas oxygen (EGO). The EGO sensor provides an indirect measurement of the air–fuel ratio. This is a highly nonlinear sensor, that essentially indicates either lean or rich air–fuel ratios.

Engine Knock. Engine knock typically is measured indirectly by some type of vibration sensor mounted on the engine block. More direct measurement using cylinder pressure is possible but not used commercially.

Vehicle Speed. An optical vehicle-speed sensor similar in operation to the engine-speed sensor is used. It is connected by a flexible cable to the driveshaft, which rotates at an angular speed proportional to the vehicle wheel speed.

Longitudinal Slip. This is the difference – normalized relative to vehicle speed – between the vehicle speed and the tire circumferential speed. It is difficult to measure directly and typically is estimated based on changes in wheel speed. The wheel speed in normal operation is limited by the inertia of the wheel but, when slip occurs, rapid changes in wheel speed take place. These changes, as obtained by differencing of the speed measurement during short time intervals, can be used as a longitudinal-slip sensor. This is the key sensor for ABS and traction-control systems.

Steering Angle. This is determined by measuring the rotation of the steering wheel. A typical approach uses photocells, a light source, and a disk with multiple codes. This is used, for example, in 4WS systems.

Vehicle Acceleration. Accelerometers can be used to measure vehicle acceleration in various directions. For lateral acceleration, which is important for ride and handling, the speed sensor and a steering-angle, d, sensor can be used to calculate the lateral acceleration from:

$$a_y \approx mu^2 d/l \tag{1.9}$$

Suspension Stroke. This typically is measured using a linear displacement sensor, such as a linear variable differential transformer (LVDT), and is important for active suspension control.

Of course, there are many other sensors; this list certainly is not exhaustive. Also, there are efforts to estimate quantities of interest (e.g., yaw rate, traction forces, and wheel slip) from indirect measurements. Following is a discussion of some important actuation devices needed for engine and vehicle-control systems.

Fuel Metering. This is accomplished using either an electronic carburetor or a fuel injector. A *fuel injector* is a solenoid-operated valve that passes fuel from a constant-pressure source when the solenoid current is on and blocks fuel flow when it is off. The fuel injector is either mounted in the throttle body (i.e., single-point fuel injection) or at each cylinder intake port near the intake valve (i.e., multipoint fuel injection). The duty cycle of the fuel injector can be varied to manipulate the air–fuel ratio.

Spark Ignition. This electronic ignition includes the coil, distributor, and spark plugs as well as the associated electronics.

Exhaust Gas. The EGR actuator is a valve that connects the intake and exhaust manifolds. This allows mixing of exhaust gases in the intake manifold.

Throttle Actuator. This adjusts the throttle angle in cruise-control applications. Typically, a control solenoid is used to actuate a pressure-control valve, which allows the vacuum from the intake manifold to exert a pull on a spring-loaded throttle lever.

Brake-Pressure Modulators. These are used in the ABS. Modulators momentarily reduce the brake pressure and, consequently, the braking force to prevent the wheels from locking.

Suspension Actuator. Semi-active suspensions use a special adjustable shock absorber in which resistance to oil flow can be adjusted by changing the size of the orifice. A fully active suspension includes an electrohydraulic actuator that applies a force between the sprung and unsprung mass in response to control signals. A fully active system typically incorporates a passive air-spring–type suspension in parallel with this actuator.

Vehicle Communication Networks

Vehicle communication networks provide the infrastructure to exchange information among vehicle electronic units. These electronic units are not only actuator and sensor components with network capabilities but also ECUs such as the engine controller.

Figure 1.8 is a configuration for common electronic components that communicate on the vehicle networks. For those components of the network for which the information exchange is less critical (e.g., automatic configuration of the seat position and/or the confirmed position), a slower and less expensive network system such as the Local Interconnect Network (LIN) (Motorola 1999) is preferred. For safety and vehicle performance, critical information exchange high-speed communication network protocols such as the Controller Area Network (CAN) (Tindell et al. 1995) or FlexRay (Makovitz and Temple 2006) are preferred.

Delay of the information can be critical for the algorithms that span multiple ECUs. In fact, characterization of the vehicle networks for a particular vehicle configuration affects algorithms run in individual control units and frequency of the information exchange. For critical safety features such as x-by-wire (e.g., drive-by-wire and steer-by-wire) applications, guaranteed delivery of the information within

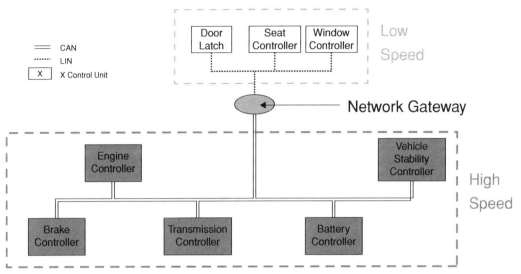

Figure 1.8. Simple vehicle-control network.

specified time boundaries is required. To satisfy the strict communication requirements, design features such as multiple physical routes and software-based message
priorities are considered (Davis et al. 2007).

Supervisory and Distributed Control Algorithms in Automotive Applications

Figure 1.9 shows powertrain components and control-oriented features for a typical
hybrid vehicle in today's market. Modern vehicles consist of many subsystems, and

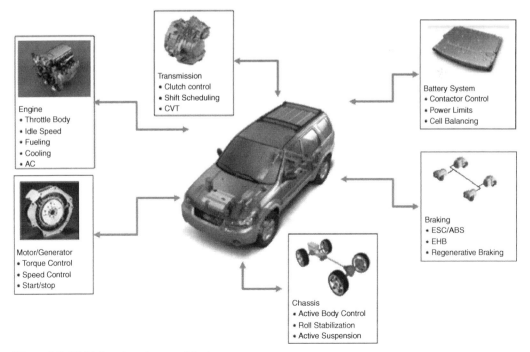

Figure 1.9. Vehicle subsystems and features.

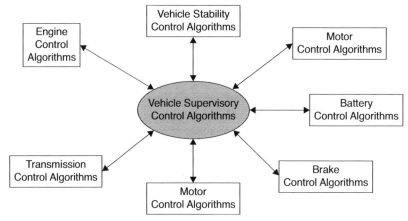

Figure 1.10. Vehicle supervisory control.

overall performance of a vehicle depends on consistent and reliable operation of all of them. Supervisory vehicle-control algorithms coordinate and monitor the operation of control algorithms located in the subsystem controller units, as shown in Figure 1.10. A typical example is the energy-management algorithms in hybrid electric vehicles. Although engine, transmission, motor, brake, and battery controller modules are individually commanded by their respective subsystems, a higher-level control algorithm is required to determine the power flow to or from the battery as well as the composition of the engine and motor torque provided to the wheels (Powers and Nicastri 2000).

With additional performance requirements for modern vehicles, the relative amount of supervisory-control software compared to the total amount is increasing. The combination of this trend with the desire to maximize the computing resource-allocation requirements and modularity for cost-reduction purposes means that supervisory control algorithms (e.g., energy management and telematics) are distributed to subsystem controllers and rely on vehicle communication networks and external information inputs (Leen and Hefferman 2002; Navet et al. 2005). Modularizing the design of components in such a networked control system to make them swappable (or "plug-n-play") can reduce development time, calibration costs, and maintenance and other costs (Çakmakcı and Ulsoy 2009; 2011).

PROBLEMS

1. From your own experience while driving and traveling in today's vehicles, provide examples of automotive control systems and discuss how they add value.

2. Table 1.1 summarizes developments in automotive electronics. Based on your reading of articles, such as (Cook et al. 2007; Jones 2002; Powers and Nicastri 2000) and others, provide additional items to those already listed. Extend the table to speculate on technologies that may appear in the future. Please cite your sources.

REFERENCES

Ashley, S., 2001, "A Low-Pollution Engine Solution," *Scientific American*, Vol. 284, June 2001, p. 90.

Bastow, D., G. Howard, and J. P. Whitehead, 2004, *Car Suspension and Handling*, SAE International.

Bosch, R., 2009, *Automotive Handbook*, Bentley Publishers.

Çakmakcı, M., and A. G. Ulsoy, 2009, "Improving Component Swapping Modularity Using Bi-Directional Communication in Networked Control Systems," *IEEE/ASME Transactions on Mechatronics*, Vol. 14, No. 3, June 2009, pp. 307–16.

Çakmakcı, M., and A. G. Ulsoy, 2011, "Modular Discrete Optimal MIMO Controller for a VCT Engine," *IEEE Transactions on Control Technology*, Vol. 19, No. 5, September 2011, pp. 1168–77.

Cook, J. A., I. Kolmanovsky, D. McNamara, E. C. Nelson, and K. V. Prasad, 2007, "Control, Computing and Communications: Technologies for the Twenty-First Century Model T," *Proceedings of the IEEE*, Vol. 95, No. 2, February 2007, pp. 334–54.

Davis, R., A. Burns, R. Bril, and J. Lukkien, 2007, "Controller Area Network (CAN) Schedulability Analysis: Refuted, Revisited and Revised," *Real-Time Systems*, Vol. 35, April 2007, pp. 239–72.

Dixon, J. C., 1992, *Tyres, Suspension and Handling*, Cambridge University Press.

Ellis, J., 1969, *Vehicle Dynamics*, Century Publishing.

Ford Motor Company, 1986, "Automotive Electronics in the Year 2000: A Ford Motor Company Perspective," *Proceedings of the CONVERGENCE '86 Conference*, October 1986.

Gillespie, T. D., 1992, *Fundamentals of Vehicle Dynamics*, SAE International.

Grizzle, J. W., K. Dobbins, and J. Cook, 1991, "Individual Cylinder Air–Fuel Ratio Control with a Single EGO Sensor," *IEEE Transactions on Vehicular Technology*, Vol. 40, No. 1, February 1991, pp. 280–6.

Hedrick, J. K., Tomizuka, M., and P. Varaiya, 1994, "Control Issues in Automated Highway Systems," *IEEE Control Systems Magazine*, December 1994.

Jones, W.D., 2002, "Building Safer Cars," *IEEE Spectrum*, January 2002, pp. 82–5.

Jurgen, R., 1995, *Automotive Electronics Handbook*, McGraw-Hill, Inc.

Leen, G. and D. Heffernan, 2002, "Expanding Automotive Electronic Systems," *Computer*, Vol. 35, 2002, pp. 88–93.

Makovitz, R. and C. Temple, 2006, "FlexRay – A Communication Network for Automotive Control Systems," *IEEE International Workshop on Factory Automation*, 2006.

Mizutani, S., 1992, *Car Electronics*, Society of Automotive Engineers.

Motorola, 1999, "LIN Specification and Press Announcement," *SAE World Congress*, Detroit, 1999.

National Highway Traffic Safety Administration (NHTSA), 1999, *Traffic Safety Facts 1998: A Compilation of Motor Vehicle Crash Data from the Fatality Analysis Reporting System and the General Estimates System*, National Center for Statistics and Analysis, U.S. Department of Transportation, Washington, D.C., October 1999.

National Highway Traffic Safety Administration (NHTSA), 2009, *Traffic Safety Facts 2008: A Compilation of Motor Vehicle Crash Data from the Fatality Analysis Reporting System and the General Estimates System*, National Center for Statistics and Analysis, U.S. Department of Transportation, Washington, D.C., October 2009.

Navet, N., Y. Song, F. Simonot-Lion and C. Wilwert, 2005, "Trends in Automotive Communication Systems, *Proceedings of the IEEE*, Vol. 93, 2005, pp. 1204–23.

Powers, W. F. and P. R. Nicastri, 2000, "Automotive Vehicle Control Challenges in the 21st Century," *Control Engineering Practice*, Vol. 8, No. 6, June 2000, pp. 605–18.

Rajamani, R., H. S. Tan, B. K. Law and W. B. Zhang, 2000, "Demonstration of Integrated Longitudinal and Lateral Control for the Operation of Automated Vehicles in Platoons," *IEEE Transactions on Control Systems Technology*, Vol. 8, No. 4, July 2000, pp. 695–708.

Ribbens, W. B., 2003, (6th edition), *Understanding Automotive Electronics*, Butterworth-Heinemann.

Segel, L., 1990, *Vehicle Dynamics*, Course Notes, Department of Mechanical Engineering, University of Michigan.

Tindell, K. W., A. Burns and A. J. Wellings, 1995, "Calculating Controller-Area Network (CAN) Message Response Times," *Control Engineering Practice*, Vol. 3, No. 8, August 1995, pp. 1163–9.

Washino, S., 1988, *Automobile Electronics*, Gordon and Breach, New York.

Wong, J. Y., 2008, *Theory of Ground Vehicles*, Wiley.

2 Automotive Control-System Design Process

2.1 Introduction

Generally, "solving" the controller design problem means finding the proper mathematical representation of a control action that meets a set of desired performance criteria. In reality, this is only one part of the solution (albeit an important part); the control-systems development process also includes steps for selecting the correct hardware – loaded with the proper software – for the controller module, which is the real end-product of this process (Figure 2.1).

The control-system development process begins by first developing the high-level system requirements, which are generally verbal and abstract and rarely point to a recognizable control design problem such as those that traditional engineering students would see in their control classes. The formal and technical requirements documents can be described as "wish lists" regarding the overall system features and performance. The result of the process is the controller module, which is to be deployed in bulk to the end product. The purpose of studying the control-systems development process is to provide a reliable, robust, and repeatable sequence of actions to develop ECUs.

In recent years, computer-aided design and analysis tools (e.g., MATLAB and Simulink) have improved the efficiency of design processes and increased the application of the model-based controller design and development process (Chrisofakis et al. 2011; Mahapatra et al. 2008; Michaels et al. 2010; Powers and Nicastri 2000). Figure 2.2 is a general outline of the model-based controller design and deployment process. The major components of this process (i.e., design, implementation, and testing) are discussed in the next section. The process outline in Figure 2.2 is based on development and testing portions that progress in parallel and continuously interact throughout the development cycle. This is, in fact, one of the most important features of model-based design, which enables debugging and validation of the current work while minimizing changes from the previous phase. Therefore, as the control development evolves, so does the testing platform for its debugging and validation.

Another important feature of model-based design and deployment is the ability to reuse models and routines (e.g., testing and data analysis) from similar past and ongoing projects. This enables engineers to compare continuously the performance

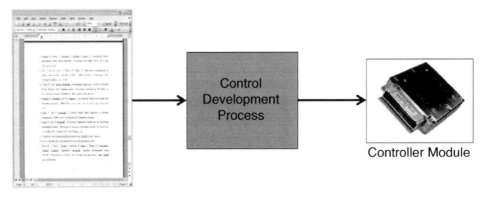

Requirement
Documents

Figure 2.1. Role of control engineering.

of their models and control algorithms. The self-documenting nature of these models eases the workload of archiving the developed tools and results of the finished project for future use.

2.2 Identifying the Control Requirements

The high-level requirements specified for a vehicle can be general and abstract, as in the following examples:

> Based on customer feedback, design a control system so that the vehicle is durable with a pleasant driving feel and good fuel economy.
> Vehicle shall meet 100,000-mile life requirements.

Today, part of a control engineer's responsibility is to determine how overall vehicle requirements affect the control problem. This may include understanding and quantifying how the vehicle-life requirements affect the way that the automatic transmission shifts gears, the engine controller operates the engine, and the battery-control module manages battery power. The formal methods for identifying control-system problems from vehicle requirements are performed by control-system engineers.

Systems Engineering for Control Development

The act of developing a control system from verbal requirements can be as simple as an engineer reading the entire document and developing the matching control design problems based on his or her experience and understanding of the document. Although this still may be acceptable for small companies (i.e., five or fewer employees), today's competitive environment – driven by the need for consistent quality and reduced warranty costs – requires a more structured and traceable approach. Probably the most common feature of all of these approaches is deliberately reviewing and documenting the results of the requirements-analysis, design, implementation, verification, and maintenance phases throughout product development. In fact, in many

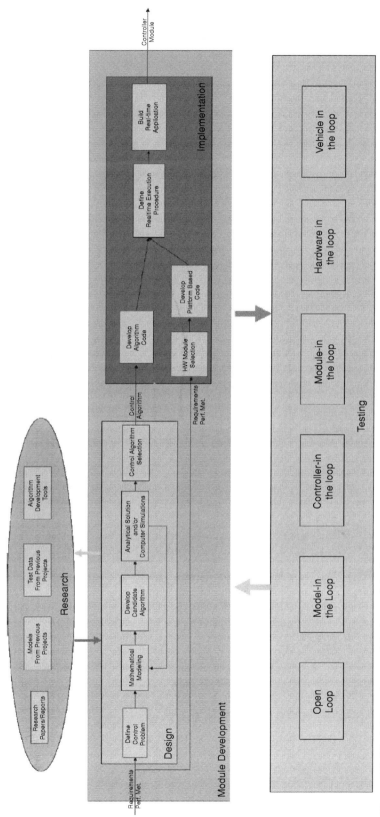

Figure 2.2. Model-based control design and deployment process.

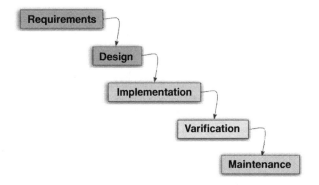

Figure 2.3. The Waterfall Model.

industries, documents related to these steps are required for regulatory and legal purposes to observe how design decisions were made. Actions taken by engineering teams at each phase and how these phases interact can be different for each product and company. The Waterfall and V-Diagram Models are two common approaches used in industry.

The Waterfall Model, illustrated in Figure 2.3, is a sequential software development process in which progress is seen as flowing steadily downward (like a waterfall) through the phases of requirements analysis, design, implementation, verification, and maintenance (Jacobson, Booch, and Rumbaugh 1999). Although this type of development model has been a widely used method for analyzing complex requirements since the early days of systems engineering, its sequential nature (i.e., no feedback) throughout the design problem is considered a weakness.

The V-Diagram Model is another development-process model that can be viewed as an extension of the Waterfall Model. Instead of moving down linearly, the process steps are redirected upward after the coding phase to form the typical V shape (Figure 2.4). The V-Diagram Model demonstrates the relationships

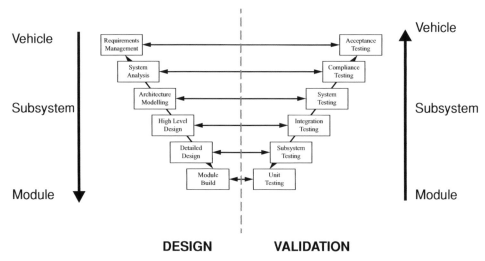

Figure 2.4. System engineering V-Diagram.

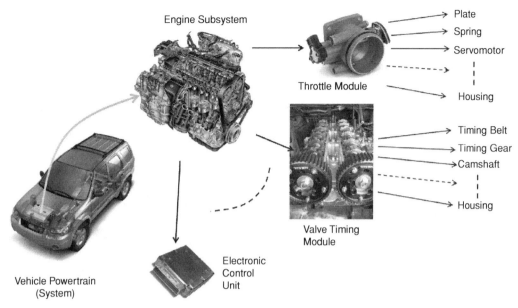

Figure 2.5. Vehicle-system decomposition.

between each phase of the development life cycle and its associated phase of testing (Jacobson, Booch, and Rumbaugh 1999; Stahl, Voelter, and Czarnecki 2006).

Figure 2.5 is a typical decomposition of a vehicle's powertrain system parts from the vehicle level to the module level for the engine subsystem. A vehicle powertrain system is composed of many subsystems, including the engine, transmission, and brakes, which in turn are composed of smaller functional modules. For example, an internal-combustion engine has a throttle module, valve-timing module, and oil-pump module. These modules can be decomposed further into parts that typically are supplied as off-the-shelf products or manufactured as single pieces. In modern automotive systems, almost every subsystem has an electronic control module to control and monitor the functional operations that communicate with the subsystem actuators, sensors, and other modules.

Developing algorithms that successfully will perform control operations can be challenging when considering the complexity of the performance requirements in today's automotive systems. The process of obtaining control-system require-ments from higher-level vehicle requirements is conducted in sequential steps that are called *requirements cascade studies* (Philips 2005). Figure 2.6 illustrates the path for developing control-algorithm solutions from a generic set of vehicle requirements.

In Figure 2.7, the method described in Figure 2.5 is shown for a specific exam-ple. The example shows how the vehicle's 100,000-mile requirement affects specific features (i.e., control problems) for a particular vehicle application. The effect of this and other requirements defines the feature control problem to be solved. The solutions obtained from all of the features represent the control algorithm for a vehicle.

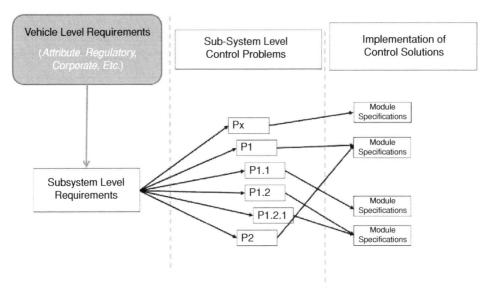

Figure 2.6. Requirement mapping and decomposition.

Algorithm Development

In the algorithm-design step (Figure 2.8), the control design problem is first formulated based on the given performance requirements and developed mathematical formulation. There is more than one control design approach to provide a solution for the control problem. By using analytical methods and/or computer simulations, the best alternative among the candidate algorithms is selected. If the control problem is similar to a previous application, development teams often prefer to start with

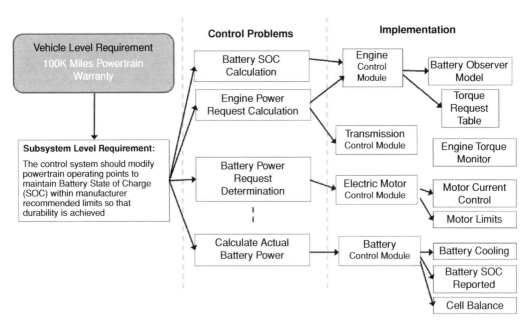

Figure 2.7. Requirement mapping and decomposition (example).

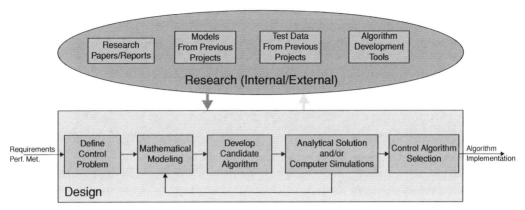

Figure 2.8. Design-phase step.

an existing control algorithm and improve the solution by building on the existing (and proven) solution.

Implementation

During the algorithm-implementation phase (Figure 2.9), the objective is to develop a real-time application to be executed in the control module using the desired control algorithm. By using an automatic code-generation tool, the target computer-language equivalent (e.g., the language C) of the control algorithm model is generated by a computer tool. This process also is known as *autocoding*. Once the algorithm block diagram model has been autocoded, it is combined with the hardware-platform-dependent code (e.g., device drivers). When the real-time application is generated, the control algorithm is ready to be downloaded and executed in the hardware module.

An alternative method to autocoding is *handcoding* – that is, developing the equivalent of the control-algorithm model by manually writing the computer code. Many developers argued in the early days of autocoding that a computer-generated code would be unnecessarily long and inefficient due to the generalized code-generation algorithms used. However, recent studies (e.g., Hodge, Ye, and Stuart 2004) show that with the appropriate initial setup, code generation can be as effective as handwritten code while requiring a fraction of the development time. The use

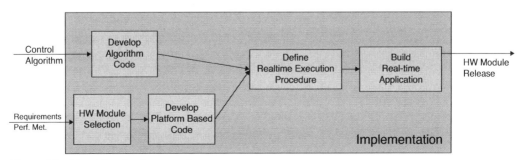

Figure 2.9. Algorithm-implementation step.

of autocoding also minimizes the effects of human error during initial development and successive modifications.

When developing the executable code, real-time constraints of the target hardware (i.e., the controller module) also should be considered. Software implementation of the algorithm should be matched to available computing resources. If there are overruns during the real-time execution, the algorithm should be simplified or new target hardware should be selected. In today's modern vehicles, controller modules also communicate with other controllers via communication networks. The effects of the loss of or limited communication with one or more contacts should be investigated and the necessary modifications implemented.

Testing and Validation

Testing in the model-based control development process starts as early as in the algorithm-development step. By testing an algorithm's open-loop (Figure 2.10a), developers can feed in simple test vectors and analyze the test output for expected functionality. These simple algorithms also can be tested against the simpler conceptual vehicle models, which are available in earlier stages of the program (Figure 2.10b). These models then are fortified with improvements based on component- and vehicle-testing data, which makes them suitable for more complex testing procedures such as module, component, and vehicle loop-type testing.

Hardware-in-the-Loop Systems Overview

During the controller module development process, combinations of hardware and simulated elements are used to evaluate system performance.

OPEN-LOOP VERSUS CLOSED-LOOP TESTING. When a developed algorithm or mathematical representation of a plant model must be verified, the quickest way that requires the least effort is to perform open-loop testing. In an open-loop-testing configuration (Figure 2.10a), algorithms or models are provided with test-vector inputs, and the outputs are compared with a set of expected results. The platform for open-loop testing can be a simulation environment (e.g., MATLAB or Simulink) or a test-bench when the controller prototype or production module is available. Perhaps the most common form of open-loop testing is vehicle road-testing. A distinctive challenge associated with open-loop testing is the generation of input vectors suitable for the test purposes. This is particularly difficult when the subject being tested is part of a larger algorithm or plant dynamics and its input/output (I/O) relationship is counterintuitive, which requires some internal states to be "staged" for testing the feature in question. Moreover, in most types of robustness testing, output-based generation of the input is required.

The type of testing in which the input to the test subject is generated based on the output from previous output is called *closed-loop testing*. These tests require the actual hardware – or its emulation – to react properly to the test output. In the simulation environment, closed-loop testing can be performed by using algorithm-in-the-loop (Figure 2.10b) and/or controller-in-the-loop testing (Figure 2.10c). The difference between the models of an algorithm and a controller is that the latter includes the algorithm model as well as the I/O processing schemes and timing

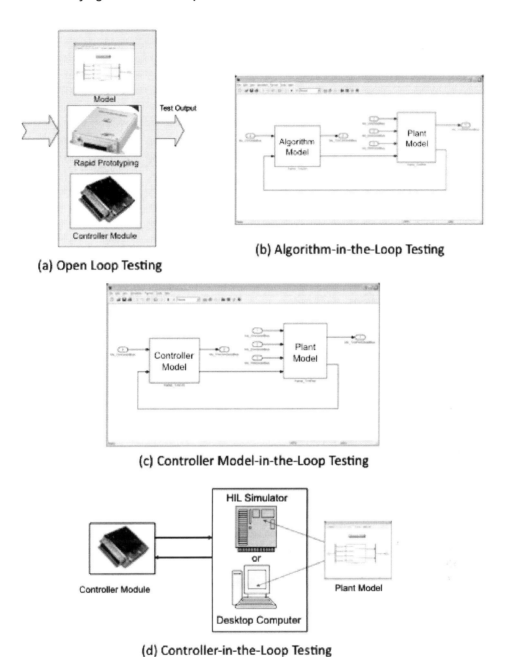

(a) Open Loop Testing

(b) Algorithm-in-the-Loop Testing

(c) Controller Model-in-the-Loop Testing

(d) Controller-in-the-Loop Testing

Figure 2.10. Different types of testing.

constraints of the targeted implementation. Generally, algorithm models are available in the development cycle as early as in the conceptual-design phase, whereas controller models begin to take shape toward the end of the implementation phase.

HARDWARE-IN-THE-LOOP TESTING AND COMMON CONFIGURATIONS. Closed-loop testing configurations in which hardware versions of one or more components exist in the test setup are referred to as *hardware-in-the-loop* (HIL) testing configurations.

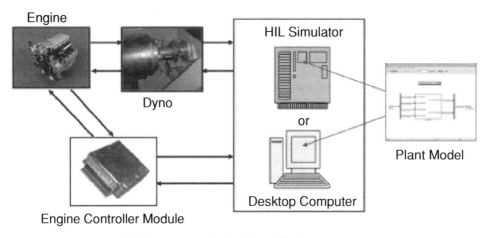

Engine

Dyno

Engine Controller Module

(a) Component-in-the-Loop Testing

(b) Vehicle-in-the-Loop Testing   (c) Vehicle Road Testing

Figure 2.11. Different types of testing (continued).

Primary components in an HIL system are the HIL simulator, host computer, and hardware being tested. An HIL simulator is the real-time–oriented test computer in which the emulation program for driving the hardware is run. The host computer is the supporting computer that communicates with the HIL simulator during the test through an HIL runtime graphic user interface (GUI), and it collects data related to testing. For some tests, secondary data-acquisition hardware such as CAN cards and calibration tools is included in the test setup for debugging and verification purposes. When physical hardware is included in the HIL setup, sensors and actuators to support its interfacing to the emulation also are included (Isermann, Schaffnit, and Sinsel 1999; Kendall and Jones 1999; Powell, Bailey, and Cikanek 1998).

The most common HIL for testing is the rapid-prototyping controller or the production-level controller module. This is the configuration in which the controller is driven by an HIL simulator that emulates the rest of the system (Figure 2.10d). More elaborate configurations of these HIL systems exist, including multiple-controllers-in-the-loop, powertrain-component-in-the-loop (Figure 2.11a), and vehicle-in-the-loop (Figure 2.11b) systems. As observed by many developers (e.g., Hatipoglu and Malik 1999; Kendall and Jones 1999), it generally is expected that as the complexity of the HIL increases, the accuracy of the testing also increases,

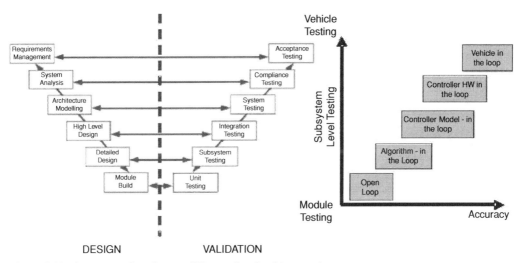

Figure 2.12. Accuracy of testing on different levels of integration.

as shown in Figure 2.12. It is shown in this figure that HIL testing fills the void between testing based on simulation models (available early) and actual vehicle testing (available later) and keeps the iterative nature of the V-Diagram Model development process intact.

PROBLEMS

1. Consider the Waterfall Model diagram in Figure 2.3. How would you modify the process described here to take advantage of some of the features of the V-Diagram Model process described in Figure 2.4?

2. Consider the high-level requirement: "Vehicle shall meet 100,000-mile life requirements." Discuss with examples what this could mean to the design of a braking system for a family sedan vehicle that will be driven primarily in city-traffic conditions.

3. Using your preferred modeling and simulation tool (e.g., MATLAB, Simulink, or SciLab):

    (a) Simulate the result of the block diagram.
    (b) Generate the c-code equivalent of the block diagram.
    (c) Discuss possible advantages and disadvantages of manually developing equivalent computer code.

4. Consider the component-testing setup in Figure 2.11a for engine testing. For which other automotive components can you develop similar testing setups? Describe the test equipment and operation of this facility using a diagram.

REFERENCES

Chrisofakis, E., et al., 2011, Simulation-Based Development of Automotive Control Software with Modelica, in *Proceedings of the 8th International Modelica Conference*, Dresden, Germany.

Hatipoglu, C., and A. Malik, 1999, "Simulation-Based ABS Algorithm Development," SAE International.

Hodge, G., J. Ye, and W. Stuart, 2004, "Multi-Target Modelling for Embedded Software Development for Automotive Applications," 2004 SAE World Congress, Detroit, MI.

Isermann, R., J. Schaffnit, and S. Sinsel, 1999, "Hardware-in-the-Loop Simulation for the Design and Testing of Engine-Control Systems," *Control Engineering Practice*, Vol. 7, 1999, pp. 643–53.

Jacobson, I., G. Booch, and J. Rumbaugh, 1999, *The Unified Software Development Process*, Pearson Education India, 1999.

Kendall, I. R., and R. P. Jones, 1999, "An Investigation into the Use of Hardware-in-the-Loop Simulation Testing for Automotive Electronic Control Systems," *Control Engineering Practice*, Vol. 7, No. 11, November 1999, pp. 1343–56.

Mahapatra, S., et al., 2008, "Model-Based Design for Hybrid Electric Vehicle Systems," SAE Paper 2008010085.

Michaels, L., et al., 2010, "Model-Based Systems Engineering and Control System Development via Virtual Hardware-in-the-Loop Simulation," SAE Technical Paper No. 01-2325.

Philips, A. M., 2005, "Technical Challenges of Hybrids," SAE Technical Symposium on Engineering Propulsion, 2005.

Powell, B. K., K. E. Bailey, and S. R. Cikanek, 1998, "Dynamic Modeling and Control of Hybrid Electric Vehicle Powertrain Systems," *IEEE Control Systems Magazine*, Vol. 18, October 1998, pp. 17–33.

Powers, W. F., and P. R. Nicastri, 2000, "Automotive Vehicle Control Challenges in the 21st Century," *Control Engineering Practice*, Vol. 8, June 2000, pp. 605–18.

Stahl, T., M. Voelter, and K. Czarnecki, 2006, *Model-Driven Software Development: Technology, Engineering, Management*, John Wiley & Sons, 2006.

3 Review of Engine Modeling

For obvious reasons, engine-control systems were among the first developed for vehicles: The engine is not only the most crucial component for automobile performance; its emission performance also significantly affects the environment. As discussed in Chapter 1, engine-control systems may include fuel-injection control (i.e., air–fuel ratio control), ignition or spark-timing control, antiknock-control systems, idle-speed control, EGR control, and transmission control. The goal of engine-control systems is to ensure that the engine operates at near-optimal conditions at all times in terms of drivability, fuel economy, and emissions.

Overall, engine-control systems are complex due to the nonlinearity of many of the components and the interactions among the several related control functions: air–fuel ratio control, idle-speed control, knock (or spark-timing) control, EGR control, and transmission control. In this chapter, each major phase of the operation of a spark-ignited gasoline engine and its dynamic modeling is discussed from the control perspective. Subsequent chapters consider specific engine-control problems (e.g., air–fuel ratio control, spark timing, EGR, and idle-speed control), as well as control problems associated with hybrid and fuel-cell vehicles.

3.1 Engine Operations

As shown in the conceptual block diagram in Figure 3.1, engine operations can be divided into several key phases. This discussion is specific to a four-stroke, spark ignition (SI), Otto gasoline engine (Figure 3.2). During each crankshaft revolution, there are two strokes of the piston and a total of four strokes, as follows (see Figure 3.2) (Heywood 1988; Stone 1994):

(a) *Induction Stroke.* The intake valve is opened and the piston travels down the cylinder and draws in a charge of air (or a charge of premixed fuel and air).

(b) *Compression Stroke.* Both valves are closed and the piston travels up the cylinder. As the piston approaches TDC, SI occurs.

(c) *Expansion (or Power) Stroke.* Combustion propagates throughout the charge, raising the pressure and temperature, thereby forcing the piston down. At the end of the power stroke, the exhaust valve opens and the irreversible expansion of the exhaust gases is termed *blow-down*.

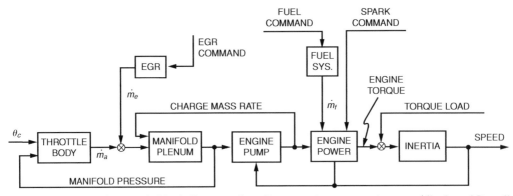

Figure 3.1. Schematic block diagram of nonlinear engine-system elements (Cook and Powell 1988).

(d) *Exhaust Stroke.* The exhaust valve remains open and as the piston travels up the cylinder. The remaining gases are expelled. At the end of the exhaust stroke, the exhaust valve closes. Some exhaust gases remain and dilute the next charge.

Because this cycle is completed only once every two crankshaft revolutions (i.e., 720 degrees), the valve (and fuel-injection) gear must be driven (usually by a camshaft) at half the engine speed. In a single-cylinder engine, power is produced only during the power stroke, which is only one quarter of the cycle. During other parts of the cycle, crankshaft rotation is maintained by power stored in a mechanical flywheel. In a multicylinder engine, the power strokes are staggered so that power is

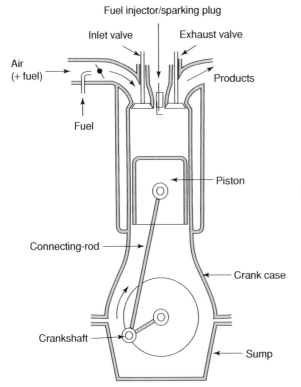

Figure 3.2. A four-stroke engine.

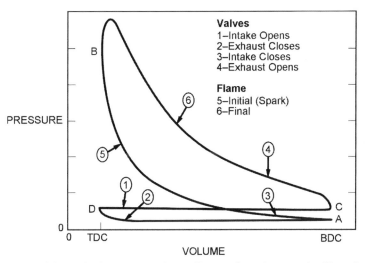

Figure 3.3. Typical pressure-volume diagram for a four-stroke SI engine.

produced during a larger fraction of the cycle than for a single-cylinder engine. For satisfactory SI and flame propagation, the air–fuel mixture must be stoichiometric (i.e., chemically balanced). This is important for emissions, as discussed in Chapter 5. Spark timing is important for performance, emissions, and prevention of engine knock (i.e., spontaneous self-ignition). Figure 3.3 is a typical pressure-volume diagram for a four-stroke engine. TDC is at B and D, and bottom dead center (BDC) is at A and C. The stroke from A to B is the compression stroke, B to C is the power stroke, C to D is the exhaust stroke, and D to A is the intake stroke. The valve openings and closures are marked 1 to 4, the spark occurs at 5, and the flame extinguishes at 6.

The following section is a brief description of each phase of engine operations, as shown in the block diagram in Figure 3.1.

Throttle Body

The mass airflow into the intake manifold is adjusted by the driver setting of the throttle input via an accelerator-pedal command. The throttle plate acts as an airflow control valve controlled by the accelerator pedal. The mass-airflow dependency on the throttle angle is nonlinear; however, for small changes about an equilibrium condition, the mass airflow rate is proportional to the change in throttle angle. The engine requires a relatively rich mixture and increased airflow to start when it is cold. This required enrichment may be provided by a choke valve, leading to different throttle dynamics when the flow is choked or unchoked.

Intake Manifold

Change in pressure in the intake manifold is proportional to the mass flow in (i.e., due to the throttle command) minus the mass flow out (i.e., from the engine pumping). Models are derived based on conservation of mass and the ideal gas law.

EGR

The EGR command adjusts the opening of an EGR valve, which directs a portion of the exhaust gas into the charge to the cylinder. This can affect the engine performance, emissions, and fuel consumption. The primary goal of EGR is to reduce NO_x emissions.

Fuel System

The fuel system controls the amount of fuel injected into the intake manifold (or cylinders). The main goal is to ensure that the ratio of the mass of air to the mass of fuel is regulated at the desired level. Under normal cruising conditions, the desired air–fuel ratio is stoichiometric (i.e., 14.7). For air–fuel ratio control, the fuel-flow rate is proportional to the airflow rate. Therefore, it is usually necessary to measure air mass flow rate. Various fuel-injection system designs exist. Typically, for best control, fuel is injected into each cylinder near each intake valve. However, fuel also can be injected by a single injector at the intake manifold.

Engine Pumping

The engine behaves like a pump to produce the airflow, EGR flow, and fuel flow out of the intake manifold and into the cylinders. An important result of this pumping mechanism is the induction-to-power (IP) delay (or lag), which can be treated as a pure delay corresponding to 360 degrees of crankshaft rotation. In other words, from the I/O perspective, the control action (e.g., change in throttle angle) takes effect after a time delay, which is engine-speed dependent.

Spark Command

The spark command ignites the air–fuel mixture in the cylinder to produce torque. The produced engine torque depends not only on the air–fuel ratio but also is affected by spark timing/advance. Spark advance is the time before TDC when a spark is initiated, and it is usually expressed in degrees of crankshaft rotation relative to TDC. The so-called MBT is used to maximize torque while maintaining a margin of safety to prevent knock (Figure 3.4).

Engine Power

Engine torque due to combustion leads to engine torque influenced by delayed pressure, delayed fuel, spark advance, and engine friction.

Engine Inertia

The powertrain rotational inertia and load torques (i.e., to drive the vehicle and accessories) must be overcome by engine torque (i.e., Newton's second law).

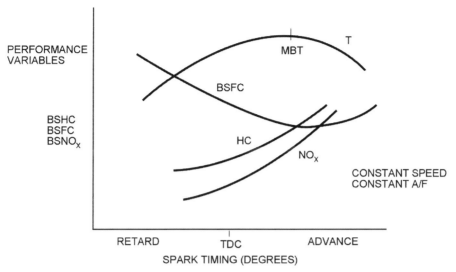

Figure 3.4. Spark-timing effect on engine performance (Ribbens 2003).

3.2 Engine Control Loops

In this section, each control loop associated with the engine operation is described.

Air–Fuel Ratio Control

As discussed previously, there are various performance metrics: emissions (i.e., CO, NO_x, and HC), fuel consumption (i.e., BSFC), and output power (i.e., torque). Most (but not all) are optimized, in conjunction with a catalytic converter, at or near stoichiometry (i.e., air–fuel ratio = 14.7). A key actuation mechanism for the air–fuel ratio function is the fuel injector(s). The throttle-angle setting from the driver determines the mass airflow rate, measured by a MAS, and the fuel flow is proportional to this rate. The MAP also is measured. The critical sensor for closed-loop air–fuel ratio control is the EGO sensor, which detects oxygen in the exhaust.

EGR

The percentage of exhaust gas in the charge is controlled by the EGR valve based on readings from the MAP sensor and engine temperature and speed. Higher percentages of EGR lower NO_x; however, other performance metrics (e.g., BSFC and HC) deteriorate with higher EGR. The EGR and air–fuel-ratio loops are highly coupled.

Spark Timing

The combustion in the cylinder is initiated by the spark-plug firing, typically a few degrees of crank angle before TDC. The so-called MBT is used to maximize torque while also maintaining a margin of safety to prevent engine knock. Advancing spark timing can increase torque and reduce fuel consumption. However, this is usually associated with increased emissions and the danger of engine knocks occurring. To achieve good spark control, the crankshaft angle must be measured or estimated accurately. Spark timing interacts with the idle-speed control and air–fuel-ratio control loops.

Idle-Speed Control

The goal is to measure and control engine speed at idle by adjusting airflow rate using the throttle or an idle-speed control valve (which provides better precision compared to the throttle). Maintaining consistent engine speed at idling despite load variations is important for perceived vehicle quality and to ensure low emissions and improved fuel economy. This control function requires measurement of the crankshaft angular position and engine speed.

Transmission Control

The main purpose of a transmission is to match the engine and vehicle speeds so that the engine can work in a more efficient region. Therefore, the gear selection for the transmission is said to be a mechanism for "engine control." Usually, a "shift map" with two independent variables is constructed, which then is used to determine up-shift and down-shift points. To implement the shift map, vehicle-speed and throttle-angle measurements are necessary. In addition to determining the gear position, it is important to ensure that shifting from one gear ratio to the next is executed smoothly. This entails precise coordination of the friction torques of various clutches.

There are several engine-operation modes that influence how each control loop operates, including the following:

(a) *Engine Crank (Start)*. The primary goal is reliable engine start-up; less emphasis is placed on fuel economy and emissions, and EGR is not used. Typically, the engine speed is low, the air–fuel ratio is low, and the spark is retarded.
(b) *Engine Warm-Up*. The primary goal is a rapid and smooth engine warm-up. Typically, the EGR is off, the air–fuel ratio is low, and fuel economy and emissions are not primary concerns.
(c) *Open-Loop Control*. The primary goal is to control the engine until the EGO sensor reaches the correct operating temperature and produces reliable output.
(d) *Closed-Loop Control*. The primary goal is tight control of performance, fuel economy, and emissions under closed-loop control using the EGO sensor.
(e) *Hard Acceleration*. The primary goal is high performance, with less emphasis on fuel economy and emissions. The air–fuel ratio is rich, EGR is off, and EGO is not in the loop.
(f) *Deceleration and Idling*. The primary goal is reduced fuel consumption and emissions. The air–fuel ratio is lean, engine speed is kept low and constant, and EGR is on.

In general, a dynamic model of the engine is a complex, nonlinear, dynamic system. There are various models suitable for simulations of a subset of the phenomena mentioned here. One example is discussed in the following subsection.

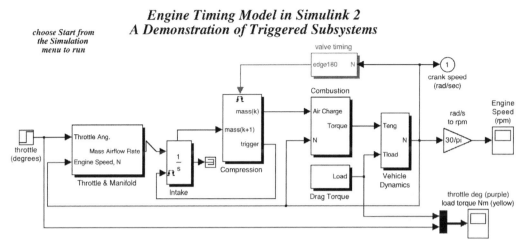

Figure 3.5. Engine example included in MATLAB/Simulink.

EXAMPLE 3.1: SIMULINK/MATLAB ENGINE MODEL. A nonlinear, three-state engine dynamics model (Crossley and Cook 1991) for a four-stroke SI engine is included as a demonstration module in MATLAB. To start the demonstration, type "sldemo_engine" at the MATLAB prompt:

$$\gg \text{sldemo\_engine}$$

To run the simulation, choose *Start* from the *Simulation* menu. Figure 3.5 shows the model in Simulink block-diagram form.

Click on the different blocks in the Simulink block diagram to see the underlying structure of each block. For example, the *throttle & manifold* block generates the mass airflow rate from throttle-angle, manifold-pressure, atmospheric-pressure, and engine-speed inputs. This block is composed of complex throttle and intake manifold blocks, as shown here:

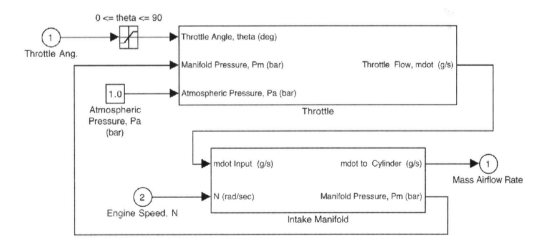

Air Intake Dynamics

The *throttle* block switches based on different cases (i.e., choked versus unchoked flow) to produce the airflow in response to the throttle-angle command. For lower manifold pressures, the flow through the throttle body is sonic and the outflow is a function of only the throttle angle, as follows:

mdot = g(Pm)*(2.821–0.05231*TA+0.10299*TA^2–0.00063*TA^3)

where:

TA: throttle angle (degrees)
Pm: manifold pressure (bar)
g(Pm) = 1 for Pm <= P(amb), otherwise
g(Pm) = (2/P(amb))*sqrt(Pm*P(amb)–Pm^2)
mdot: mass flow rate of air (g/s)

This is shown in the following Simulink block diagram:

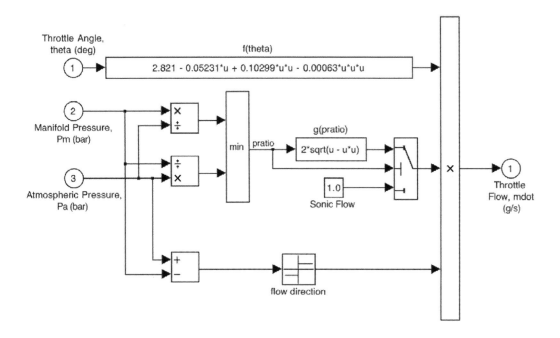

Throttle Flow vs. Valve Angle and Pressure

The *intake manifold* block includes manifold-pressure dynamics and calculates the manifold pressure and produces the mass airflow rate to the cylinder. The difference in the inflow and outflow mass rates is multiplied by a gain and integrated to obtain the manifold pressure based on the ideal gas law and homogeneous temperatures and pressures in the air–fuel mixture. It is assumed that there is no EGR, but it can be added easily. The mass flow rate of the air–fuel mixture pumped into the cylinders is given by an empirically derived equation as a function of the manifold pressure and engine speed. This is shown in the following block diagram:

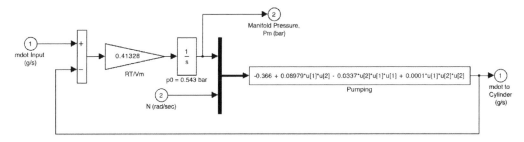

Intake Manifold Vacuum

where:

R: specific gas constant
T: temperature (K)
Vm: manifold volume (m^3)
gamma: ratio of specific heats (1.4)
mdot: mass flow rate of air (g/s)
Pdot_m: rate of change of manifold pressure, Pm
N: engine speed (rad/s)
Pm: manifold pressure (bar)

The *combustion* block generates the engine torque using an empirical curve fit to the input air charge, fuel flow, spark advance, and engine speed, as follows:

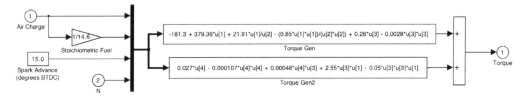

Engine Torque

The *vehicle dynamics* block is a simple rotational inertia (J) with load and engine-torque inputs, as follows:

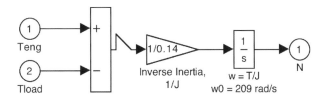

There are additional blocks for *intake*, *compression*, *valve timing*, and *drag torque* in this Simulink model. The intake-to-power-stroke delay is given by $Delay = \pi/N$, where N is the engine speed in rad/s. This assumes that the delay is one quarter of the total engine cycle (i.e., intake, compression, power and exhaust strokes) for a four-stroke engine.

The output of this simulation model is a plot of the engine speed (revolutions per minute [RPM]) versus time (seconds), for a simulation time of 10 seconds, and a plot of the throttle input (degrees) and load torques (N · m) versus time as shown here. The first plot shows the system inputs, the throttle angle (lower curve), and the load torque versus time in seconds:

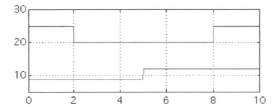

The second plot shows the system response – that is, the engine speed in RPM versus time in seconds.

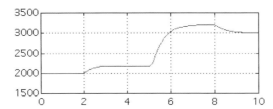

This Simulink engine model can be used as a standalone simulation model or combined as part of a larger powertrain simulation model. For example, this model could be used as part of an integrated vehicle and powertrain simulation for the development of a traction-control system.

3.3 Control-Oriented Engine Modeling

Many different models for engine control are reported in the literature (e.g., Cassidy et al. 1980; Cho and Hedrick 1989; Dobner 1983; Hendricks 1990; Kamei et al. 1987; and Powell 1979). In this section, the dynamic modeling of an engine (see Figure 3.1) is discussed for purposes of control-system design and evaluation. First, a simple linearized model of engine dynamics is derived and presented. Then, a more complex nonlinear engine model, suitable for engine diagnostics, is provided.

Linearized Engine Dynamics Model

First, we model the throttle following the block diagram shown in Figure 3.1. A linearized throttle model, which represents the mass airflow rate as proportional to change in throttle angle, is given by:

$$\dot{m}_a = K_\theta \Delta\theta$$

where θ is the throttle angle, $\dot{m}_a$ is the air mass inflow rate, K_θ is the linearized airflow rate sensitivity, and $\Delta\theta$ represents the change in throttle angle from the steady condition about which the system is linearized. In general, the symbol Δ is used to represent incremental variables that give changes about the steady values for which linearization was performed.

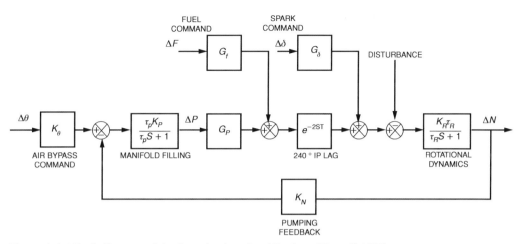

Figure 3.6. Block diagram of the linearized engine (Cook and Powell 1988).

The manifold-filling dynamics is based on the Ideal Gas Law and expresses the change in intake-manifold pressure due to mass airflow into the manifold minus mass airflow out of the manifold:

$$\dot{P} = \frac{RT}{V}(\dot{m}_a - \dot{M}) \tag{3.1}$$

where P is the manifold pressure, R is the ideal-gas constant, T is the air temperature, V is the manifold volume, $\dot{m}_a$ is the mass flow rate of air into the manifold (see Eq. (3.1)), and $\dot{M}$ is the mass flow of air out of the manifold due to engine pumping. After linearization, this equation becomes:

$$\Delta\dot{P} = K_P\left(\frac{\partial\dot{m}_a}{\partial P} - \frac{\partial\dot{M}}{\partial P}\right)\Delta P + K_P K_\theta \Delta\theta - K_P K_N \Delta N \tag{3.2}$$

where N is the engine speed, $K_P = \frac{\dot{P}}{\dot{m}_a - \dot{M}}$, and $K_N = \frac{\partial\dot{M}}{\partial N}$. Defining $\tau_p = \frac{-1}{K_P\left(\frac{\partial\dot{m}_a}{\partial P} - \frac{\partial\dot{M}}{\partial P}\right)}$ results in the transfer function shown in Figure 3.6. Next, define AM as an estimate of the mass flow rate, with $\dot{M}$ obtained from a speed density air-sensing system:

$$AM = cPN \tag{3.3}$$

where c is a proportionality constant. Linearizing Eq. (3.3) yields:

$$\Delta AM = c(P_o\Delta N + N_o\Delta P) \tag{3.4}$$

Figure 3.7 is an engine-induction map that enables calculation of the various coefficients in Eqs. (3.1), (3.2), and (3.4) at various operating conditions.

For control to a specific air–fuel ratio, the engine fuel-flow rate is proportional to the airflow rate. The amount of fuel injected in any one event is proportional to the airflow rate divided by engine speed (which is assumed proportional to the air charge). Two possible injection-timing strategies are sequential and bank to bank. Sequential injection-timing meters fuel individually to each cylinder during the appropriate portion of the engine cycle (e.g., immediately before the intake valve opens in each cylinder). Thus, each cylinder receives a fuel charge delayed by

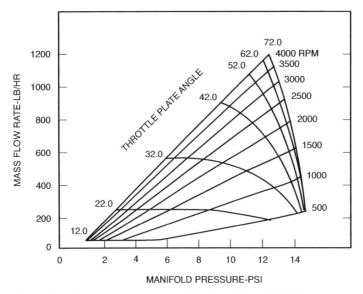

Figure 3.7. Engine induction map (Cook and Powell 1988).

the same amount from the time of injection. In bank-to-bank fuel injection (e.g., for a six-cylinder engine), the amount of fuel to be injected is calculated once per engine revolution based on the speed-density airflow. For simplicity and economy, the injectors are "slaved" in groups of three and fired alternately at 360-degree crank-angle increments delayed by 240 degrees from the fuel-metering calculation.

An estimate of the air, fuel, and EGR mass flow rates out of the manifold and into the cylinders can be obtained by treating the engine as a pump. A pure transport delay, which is referred to as the IP stroke lag, is shown in Figure 3.6. Mass-flow-rate samples for each cylinder eventually produce torque via the combustion process. This delay is 180 to 360 degrees of crankshaft rotation. The sampling in the system is engine-speed–based rather than time-based. Thus, the IP lag can be represented by a transport delay of two sampling periods, T. For a six-cylinder engine, the control sampling time $T = 720/6 = 120$ degrees.

Torque is produced from the combustion process and depends on several variables in the system. For the linear model, the following functional dependence is assumed for the engine brake or output torque:

$$T_e = F(M_d, F_d, \delta, N) \tag{3.5}$$

where M_d is the mass charge delayed by the IP lag, F_d is the fuel delayed by the IP lag, and δ is the ignition timing in degrees before TDC. The mass charge is a function of manifold pressure and engine speed; therefore, a linearized relationship can be given as:

$$\Delta T_e = G_p \Delta P_d + G_f \Delta F_d + G_\delta \Delta \delta + F_N \Delta N \tag{3.6}$$

where G_p is the influence of delayed pressure on torque, G_f is the influence of delayed fuel on torque, G_δ is the influence of spark advance on torque, and F_N is the engine friction. The first three terms define what is usually called the combustion torque, T_c.

Table 3.1. *Six-cylinder-engine model
parameters at N = 600 RPM (Cook and
Powell 1988)*

Parameter	Unit	Value
K_θ	(lb/hr)/deg	20.000
t_P	sec	0.210
K_P	lbf-h/(lbm-in^2-sec)	0.776
G_P	ft-lbf/psi	13.370
G_δ	ft-lbf/deg	10.000
T	sec	0.033
t_R	sec	3.980
K_R	rpm/(ft-lbf-sec)	67.200
K_N	lbm/(rpm-hr)	0.080
G_f	ft-lbf/lbm	36.600

A powertrain rotational dynamics model is needed to complete the model in Figure 3.2. Applying Newton's second law to the crankshaft rotation gives:

$$J_e \dot{N} = \frac{30}{\pi} T_e - \frac{30}{\pi} T_L \tag{3.7}$$

where J_e is the engine rotational inertia and T_L is the external (i.e., disturbance or load) torque that the engine must overcome. For a vehicle with an automatic transmission, T_L consists of the load applied by the torque converter, and it loads from the driven accessories (e.g., air-conditioning compressor). Linearizing Eq. (3.7) and substituting from Eq. (3.6) gives:

$$\Delta\dot{N} + \frac{30}{J_e\pi}\left(\frac{2N}{K_i^2} - F_N\right)\Delta N = \frac{30}{J_e\pi}(G_P\Delta P + G_f\Delta F_d + G_\delta\Delta\delta - \Delta T_d) \tag{3.8}$$

where the transfer function shown in Figure 3.6 is obtained by defining the time constant $\tau_R = 1/\frac{30}{J_e\pi}(\frac{2N}{K_i^2} - F_N)$, $K_R = \frac{30}{J_e\pi}$, and ΔT_d represents the incremental disturbance torque. The model parameters must be determined by testing, and typical values of the model parameters are listed in Table 3.1 for a six-cylinder engine at $N = 600$ RPM. Figure 3.8 shows results from validation studies for the linearized engine model given in Eqs. (3.1), (3.2), (3.4), and (3.8).

EXAMPLE 3.2: LINEARIZED ENGINE DYNAMICS MODEL. Consider the linearized engine model given in Eqs. (3.1)–(3.8) and express these equations in standard-state equation form:

$$\dot{\mathbf{x}} = \mathbf{A}\mathbf{x} + \mathbf{B_u}\mathbf{u} + \mathbf{b_v}v \tag{3.9}$$

$$\mathbf{y} = \mathbf{C}\mathbf{x} + \mathbf{D}\mathbf{u} \tag{3.10}$$

where the states of the model are the mass airflow, the change in manifold pressure, and the change in engine speed:

$$\mathbf{x} = \left\{\begin{array}{c} m_a \\ \Delta P \\ \Delta N \end{array}\right\} \quad \mathbf{u} = \left\{\begin{array}{c} \Delta\theta \\ \Delta\delta \\ \Delta F_d \end{array}\right\} \quad \text{and } v = \Delta T_d \tag{3.11}$$

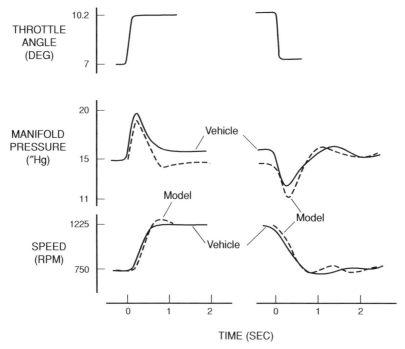

Figure 3.8. Sample transient response for validation (Cook and Powell 1988).

The inputs are the change in throttle angle, the change in spark timing, the change in delayed fuel, and the disturbance torque. The coefficient matrices for the state equations in Eq. (3.9) are:

$$\mathbf{A} = \begin{bmatrix} 0 & 0 & 0 \\ 0 & -1/\tau_P & -K_P K_N \\ 0 & K_R G_P & -1/\tau_R \end{bmatrix} \quad \mathbf{B_u} = \begin{bmatrix} K_\theta & 0 & 0 \\ K_P K_\theta & 0 & 0 \\ 0 & K_R G_\delta & K_R G_f \end{bmatrix} \quad \mathbf{b_v} = \left\{ \begin{array}{c} 0 \\ 0 \\ -K_R \end{array} \right\}$$

$$(3.12)$$

If we let $\mathbf{y} = \mathbf{x}$, then $\mathbf{C} = \mathbf{I}$ and $\mathbf{D} = \mathbf{0}$ in Eq. (3.10). These equations now can be used as the basis for various open-loop analyses and simulation studies. Parameter values needed for the linearized model of a six-cylinder engine at $N = 600$ RPM are listed in Table 3.1. The value of $K_\theta = 20$ (lb/hr)/deg can be estimated from Figure 3.7, and the value of $G_\delta = 10$ (ft-lbf/deg) can be estimated from other data. Recall that the incremental fuel-flow input ΔF_d is delayed by approximately two sampling periods, T, which also is given in Table 3.1.

Nonlinear Engine Dynamics Model

Whereas the linearized model in the previous section may be adequate for a basic understanding of the engine air-intake–fueling–combustion dynamics, achieving levels of control and diagnostic performance that are mandated by current and future U.S. Environmental Protection Agency (EPA) regulations requires the use of a considerably more precise model. The derivation of such a model is presented here,

Table 3.2. *List of symbols for engine model*

No.	Symbols	Description [Units]
1.	p_m	intake manifold pressure [Pa]
2.	p_a	ambient pressure [Pa]
3.	T_a	ambient temperature [K]
4.	$C_{d,th}$	throttle discharge coefficient
5.	$\dot{m}_{a,th}$	mass flow rate of air at throttle [Kg/sec]
6.	R	ideal gas constant [J/(kg · K)]
7.	$\dot{m}_{a,cyl}$	mass flow rate of air into cylinder [kg/sec]
8.	V_m	intake manifold volume [m^3]
9.	V_d	displacement volume [m^3]
10.	η_v	volumetric efficiency
11.	$\dot{m}_{ff}$	fuel flow rate from fuel puddle [kg/sec]
12.	$\dot{m}_{fi}$	fuel flow rate from injector [kg/sec]
13.	τ_f	fuel evaporation time constant [sec]
14.	X	fraction of injected fuel entering puddle
15.	$\dot{m}_{fc}$	mass flow rate of fuel entering the cylinder
16.	τ_m	fuel flow rate from film [kg/sec]
17.	t_c	cycle delay [sec]
18.	t_t	transportation delay [sec]
19.	ϕ_m	measured equivalence ratio
20.	θ_s	crank-angle sampling interval (radians)

based on (Krishnaswamy et al. 1996). This model is developed in the crank-angle domain and incorporates sufficient detail to capture the essential characteristics of powertrain-system behavior while also remaining simple enough to be implementable onboard a vehicle (e.g., for engine diagnostics).

Toward this end, a hybrid identification approach was taken for model development. That is, the basic equations describing the system dynamics were derived from consideration of physical principles, but the equations then were parameterized by constants (e.g., discharge coefficient for the throttle) that were identified using empirical identification techniques from experimental data. This approach enables the development of reasonably accurate models that still accommodate physical intuition, which allows inferences to be made about system operation through the monitoring of these physically based constants and system variables. Table 3.2 lists all of the variables used in the model.

Air-Intake Model

The air-intake model incorporates a nonlinear airflow model that models the flow past the throttle, taking into account the different dynamics under both choked and unchoked flow conditions, the manifold-filling dynamics, and the variation in engine volumetric efficiency as a function of the engine operating condition. The mass flow rate of air at throttle depends on the following flow conditions:

For unchoked flow:

$$\frac{p_m}{p_a} > \left[\frac{2}{\gamma + 1}\right]^{\gamma/(\gamma-1)} \tag{3.13}$$

$$\dot{m}_{a,th} = \frac{C_{D,th}A_{th}p_a}{\sqrt{RT_a}}\left(\frac{p_m}{p_a}\right)^{1/\gamma}\left\{\frac{2\gamma}{\gamma-1}\left[1-\left(\frac{p_m}{p_a}\right)^{(\gamma-1)/\gamma}\right]\right\}^{1/2} \qquad (3.14)$$

For choked flow:

$$\frac{p_m}{p_a} \le \left[\frac{2}{\gamma+1}\right]^{\gamma/(\gamma-1)} \qquad (3.15)$$

$$\dot{m}_{a,th} = \frac{C_{D,th}A_{th}p_a}{\sqrt{RT_a}}\gamma^{1/2}\left(\frac{2}{\gamma+1}\right)^{(\gamma+1)/2(\gamma-1)} \qquad (3.16)$$

Using the conservation of mass in the intake manifold, the manifold dynamics can be described as:

$$\frac{dm_{a,m}}{dt} = \dot{m}_{a,th} - \dot{m}_{a,cyl} \qquad (3.17)$$

Using the Ideal Gas Law, we can write:

$$\frac{dp_m}{dt} + \frac{\eta_v V_d \omega}{4\pi V_m}p_m = \dot{m}_{a,th}\frac{RT_a}{V_m} \qquad (3.18)$$

In the crank-angle domain:

$$\omega\frac{dp_m}{d\theta} + \frac{\eta_v V_d \omega}{4\pi V_m}p_m = \dot{m}_{a,th}\frac{RT_a}{V_m} \qquad (3.19)$$

The actual mass flow rate of air into a cylinder can be computed using the volumetric efficiency (i.e., speed-density equation):

$$\dot{m}_{a,cyl} = \frac{\eta_v P_m V_d \omega}{4\pi RT_a} \qquad (3.20)$$

The volumetric efficiency was modeled as a polynomial function of engine speed, manifold pressure, and throttle opening:

$$\eta_v = a_0 + a_1\omega + a_2\omega^2 + a_3 P_m + a_4\alpha + a_5\alpha^2 + a_6\alpha^3 \qquad (3.21)$$

where the coefficients a_i's usually are determined empirically.

Fuel Dynamics (Wall Wetting Model)

The physics of the process by which fuel that is sprayed into the cylinder by the injectors vaporizes and participates in combustion is complex. In simplified form, it can be considered that part of the fuel vaporizes quickly enough (or is sufficiently finely atomized) to participate directly in combustion, whereas the rest of the fuel spray impinges on the cylinder wall, where it combines with the fuel "puddle" from earlier injections and then evaporates from the puddle to form part of the air–fuel mixture. Thus, the dynamics of the fueling system can be modeled as follows:

$$\frac{d\dot{m}_{ff}}{dt} = -\frac{1}{\tau_f}\dot{m}_{ff} + X\dot{m}_{fi}$$

$$\dot{m}_{fc} = \frac{1}{\tau_f}\dot{m}_{ff} + (1-X)\dot{m}_{fi} \qquad (3.22)$$

Table 3.3. *Empirically determined engine parameters*

No.	Parameter	Description
1.	$C_{d,th}$	Throttle-discharge coefficient
2.	η_v	Volumetric efficiency
3.	τ_f	Fuel-evaporation constant
4.	X	Direct entry fraction of fuel
5.	$\theta_c + \theta_t$	Transport + EGO sensor delay
6.	τ_m	EGO sensor time constant

where the variables are as defined in Table 3.2. Again, in the crank-angle domain, the model becomes:

$$\frac{d\dot{m}_{ff}}{d\theta} = -\frac{1}{\tau_f \omega}\dot{m}_{ff} + \frac{X}{\omega}\dot{m}_{fi}$$

$$\dot{m}_{fc} = \frac{1}{\tau_f}\dot{m}_{ff} + (1-X)\dot{m}_{fi}$$

(3.23)

Air–Fuel-Ratio Dynamics

Using the mass flow rate of air and fuel from Eqs. (3.20) and (3.23), the air–fuel ratio in a cylinder can be expressed as:

$$\frac{\dot{m}_{fc}}{\dot{m}_{ac}} = \frac{1}{\tau_f}\frac{\dot{m}_{ff}}{\dot{m}_{ac}} + (1-X)\frac{\dot{m}_{fi}}{\dot{m}_{ac}}$$

(3.24)

Because $AF = \dot{m}_{ac}/\dot{m}_{fc}$,

$$\frac{1}{AF} = \frac{1}{\tau_f}\frac{\dot{m}_{ff}}{\dot{m}_{ac}} + (1-X)\frac{\dot{m}_{fi}}{\dot{m}_{ac}}$$

(3.25)

which introduces the equivalence ratio $\phi = 1/AF$,

$$\phi = \frac{1}{\tau_f}\frac{\dot{m}_{ff}}{\dot{m}_{ac}} + (1-X)\frac{\dot{m}_{fi}}{\dot{m}_{ac}}$$

(3.26)

Exhaust Transport Delay and Sensor Dynamics

The delay of the exhaust gas is the sum of cycle delay and transport delay. Therefore, sensor dynamics combined with delay can be expressed as:

$$\tau_m \frac{d\phi_m}{dt} + \phi_m = \phi(t - t_c - t_t)$$

(3.27)

In the crank-angle domain:

$$\tau_m \omega \frac{d\phi_m}{d\theta} + \phi_m = \phi(\theta - \theta_c - \theta_t)$$

(3.28)

Table 3.3 lists the parameters in these equations that were determined empirically. Determination of the parameters is not a trivial task and involves the resolution of many issues, including the performance of both dynamic and static engine tests – with suitably designed inputs that excite the relevant dynamics of the engine – and the means of analysis of the data (i.e., both linear and nonlinear least squares techniques). However, a discussion of the identification procedures is beyond the scope of this chapter.

PROBLEMS

1. Run in Simulink the open-loop-engine demonstration simulation described in this chapter. Indicate the operating conditions selected for your simulation and present the output plots from it.

2. *Open-loop engine linear dynamics simulation.* Use the linearized engine model given in Eqs. (3.1)–(3.8) and the parameter values given in Table 3.1 and Example 3.1 to simulate the dynamics of the engine to the following inputs:

> (a) Unit step disturbance torque (qualitatively compare to top trace in Figure 8.4 and to the results in Example 3.1).
> (b) Separate unit step inputs for $\Delta\theta$, $\Delta\delta$, and ΔF_d (qualitatively compare to the results in Example 3.1 for the input $\Delta\theta$).

Plot the responses **y** versus time.

3. Repeat the simulations in Problem 2; however, include the IP delay. That is, in your simulation, the inputs with subscript d should be delayed by the time $2T$, and $T = 0.033$ seconds, as given in Table 3.1.

4. You are an engineer at an automotive company, which has determined the parameters for the open-loop engine model. The state-space model for the engine is given by:

$$\mathbf{x} = \mathbf{A}\mathbf{x} + \mathbf{B_u}\mathbf{u} + \mathbf{b_v}v$$

$$\mathbf{y} = \mathbf{C}\mathbf{x} + \mathbf{D}\mathbf{u}$$

where the states of the model are the mass airflow, the change in the manifold pressure, and the change in engine speed; the inputs to the open-loop model are throttle angle, spark-advance command, and fuel injected to the system. That is,

$$\mathbf{x} = \left\{ \begin{array}{c} m_a \\ \Delta P \\ \Delta N \end{array} \right\}, \quad \mathbf{u} = \left\{ \begin{array}{c} \Delta\theta \\ \Delta\delta \\ \Delta F_d \end{array} \right\}$$

v is the torque disturbance and $v = \Delta T_d$. The coefficient matrices for the state equation are:

$$\mathbf{A} = \begin{bmatrix} 0 & 0 & 0 \\ 0 & 1/\tau_P & -K_P K_N \\ 0 & K_R G_P & -1/\tau_R \end{bmatrix} \quad \mathbf{B_u} = \begin{bmatrix} K_\theta & 0 & 0 \\ K_P K_\theta & 0 & 0 \\ 0 & K_R G_\delta & K_R G_f \end{bmatrix} \quad \mathbf{b_v} = \begin{bmatrix} 0 \\ 0 \\ -K_R \end{bmatrix}$$

Your supervisor wants you to provide a quick estimate for the step response to the throttle angle for the closed-loop engine model. From experience, you know that the fuel injected must be kept close to stoichiometry – thus, $\Delta F_d = K_{AF}\dot{m}_a$ – and that as a general rule, spark timing must be advanced as engine speed increases. Therefore, $\Delta\delta = K_{SA}\Delta N$. You may neglect further the effects of torque disturbance assuming smooth driving conditions.

Please answer the following:

(a) Find the simplified state-space model with $\Delta\theta$ as the only input.
(b) Find the transfer functions from $\Delta\theta$ to each of the three states.
(c) Find the steady-state value of ΔN and m_a for a unit step in $\Delta\theta$. Will both of them be stable? Explain.

5. An engineer was transferred to the engine-test department and he needs to model the breathing dynamics of a four-stroke, six-cylinder SI engine. The notebook of the previous engineer contains the manifold-filling equation:

$$\frac{d}{dt}P_m = 0.5(\dot{m}_\theta - \dot{m}_{cyl})$$

where $(P_m,$ bar) is the intake manifold pressure, $(\dot{m}_\theta,$ g/s) is the mass airflow rate into the manifold through the throttle body, and $(\dot{m}_{cyl},$ g/s) is the pumping mass airflow rate into the cylinders. In the next page of the notebook, the new engineer finds this equation:

$$\dot{m}_\theta = k_\theta\theta + k_1 P_m$$

where k_q is either $(10/3)$ or -3. Please answer the following questions:

(a) Which value (i.e., $10/3$ or -3) would you choose for k_q? (*Hint:* Recall that the flow rate is proportional to the pressure difference and that the intake-manifold pressure typically is less than the pressure upstream in the throttle body.)

The next page in the notebook contains the following experiments:

Experiment 1: $\theta = 30°, P_m = 0.0$ bar, $\dot{m}_\theta = 100$ g/s
Experiment 2: $\theta = 30°, P_m = 0.5$ bar, $\dot{m}_\theta = 45$ g/s

(b) Can you determine the constant k_1?

Finally, the linear function that defines the engine-pumping rate is found to be:

$$\dot{m}_{cyl} = \frac{N}{500} + k_2 P_m$$

where N in RPM is the engine speed. Two sets of experiments were conducted to identify k_2. The notebook indicates manifold-pressure measurements of 0.5 and 0.9 bar and pumping rates of 50 and 10 g/s without pairing the pressure and pumping rate of each experiment.

(c) Can you guess the pairs of pressure and cylinder flow?
(d) Based on your answer to (c), find the value of the constant k_2.

REFERENCES

Ashley, S., 2001, "A Low-Pollution Engine Solution," *Scientific American*, June 2001, pp. 91–5.

Bidan, P., S. Boverie, and V. Chaumerliac, 1995, "Nonlinear Control of a Spark-Ignition Engine," *IEEE Transactions on Control Systems Technology*, Vol. 3, No.1, March 1995.

Bosch, R., 2008, *Automotive Handbook*, Wiley & Sons.

Cassidy, J., M. Athans, and W. H. Lee, 1980, "On the Design of Electronic Automotive Engine Controls Using Linear Quadratic Control Theory," *IEEE Transactions on Automatic Control*, Vol. 25, No. 5, pp. 901–12.

Cho, D., 1991, "Research and Development Needs for Engine and Powertrain Control Systems," in S. A. Velinsky, R. H. Fries, I. Haque, and D. Wang (eds.), 1991, *Advanced Automotive Technologies-1991*, ASME DE–Vol. 40, New York, pp. 23–34.

Cho, D., and J. K. Hedrick, 1989, "Automotive Powertrain Modeling for Control," *ASME Journal of Dynamic Systems, Measurement and Control*, Vol. 111, December 1989, pp. 568–76.

Connoly, F., and G. Rizzoni, 1994, "Real-ime Estimation of Engine Torque for the Detection of Engine Misfires," *ASME Journal of Dynamic Systems, Measurement and Control*, pp. 675–86.

Cook, J. A., and B. K. Powell, 1988, "Modeling of an Internal Combustion Engine for Control Analysis," *IEEE Control Systems Magazine*, August 1988, pp. 20–6.

Cook, J. A., J. Sun, J. H. Buckland, I. V. Kolmanovsky, H. Peng, and J. W. Grizzle, 2006, "Automotive Powertrain Control: A Survey," *Asian Journal of Control*, Vol. 8, No. 3, September 2006, pp. 237–60.

Crossley, P. R., and J. A. Cook, 1991, *Proceedings of the IEEE International Control 91 Conference*, Vol. 2, pp. 921–25, March 1991, Edinburgh, Scotland, UK.

Crouse, W. H., and D. Anglin, 1977, *Automotive Emission Control*, McGraw-Hill.

Crouse, W. H., and D. L. Anglin, 1986, *Automotive Engines*, seventh edition, McGraw-Hill.

Dobner, D. J., 1983, "Dynamic Engine Models for Control Development – Part I: Non-Linear and Linear Model Formulation," *Application of Control Theory in the Automotive Industry*, Inderscience Publishers.

Green, J. H., and J. K. Hedrick, 1990, "Nonlinear Speed Control for Automotive Engines," *Proceedings of the American Control Conference*, San Diego, CA, May 1990, pp. 2891–7.

Guzzella, L., and A. Sciarretta, *Vehicle Propulsion Systems: Introduction to Modeling and Optimization*, Springer Verlag, 2007 (second edition).

Hebbale, K. V., and Y. A. Ghoneim, 1991, "A Speed and Acceleration Estimation Algorithm for Powertrain Control," *Proceedings of the American Control Conference*, Boston, MA, June 1991, pp. 415–20.

Hendricks, E., 1990, "Mean Value SI Engine Model for Control Studies," *Proceedings of the American Control Conference*, San Diego, CA, May 1990, pp. 1882–6.

Heywood, J., 1988, *Internal Combustion Engine Fundamentals*, McGraw-Hill.

Hrovat, D., and W. F. Powers, 1988, "Computer Control Systems for Automotive Power Trains," *IEEE Control Systems Magazine*, August 1988, pp. 3–10.

Jankovic, M., and I. Kolmanovsky, 2009, "Developments in Control of Time-Delay Systems for Automotive Powertrain Applications," in B. Balachandran et al. (eds.), *Delay Differential Equations: Recent Advances and New Directions*, Springer.

Kamei, E., H. Namba, K. Osaki, and M. Ohba, 1987, "Application of Reduced Order Model to Automotive Engine Control System," *ASME Journal of Dynamic Systems, Measurement and Control*, Vol. 109, September 1987, pp. 232–7.

Kiencke, U., and L. Nielsen, 2005, *Automotive Control Systems for Engine, Driveline and Vehicle*, Springer.

Knowles, D., 1989, *Automotive Emission Control & Computer Systems*, second edition, Prentice-Hall.

Krishnaswamy, V., C. Siviero, F. Carbognani, G. Rizzoni, and V. Utkin, 1996, "Application of Sliding Mode Observers to Automobile Powertrain Diagnostics," *Proceedings of the IEEE Conference on Control Applications*, Dearborn, MI, CCA 96, pp. 355–60.

Melgaard, H., E. Hendricks, and H. Madsen, 1990, "Continuous Identification of a Four-Stroke SI Engine," *Proceedings of the American Control Conference*, San Diego, CA, May 1990, pp. 1876–81.

Moskwa, J. J., and J. K. Hedrick, 1990, "Nonlinear Algorithms for Automotive Engine Control," *IEEE Control Systems Magazine*, April 1990, pp. 88–93.

Nesbit, C., and J. K. Hedrick, 1991, "Adaptive Engine Control," *Proceedings of the American Control Conference*, Boston, MA, June 1991, pp. 2072–6.

Powell, B. K., 1979, "A Dynamic Model for Automotive Engine Control Analysis," *Proceedings of the IEEE Conference on Decision and Control*, pp. 120–6.

Ribbens, W. B., 2003, *Understanding Automotive Electronics* (6th edition), Butterworth-Heinemann.

Shiao, Y., and J. J. Moskwa, "Cylinder Pressure and Combustion Heat Release Estimation for SI Engine Diagnostics Using Nonlinear Sliding Observers," *IEEE Transactions on Control Systems Technology*, Vol. 3, No. 1, March 1995.

Shiga, H., and S. Mizutani, 1988, *Car Electronics*, Nippondenso Co.

Stone, Richard, 1994, *Introduction to Internal Combustion Engines*, SAE International, second edition, 1994.

Sweet, L. M., 1981, "Control Systems for Automotive Vehicle Fuel Economy: A Literature Review," *ASME Journal of Dynamic Systems, Measurement, and Control*, Vol. 103, September 1981, pp. 173–80.

Tabe, T., M. Ohba, E. Kamei, and H. Namba, 1987, "On the Application of Modern Control Theory to Automotive Engine Control," *IEEE Transactions on Industrial Electronics*, Vol. 34, No. 1, February 1987, pp. 35–9.

Washino, S., 1989, *Automobile Electronics*, Gordon and Breach Science Publishers, New York.

4 Review of Vehicle Dynamics

Design of control systems for ground vehicles must start from an adequate understanding of their dynamic behavior. Although a detailed discussion of vehicle dynamics is beyond the scope of this chapter, simple dynamic models suitable for controller design are necessary for control studies and are developed and presented herein. These simple models are used in subsequent chapters as the basis for controller designs (e.g., cruise control, antilock brakes, traction control, steering control, and active suspensions). More complex (i.e., nonlinear, high-order, and fully coupled) models for vehicle dynamics often are needed to evaluate, using simulation studies, the controllers that are designed using simple control-design models. Such complex models are discussed in detail in the literature (Ellis 1966; Gillespie 1992; Segel 1990; Venhovens 1993; Wong 2008) and also can be implemented in commercial dynamic simulation software (e.g., ADAMS and CARSIM).

First, the standard notation and terminology for vehicle dynamics is introduced with definitions of reference frames and coordinates used to describe vehicle motion. Next, the longitudinal motion of the vehicle, including braking and acceleration, is presented. Then, lateral-motion dynamics, or vehicle steering or handling, is described. Finally, the vertical motion of vehicles is discussed.

4.1 Coordinates and Notation for Vehicle Dynamics

Lumped parameter models concentrate the distributed mechanical properties of mass (kg), stiffness (N/m), and damping (Ns/m) at imagined physical locations. In particular, many elementary analyses of vehicles are based on treating the vehicle as one or more concentrated (lumped) masses located at the center of gravity (CG) of their respective rigid bodies. The Society of Automotive Engineers (SAE) has introduced standard coordinates and notation for describing vehicle dynamics that are widely used (SAE 1976).[1]

---

[1] The ISO coordinates are also frequently used and are the same as the SAE coordinates in Figure 4.1 except that the vehicle-fixed axis z points up rather than down; therefore, following the right-hand rule, the y axis points to the left of the driver rather than to the right. Because this difference can cause confusion when reading the literature on vehicle dynamics and control, readers should be sure to establish which coordinate system is being used.

Table 4.1. *SAE body fixed vehicle axis system: symbols and definitions*

Axis	Translational velocity	Angular displacement	Angular velocity	Force component	Moment component
x	u (forward)	ϕ	p or $\dot{\phi}$ (roll)	F_x	M_x
y	v (lateral)	θ	q or $\dot{\theta}$ (pitch)	F_y	M_y
z	w (vertical)	ψ	r or $\dot{\psi}$ (yaw)	F_z	M_z

Figure 4.1 is the vehicle-fixed x, y, z coordinate system for describing the motion of a vehicle treated as a single lumped mass concentrated at the CG of the vehicle. This body-fixed coordinate system moves with the vehicle, which is assumed to be rigid. This coordinate system – a right-hand rule, Cartesian coordinate system – is summarized in Table 4.1. Vehicle motion typically is described in terms of the velocities (i.e., forward, lateral, vertical, roll, pitch, and yaw) in the vehicle-fixed coordinate system as referenced to an earth-fixed (i.e., inertial) reference frame.

Figure 4.2 shows an earth-fixed, Cartesian reference frame with coordinates X, Y, and Z (i.e., vertical travel, positive downward) and defines the heading angle (i.e., between x and X in ground plane) ψ; the course angle (i.e., between vehicle-velocity vector and X axis), v; and the sideslip angle (i.e., between x axis and the vehicle-forward-velocity vector, $V = u\,e_x + v\,e_y$), β. Note that the course angle $v\psi + \beta$. The sideslip angle is a result of the compliance of the pneumatic tire, and it has a significant effect on vehicle dynamics.

Figure 4.3 illustrates the tire coordinate system, with coordinates X', Y', and Z'. The forces (F_x, F_y, F_z) and moments (M_x, M_y, M_z) associated with the individual tire are defined in the axis directions. The wheel torque is usually denoted as T and wheel-spin velocity as ω. Therefore, the steer angle, δ, is between the direction of the wheel heading (X') and the vehicle heading.

Equations of motion typically are obtained by the application of Newton's second law, which for the translational motion of a rigid vehicle of mass m is:

$$\mathbf{F} = m\mathbf{a} \tag{4.1}$$

and for rotational motion (about the center of mass, a fixed point, or the instant center) is:

$$\mathbf{M} = \dot{\mathbf{H}} \tag{4.2}$$

Figure 4.1. Vehicle-fixed coordinate system.

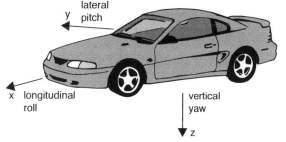

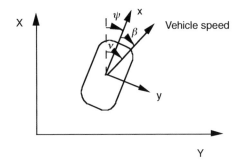

Figure 4.2. Earth-fixed and vehicle-fixed coordinate systems.

where F is the vector sum of the external forces acting on the vehicle mass m, a is the acceleration vector, M is the vector sum of the external moments acting on the vehicle, and $\dot{H}$ is the time rate of change of the moment of momentum of the rigid body about the mass center or a fixed point in an inertial reference frame. When a or $\dot{H}$ is zero, Eqs. (4.1) and (4.2) represent static equilibrium. Several examples of the application of Newton's second law to vehicle static and dynamic calculations are provided in the following sections. Instead of using Newton's laws, the equations of motion also can be derived using Lagrange's method based on energy concepts.

A two-step procedure is used when applying Newton's laws to derive vehicle-dynamic equations. First, *kinematics* (i.e., the geometry of motion) is used to derive the acceleration term represented by the vector a or $\dot{H}$ using the coordinates that are defined in Table 4.1 and Figures 4.1 through 4.3. Next, the force or moment terms, represented by the vectors F or M, must be obtained. Depending on the

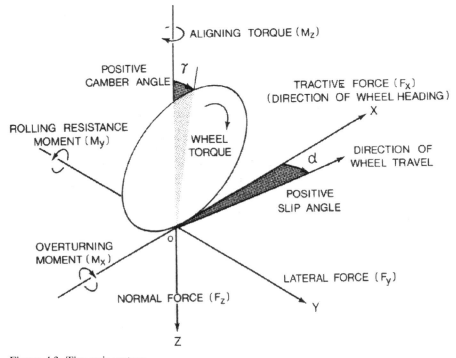

Figure 4.3. Tire-axis system.

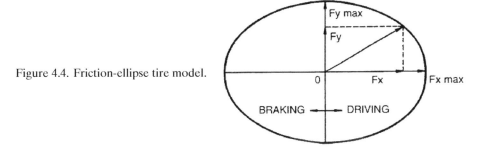

Figure 4.4. Friction-ellipse tire model.

particular situation being modeled, they include the vehicle weight (i.e., due to gravity), dissipative forces (e.g., aerodynamic drag and rolling resistance), and – most important, – the forces and moments that arise due to road–tire contact (e.g., braking and tractive and handling forces). The pneumatic tires have an important role in the dynamic behavior of a vehicle, and a brief background on this topic is provided here (Clark 1971; Wong 1993). Referring to the tire-axis system in Figure 4.3, the longitudinal and lateral tire forces are obtained by integrating the longitudinal and lateral shear stresses over the tire-contact patch.

The forces created at the tire–road interface are friction limited and depend on a tire–road friction coefficient, μ. The tire–road friction coefficient can vary significantly with road conditions (e.g., paved, gravel, wet, or icy). Typically, we are interested in the tire longitudinal, F_x, and lateral, F_y, forces and the self-aligning torque, M_z. When local shear forces are below the friction limit and the tire elements adhere to the road surface, the tire forces and moments, based on a quasistatic assumption, can be represented as follows:

$$\begin{Bmatrix} F_x \\ F_y \\ M_z \end{Bmatrix} = f(r_w, \gamma, r, \lambda, \alpha)$$

where λ is the longitudinal slip, $\tan\alpha = (v/u)$ is the lateral slip in terms of tire lateral and longitudinal velocities v and u, γ is the camber angle, and r_w is the wheel radius. When the excitation is large enough to cause loss of tire–road adhesion (i.e., the no-sliding assumption is violated), then:

$$\begin{Bmatrix} F_x \\ F_y \\ M_z \end{Bmatrix} = f(r_w, \gamma, r, \lambda, \alpha, \mu)$$

Tire forces and moments are highly nonlinear and difficult to model. Simplified models such as the Brush or Pacejka "Magic Formula" models (Bakker et al. 1987, 1989) were developed for use in simulation studies (Venhovens 1993).

A useful concept for combined lateral and longitudinal slip, as shown in Figure 4.4, is the friction-ellipse tire model. The basic idea for combined lateral and longitudinal tire forces is that when the resultant shear force exceeds the local friction limit, sliding occurs. This model is more useful conceptually than computationally.

The next section considers the longitudinal motion of the vehicle, which is required to understand the dynamic behavior in terms of braking and acceleration.

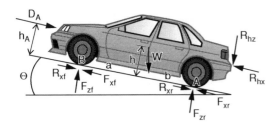

Figure 4.5. Forces acting on a vehicle.

4.2 Longitudinal Vehicle Motion

Figure 4.5 is a free-body diagram of a vehicle with mass m on an incline with angle Θ. The diagram includes the significant forces acting on the vehicle: g is the gravitational constant; D_A is the aerodynamic force; R_h is a drawbar force; $W = mg$ is the weight of the vehicle; F_x is the tractive force; R_x is the rolling-resistance force; and ma_x, an equivalent inertial force, acts at the CG. The subscripts f and r refer to the front (at B) and rear (at A) tire-reaction forces, respectively.

Application of Newton's second law for the z direction (i.e., no vertical acceleration) gives:

$$0 = W \cos \Theta - F_{zf} - F_{zr} + R_{hz} \tag{4.3}$$

and in the x direction:

$$ma_x = (W/g)a_x = m \frac{du}{dt} = F_{xr} + F_{xf} - W \sin \Theta - R_{xr} - R_{xf} - D_A + R_{hx} \tag{4.4}$$

The aerodynamic-drag force depends on the relative velocity between the vehicle and the surrounding air and is given by the semi-empirical relationship:

$$D_A = 0.5\rho \, C_d \, A(u + u_w)^2 \tag{4.5}$$

where ρ is the air density ($= 1.202$ kg/m^3 at an altitude of 200 m), C_d is the drag coefficient, A is the maximum vehicle cross-sectional area (≈ 0.9)(track) (height) for passenger vehicles), u is the vehicle-forward velocity, and u_w is the wind velocity (i.e., positive for a headwind and negative for a tailwind). The drag coefficient for vehicles ranges from about 0.2 (i.e., streamlined passenger vehicles with underbody cover) to 1.5 (i.e., trucks); 0.4 is a typical value for passenger cars (Bosch 2009; Gillespie 1992).

The rolling resistance arises due to the work of deformation on the tire and the road surface, and it is roughly proportional to the normal force on the tire:

$$R_x = R_{xf} + R_{xr} = f(F_{zf} + F_{zr}) \tag{4.6}$$

where f is the rolling-resistance coefficient in the range of about 0.01 to 0.4, with 0.015 as a typical value for passenger vehicles.

EXAMPLE 4.1: DETERMINE AERODYNAMIC DRAG AND ROLLING RESISTANCE. To determine the aerodynamic-drag and the rolling-resistance coefficients, measurements from two coast-down tests can be used: one at a high speed and the other at a lower speed.

Table 4.2. *Measurements of two coast-down tests*

	High-speed test	Low-speed test
Initial speed	V_{i1}	V_{i2}
Final speed	V_{f1}	V_{f2}
Time duration	t_1	t_2
Average speed	$V_1 = \dfrac{V_{i1} + V_{f1}}{2}$	$V_2 = \dfrac{V_{i2} + V_{f2}}{2}$
Average deceleration	$a_1 = \dfrac{V_{i1} - V_{f1}}{t_1}$	$a_2 = \dfrac{V_{i2} - V_{f2}}{t_2}$

Using the example results in Table 4.2, we obtain:

$$D_{A1} + R_x = 0.5\rho C_d A V_1^2 + fmg = ma_1$$
$$D_{A2} + R_x = 0.5\rho C_d A V_2^2 + fmg = ma_2$$

Coefficients of aerodynamic drag and rolling resistance then can be obtained:

$$C_d = \frac{m(a_1 - a_2)}{0.5\rho A \left(V_1^2 - V_2^2\right)} \qquad f = \frac{a_1 V_2^2 - a_2 V_1^2}{g\left(V_2^2 - V_1^2\right)}$$

EXAMPLE 4.2: AXLE LOADS. As a first step in the analysis of braking and acceleration performance, it is necessary to determine axle loads (see the free-body diagram in Figure 4.5). The loads at each axle consist of a static component plus load transferred from front to rear (or vice versa) due to the forces acting on the vehicle. The load on the front axle is found by summing moments about Point A and on the rear axle by summing moments about Point B. Assuming a constant value of a_x (i.e., constant acceleration or deceleration in the longitudinal direction), $R_h = 0$, and the vehicle has constant pitch, then these moments must sum to zero:

$$F_{zf} L + D_A h_A + (W/g)a_x h + W\, h \sin \Theta - W\, b \cos \Theta = 0 \qquad (4.7)$$
$$F_{zr} L - D_A h_A - (W/g)a_x h - W\, h \sin \Theta - W\, a \cos \Theta = 0 \qquad (4.8)$$

Special cases can be obtained; for example, on level ground, the cosine term is one and the sine term is zero. Grades usually are given as slopes in percentage and correspond to tan Θ. Grades on interstate highways typically are restricted to less that 4 percent and rarely exceed 10 to 12 percent on primary and secondary roads. Thus, in most cases, the approximations $\sin \Theta \approx \Theta$ and $\cos \Theta \approx 1$ can be used. The aerodynamic-drag force is proportional to the square of the vehicle-forward velocity and typically is negligible for low speeds. For forward acceleration, as shown in Figure 4.6, the load is transferred from the front axle to the rear axle in proportion to the normalized acceleration (a_x/g) and the normalized CG height (h/L).

Next, we consider the forces F_{xf} and F_{xr} during braking. Braking forces – as long as all of the wheels are rolling – can be described by the equation:

$$F_b = (T_b + I_w \alpha_w)/r_w \qquad (4.9)$$

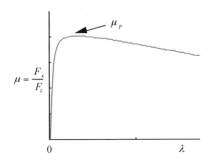

Figure 4.6. Braking coefficient versus slip.

where F_b is the brake force at a particular wheel, T_b is the brake torque produced at that wheel by application of the brake, I_w is the total rotational inertia of the wheel and drive components, α_w is the rotational deceleration of the wheel, and r_w is the wheel radius. Except when the wheel is locked up and slipping, $\alpha_w = A_x/r_w$, which can be used when Eq. (4.9) is combined with Eq. (4.4). The brake force in Eq. (4.9) is limited by the frictional coupling between the tire and the road, and it depends on the slip – caused by deformation – between the tire and the road surface. The slip ratio λ is defined as the ratio of longitudinal-slip velocity in the contact patch (i.e., tire-circumferential speed–vehicle-forward speed) to vehicle-forward speed:

$$\lambda = \frac{r_w\omega - u}{u} \quad \text{during braking} \tag{4.10}$$

$$= \frac{r_w\omega - u}{r_w\omega} \quad \text{during acceleration} \tag{4.11}$$

where ω is the tire-rotational speed. However, it is common practice to ignore the sign of λ when presenting tire-force characteristics. As shown in Figure 4.6, except at very low tire–road friction, the braking force reaches a peak value at $\lambda \approx 0.15$, or 15 percent. The braking coefficient μ is defined as the ratio between braking force and tire normal force, and it represents a normalized measure of braking forces. The peak value of $\mu = \mu_p$ is a key property because it determines the maximum braking force for a particular tire–road combination. At higher values of slip, the coefficient diminishes continuously.

EXAMPLE 4.3: VEHICLE LONGITUDINAL MOTION. Equation (4.4), a simple model of the longitudinal motion of a vehicle, can be used to determine changes in the vehicle-forward motion due to grades, braking, acceleration, and so forth. With $R_{hx} = 0$, $F_{xr} = 0$ (FWD), and $a_x = du/dt$ and using Eqs. (4.5) and (4.6) in Eq. (4.4), we obtain:

$$m(du/dt) = F_{xf} - W \sin \Theta - fW \cos \Theta - 0.5\rho C_d A (u + u_w)^2 \tag{4.12}$$

Equation (4.12) can be used in a variety of simple analyses. Assume that in a braking maneuver, all of the forces on the right-hand side are constant and equal to $-F_x$; the equation then becomes:

$$m \, du = -F_x \, dt \tag{4.13}$$

which can be integrated to give:

$$u_i - u_f = (F_x/m)(t_f - t_i) \tag{4.14}$$

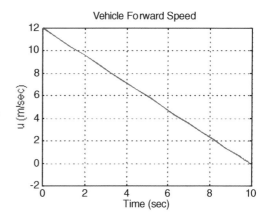

Figure 4.7. Simulation results of Example 4.3.

where the subscripts i and f denote initial and final values, respectively, and t denotes time. Using the fact that $u = dx/dt$ and that for the vehicle to stop, $u_f = 0$, the stopping distance can be found as:

$$(x_f - x_i) = mu_i^2/(2F_x) \qquad (4.15)$$

and the stopping time as:

$$(t_f - t_i) = mu_i/F_x \qquad (4.16)$$

Based on this simplified analysis, the stopping time is proportional to the vehicle-forward velocity at the time that the brakes are applied, whereas the stopping distance is proportional to the square of that velocity. More accurate results for the vehicle-forward velocity $u(t)$ in various braking or acceleration cases are obtained by numerically integrating the nonlinear ordinary differential equation, Eq. (4.12). A simple MATLAB program for this purpose is provided herein. Notice that the results are similar to the prediction based on Eqs. (4.14)–(4.16). With $m = 2,000$kg, $u_i = 12$ m/s, the average velocity $u_{av} = (u_i + u_f)/2 = 6$ m/s, and $F_x = 2,000 + (0.02)(2,000)(9.8) + (0.5)(1.202)(0.4)(2)(36)$ N, Eq. (4.16) predicts a stopping time of 9.89 seconds, which agrees closely with the simulation results shown in Figure 4.7.

```
% Ex4_3.m
% Init. time, final time, and initial
% values of the variable x are:
  ti=0.0; tf=10.0; ui = [12.0];
% Tol and trace are used
% by the integration routine ode23:
  tol = 1.0E-4; trace = 1;
% Perform integration and store
% the results in x
  [t,u] = ode23('Ex3_3a',ti,tf,ui,tol,trace);
% Plot the results
plot(t,u,'r')
title('Vehicle Forward Speed');
```

```
xlabel('Time (sec)')
ylabel('u (m/sec)'); grid;

function udot = Ex3_3a(t,u);
% Equations of longitudinal motion
% for a vehicle.
% The parameters are:
% u - vehicle forward velocity
% m - vehicle mass
% Fx - tractive or braking force
% W - vehicle weight, W = m*g
% Theta - road grade angle in radians
% f - rolling friction coefficient
% rho - density of air
% Cd - aerodynamic drag coefficient
% A - cross sectional area of vehicle
% uw - wind velocity
%
% Parameter values):
m=2000; g=9.8; W=m*g;
Fx=-2000; Theta=0.0; f=0.02;
rho=1.202; Cd=0.4; A=2; uw=0.0;
%
if u > 0
    udot = [(1/m)*(Fx-W*sin(Theta)
         - f*W*cos(Theta)-0.5*rho*Cd*A*(u+uw)^2)];
else
    udot=0;
end
```

EXAMPLE 4.4: POWER LIMITS ON LONGITUDINAL ACCELERATION. The traction forces F_{xf} and/or F_{xr} are required to move a vehicle forward. They are produced at the tire–road interface due to rotation of the wheels by the engine and drivetrain. The engine torque (as well as power and specific fuel consumption) produced is a function of speed. Using two assumptions – manual transmission and small (no) tire slip – the tractive force available from the engine to overcome load forces and to accelerate a vehicle is shown to be as follows (Gillespie 1992):

$$F_x = T_e N_t N_f (\eta_{tf}/r_w) - \left\{ (I_e + I_t)N_t^2 N_f^2 + I_d N_f^2 + I_w \right\} (a_x/r_w^2) \qquad (4.17)$$

where T_e is the engine torque, N_t is the gear ratio of the transmission, N_f is the gear ratio of the final drive, η_{tf} (<1) is a correction factor to represent the efficiency of the overall drive system, r_w is the radius of the wheel, I_w is the rotational inertia of the wheels and axle shafts, and I_d is the rotational inertia of the driveshaft.

This tractive-force expression has two parts. The first term on the right-hand side is the engine torque times the overall gear ratio and efficiency of the drive

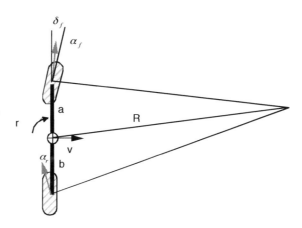

Figure 4.8. Schematic of a "bicycle model."

system. The second term on the right-hand side represents the "loss" of tractive force due to the inertia of the drive-system components. Equation (4.17) can be substituted in Eq. (4.4) to compute the vehicle motion under acceleration using computational methods.

Maximum performance in longitudinal acceleration can be limited by engine power (at high speeds) or by the traction limits of the drive wheels (e.g., at low speeds or on a slippery road surface). The engine provides the propulsion force needed to accelerate the vehicle and typically is characterized by torque and power curves as a function of speed (Figure 4.8). Power is given by the product of torque and speed, as follows:

$$P = T\Omega \tag{4.18}$$

where P is power in watt (or ft-lb/sec), T is engine torque in N-m (or ft-lb), and Ω is engine speed in rad/sec. Also note that 1 horsepower (HP) = 550 ft.lb/sec, or 0.736 kW.

Neglecting all resistance forces (i.e., at low to moderate speeds), a very rough upper limit on forward acceleration is obtained from Newton's second law:

$$ma_x = F_x \tag{4.19}$$

Because the drive power is the product of tractive force F_x times the forward speed u (ft/sec), the acceleration can be rewritten as:

$$a_x = (F_x/m) = (P/um) = (Pg/uW) \tag{4.20}$$

where W is the weight of the automobile (lb) and g is the gravitational constant (32.2 ft/sec^2). Thus, the maximum acceleration is inversely proportional to the forward speed and the weight of the vehicle.

A more detailed analysis of vehicle acceleration, or tractive-force limit, can be obtained by considering the engine, torque converter, transmission, and tire characteristics as well as the losses due to inertia and friction in the drivetrain. At times, engine dynamics is approximately included in longitudinal dynamics studies (e.g., cruise control and platooning) as a first-order dynamic lag between throttle command and the generation of the engine-drive torque.

Assuming adequate power from the engine, the acceleration (or tractive force) may be limited by the coefficient of friction between the tire and the road:

$$F_x = \mu F_z \tag{4.21}$$

where F_x is the traction force, m is the coefficient of friction, and F_z is the normal force on the drive wheels. Similar to the μ-λ curve in Figure 4.6, the coefficient of friction and the traction force depend on the longitudinal slip between the tire and the road surface. The calculation of F_z must consider not only fore–aft weight shift due to acceleration and braking (see Example 4.1) but also transverse weight shift due to cornering. The true vehicle-acceleration performance under the combination of all of these possible limiting factors usually is obtained from simulations. Under simplifying assumptions, we may be able to study analytically the meaningful vehicle performance. For example, assuming (1) small (i.e., no) lateral and yaw motions, and (2) a slippery road (i.e., tire friction is the performance limit), we can study the acceleration performance of FWD versus rear-wheel drive (RWD) vehicles, or the "optimal" brake (or drive) distribution ratio of front versus rear axles for maximum deceleration (or acceleration) performance.

4.3 Lateral Vehicle Motion

This section presents both two and three degree-of-freedom models for vehicle handling.

Two-Degree-of-Freedom Lateral Model ("Bicycle Model") in Steady Cornering

The lateral motion of a vehicle is discussed based on the so-called linear bicycle model illustrated in Figure 4.8. The term *bicycle model* is a misnomer in that this model typically is not used to model bicycle-handling dynamics; rather, it has arisen in the automotive literature because in this model, the right and left wheels are collapsed into one (see Figure 4.8). The steady turning (i.e., cornering) behavior, at constant forward speed $u0$, due to a small and steady steer displacement, δ_f, at the front wheels is considered first. For the analysis presented here to be valid, the following assumptions must hold: (1) the radius of the turn, R, must be large compared to the vehicle wheelbase, $L = a + b$, and the vehicle track, t; (2) the left and right steer angles of the front wheels must be approximately the same ($= \delta_f$); (3) the sideslip angles of the front wheels, α_f, are equal, as are the sideslip angles of the rear wheels, α_r; (4) the sideslip angle at the CG is $\beta = \tan^{-1} \frac{v}{u} \approx \frac{v}{u}$; and (5) the radius, R; sideslip angle, β; and the yaw velocity, $r = \dot{\psi}$ are fixed in a steady turn so that the instantaneous speed tangent to the path at the CG is $u = rR$.

Before applying Newton's second law, consider first the necessary geometrical relationships implied in Figure 4.8. Assuming a positive steer angle causing a turn to the right and resulting in positive slip angles being established at the front and rear tires (i.e., the tire-slip angle is defined as the angle from tire speed to tire-orientation direction), we obtain:

$$\alpha_f = \delta_f - \frac{v + ar}{u} \quad \text{and} \quad \alpha_r = -\frac{v - br}{u} \tag{4.22}$$

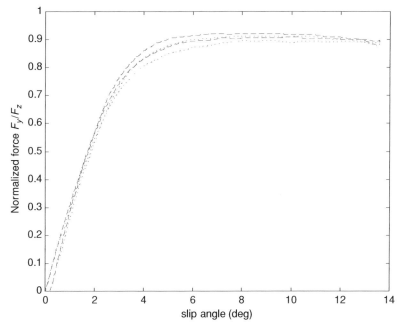

Figure 4.9. Tire lateral force, F_y, versus slip angle, α.

Therefore,

$$\alpha_f - \alpha_r = \delta_f - \frac{L}{R} \tag{4.23}$$

or

$$\delta_f = \frac{L}{R} + \alpha_f - \alpha_r \tag{4.24}$$

This implies that the steering angle, δ_f, which is necessary to negotiate a curve, consists of two parts: the static part is equal to L/R (also known as the *Ackermann angle*) and the dynamic part, which is equal to the difference between the front- and rear-tire slip angles. If the front tire-slip angle is larger than the rear slip angle, this condition is termed *understeer*. This implies that the steering angle must be larger than the Ackermann angle to maintain a constant-radius turn at nonzero speed. If the rear slip angle is greater than the front slip angle, the front steering angle is less than the Ackerman angle, a condition termed *oversteer*. Finally, if the front and rear slip angles are equal, the steering angle is equal to the Ackerman angle and the condition is termed *neutral steer*.

The side forces (i.e., cornering forces) acting at the vehicle tires, F_{yf} and F_{yr}, are related to the slip angles, α_f and α_r, as shown in Figure 4.9. Although the relationship between F_y and α is nonlinear, for small slip angles, we can use the approximation:

$$F_y = C_\alpha \alpha \tag{4.25}$$

where the cornering stiffness, C_α, is defined as the slope of the F_y versus α curve at $\alpha = 0$ and therefore is positive. The cornering stiffness depends on several tire geometric and material properties; for a given tire, it depends on the vertical load, F_z, and the inflation pressure. It should be emphasized that the force and cornering stiffness for bicycle models are defined for vehicle axles. The cornering-stiffness value, therefore, is about twice the value of tire-cornering stiffness encountered in the literature when the tire force of individual tires is analyzed.

Now we can derive the equations of motion applicable to the situation shown in Figure 4.8. Applying Newton's second law to translation in the y direction gives:

$$mu^2/R = F_{yf} + F_{yr} \qquad (4.26)$$

where the acceleration in the y direction, $a_y = ur = u^2/R$, is due to the centripetal acceleration of a vehicle in a steady turn. Also requiring the sum of the moments about the CG to be zero in the steady turn gives:

$$0 = F_{yf}a - F_{yr}b \qquad (4.27)$$

Thus, the cornering forces at the front and rear tires, respectively, are:

$$F_{yr} = m(a/L)(u^2/R) \qquad (4.28)$$
$$F_{yf} = m(b/L)(u^2/R) \qquad (4.29)$$

Substituting Eqs. (4.28) and (4.29) in Eq. (4.25) and solving for the slip angles gives:

$$\alpha_f = \frac{mu^2}{R}\frac{b}{L}\frac{1}{C_{\alpha f}} \qquad (4.30)$$

$$\alpha_r = \frac{mu^2}{R}\frac{a}{L}\frac{1}{C_{\alpha r}} \qquad (4.31)$$

Now, substitute these slip-angle expressions into Eq. (4.23) to obtain the steering angle:

$$\delta_f = \frac{L}{R} + \left(\frac{mb}{LC_{\alpha f}} - \frac{ma}{LC_{\alpha r}}\right)\frac{u^2}{R} \equiv \frac{L}{R} + K_{us} \cdot a_y \qquad (4.32)$$

where K_{us}, defined in Eq. (4.32), is termed the *understeer coefficient* and has the units of (rad/(m/s^2)).

When $K_{us} > 0$, the vehicle is termed *understeer*; when $K_{us} < 0$, the vehicle is termed *oversteer*; and when $K_{us} = 0$; the vehicle is termed *neutral steer*. For a neutral-steer vehicle, no change in steering angle with speed will be required to maintain a constant-radius turn. For an understeer vehicle, the steer angle must increase with speed (i.e., with the square of the speed) to maintain a constant-radius turn. An oversteer vehicle in a constant-radius turn must decrease the steer angle with increasing speed. Notice also that an understeer vehicle develops greater sideslip angles at the front wheels than at the rear (i.e., $|\alpha_f| > |\alpha_r|$), whereas the opposite is true for oversteered vehicles. For an understeer vehicle, the *characteristic speed* is defined as the speed at which the steer angle is twice the Ackermann angle (i.e., $K_{us} \cdot a_y = \frac{L}{R}$). Thus, the characteristic speed is given by:

$$u_{char} = \sqrt{\frac{L}{K_{us}}} \qquad (4.33)$$

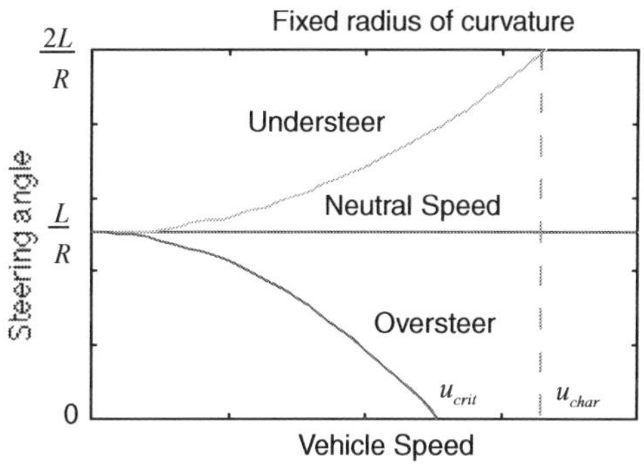

Figure 4.10. Effect of vehicle speed on steering angle.

In the oversteer case, there is a *critical speed* above which the vehicle becomes unstable:

$$u_{crit} = \sqrt{\frac{-L}{K_{us}}} \tag{4.34}$$

where, in the oversteer case, $K_{us} < 0$ by definition.

The effects of speed on steering angle in a constant-radius turn are summarized in Figure 4.10. These results can be interpreted further in terms of the lateral-acceleration gain:

$$G_a = \frac{a_y}{\delta_f} = \frac{\dfrac{u^2}{R}}{\delta_f} = \frac{\dfrac{u^2}{L}}{\delta_f \cdot \dfrac{R}{L}} = \frac{\dfrac{u^2}{L}}{1 + K_{us}\dfrac{u^2}{L}} = \frac{u^2}{L + K_{us}u^2} \tag{4.35}$$

and the yaw-rate (or yaw-velocity) gain:

$$G_r = \frac{r}{\delta_f} = \frac{\dfrac{u}{R}}{\delta_f} = \frac{u}{L + K_{us}u^2} \tag{4.36}$$

From Eqs. (4.35) and (4.36), it is clear that when $K_{us} < 0$ (i.e., oversteer), these gains become unbounded at the critical-speed value defined in Eq. (4.34). Also note that the vertical forces at the front and rear axles, F_{zf} and F_{zr}, can be expressed (by applying Newton's second law in the z direction) as follows:

$$F_{zf} = (mg)(b/L) \tag{4.37}$$

and

$$F_{zr} = (mg)(a/L) \tag{4.38}$$

Consequently, Eq. (4.32) can be rewritten as follows:

$$\delta_f = \frac{L}{R} + \left(\frac{F_{zf}}{C_{\alpha f}} - \frac{F_{zr}}{C_{\alpha r}}\right)\frac{u^2}{Rg} = \frac{L}{R} + K'_{us}\frac{u^2}{Rg} \tag{4.39}$$

where (u^2/Rg) is a nondimensional lateral acceleration in g units and the ratios (F_z/C_α) are the "cornering-compliance coefficients" in (Nz/Ny/deg) for the front and rear wheels. Because the cornering stiffness, C_α, is strongly dependent on the normal (i.e., vertical) load at the tire, data often are given for the cornering-compliance coefficient, (F_z/C_α), or its inverse, the cornering-stiffness coefficient. The nondimensional understeer gradient K'_{us} has units of rad/g rather than rad/(m/s^2).

EXAMPLE 4.5: STEADY TURNING BEHAVIOR AND UNDERSTEER GRADIENT. Consider an automobile with a weight of 8,000N on the front axle and 6,000N on the rear axle, a wheel base of 2.5 meters, and the following tire cornering-stiffness characteristics (i.e., for one tire):

Vertical Load (N)	Cornering Stiffness (N/deg)	Cornering Compliance Coefficient (N/(N/deg))
1,500	502	2.987
2,000	656	3.050
2,500	787	3.178
3,000	903	3.323
3,500	1,014	3.450
4,000	1,100	3.636

(a) Determine the Ackermann angles for a turn-radius value of $R = 200$ meters. This can be calculated easily from $\delta_f = \frac{L}{R}$ with the correct units:

$$\delta_f = \frac{2.5(m)}{200(m)} = 0.0125(rad) = 0.716(deg)$$

(b) Determine the understeer gradient, K_{us}, using Eq. (4.39). This requires that we know the cornering stiffness of the tires at the prevailing loads. On the front axle, we have a load of 4,000N per tire. From the data in the table, we obtain a cornering stiffness of 1,100 N/deg. Similarly, for the rear axle, the load is 3,000N per tire and the cornering stiffness is 903 N/deg. Although the units of K_{us} are degrees, the unit "deg/g" is written as a reminder. Thus:

$$K'_{us}\left(\frac{8000N}{1100*2N/\deg} - \frac{6000N}{903*2N/\deg}\right) = (3.6363 - 3.3222)\,deg/g$$
$$= 0.3141\,deg/g$$

(c) Determine the characteristic speed using Eq. (4.33):

$$K_{us} = K'_{us}/9.81 = 0.032\,deg/(m/sec^2) = 0.0005588\,rad/(m/sec^2)$$
$$u_{char} = \sqrt{L/K_{us}} = 66.9\,m/sec$$

(d) Determine the lateral-acceleration gain using Eq. (4.35) for $u = 55$ mph $= 24.56$ m/sec:

$$G_a = \frac{u^2}{L + K_{us}u^2} = 212.7\left(\frac{m/sec^2}{rad}\right) = 3.71\left(\frac{m/sec^2}{deg}\right) \quad (4.40)$$

(e) Determine the yaw-velocity gain using Eq. (4.36) for $u = 55$ mph $= 24.56$ m/sec:

$$G_r = \frac{u}{L + K_{us}u^2} = 8.6586\left(\frac{rad/sec}{rad}\right) = 8.6586\left(\frac{deg/sec}{deg}\right) \quad (4.41)$$

The calculated understeer gradient shows that this vehicle is close to being neutral steer. This is a consequence of only the tire properties; the steering

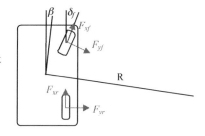

Figure 4.11. Projection of forces onto the direction of the instant center.

and suspension systems also contribute to the actual value of the understeer gradient. Also note the units in these calculations, especially those in Eqs. (4.40) and (4.41).

Finally, we consider the effect of tractive forces, F_{xf} and F_{xr}, on the under-steer/oversteer behavior of a vehicle. For a vehicle in a steady turn, the application of Newton's second law in the lateral direction now gives (instead of Eq. [4.26]) the following equations at each axle:

$$F_{yf}\cos(\delta_f - \beta) + F_{xf}\sin(\delta_f - \beta) = F_{yf}\cos\left(\frac{a}{R} + \alpha_f\right) + F_{xf}\sin\left(\frac{a}{R} + \alpha_f\right) = m\frac{b}{L}\frac{u^2}{R} \tag{4.42}$$

and

$$F_{yr}\cos(\beta) - F_{xr}\sin(\beta) = F_{yr}\cos\left(\frac{b}{R} - \alpha_r\right) - F_{xr}\sin\left(\frac{b}{R} - \alpha_r\right) = m\frac{a}{L}\frac{u^2}{R} \tag{4.43}$$

which can be obtained easily by projecting all of the forces onto the direction of the instantaneous rotation center, as shown in Figure 4.11.

Assuming small slip angles such that $\sin\alpha = \alpha$ and $\cos\alpha = 1$ and a linear-tire assumption as in Eq. (4.25), we obtain:

$$\alpha_f = \frac{m\dfrac{b}{L}\dfrac{u_o^2}{R} - F_{xf}\dfrac{a}{R}}{C_{\alpha f} + F_{xf}} \qquad \alpha_r = \frac{m\dfrac{a}{L}\dfrac{u_o^2}{R} + F_{xr}\dfrac{b}{R}}{C_{\alpha r} + F_{xr}}$$

A modified form of Eq. (4.32) to include the effects of tractive forces thus can be obtained:

$$\delta_f = \frac{L}{R} + (\alpha_f - \alpha_r) = \frac{L}{R} + \frac{m\dfrac{b}{L}\dfrac{u^2}{R} - F_{xf}\dfrac{a}{R}}{C_{\alpha f} + F_{xf}} - \frac{m\dfrac{a}{L}\dfrac{u^2}{R} + F_{xr}\dfrac{b}{R}}{C_{\alpha r} + F_{xr}}$$

$$= \frac{L}{R} - \frac{F_{xf}}{C_{\alpha f} + F_{xf}}\frac{a}{R} - \frac{F_{xr}}{C_{\alpha r} + F_{xr}}\frac{b}{R} + \left[\frac{mb}{L(C_{\alpha f} + F_{xf})} - \frac{ma}{L(C_{\alpha r} + F_{xr})}\right]\frac{u^2}{R} \tag{4.44}$$

Although it is much more complex than Eq. (4.32), this equation also has the same form. The Ackermann angle now is modified by the presence of the tractive forces, as is the expression for the understeer coefficient, K_{us}. For FWD vehicles, the Acker-mann angle is reduced and the understeer coefficient, K_{us}, increases (i.e., the vehicle becomes more understeer).

Two- and Three-Degree-of-Freedom Lateral Dynamic Models

In the previous analyses based on the two-degree-of-freedom (DOF) bicycle model, it was assumed that a vehicle is in a steady turn and the rolling motion of the vehicle sprung mass is negligible. Transient linear models (i.e., not cornering in a steady turn) now are considered for both the two- and three-DOF cases. The three-DOF dynamic model includes a roll degree of freedom, ϕ, in addition to the lateral motion y and the yaw motion ψ.

From Newton's second law, we obtain:

$$F_{yf} + F_{yr} = ma_y = m\ddot{y} = m(u_o r + \dot{v}) = m(u_o r + u_o \dot{\beta}) \tag{4.45}$$

$$aF_{yf} - bF_{yr} = I_z \dot{r} \tag{4.46}$$

where $\tan \beta = \frac{v}{u_0}$ or $v \approx \beta u_0$ was used. If we further assume the front and rear tires (or, more precisely, axles) are linear, we have:

$$F_{yf} = C_{\alpha f}\alpha_f = C_{\alpha f}\left(\delta - \left(\frac{v + ar}{u_o}\right)\right) = C_{\alpha f}\left(\delta - \beta - \frac{ar}{u_o}\right)$$

$$F_{yr} = C_{\alpha r}\alpha_r = C_{\alpha r}\left(\frac{br - v}{u_o}\right) = C_{\alpha r}\left(\frac{br}{u_o} - \beta\right)$$

Combining with Eqs. (4.45) and (4.46), we obtain:

$$mu_o\dot{\beta} + mu_o r = -(C_{\alpha f} + C_{\alpha r})\beta + \left(\frac{C_{\alpha r}b - C_{\alpha f}a}{u_o}\right)r + C_{\alpha f}\delta \tag{4.47}$$

$$I_z\dot{r} = (bC_{\alpha r} - aC_{\alpha f})\beta - \left(\frac{C_{\alpha f}a^2 + C_{\alpha r}b^2}{u_o}\right)r + aC_{\alpha f}\delta \tag{4.48}$$

which is the two-DOF linear model for lateral dynamics. In state-space form, $\dot{\mathbf{x}} = \mathbf{A}\mathbf{x} + \mathbf{B}\mathbf{u}$, where $\mathbf{x} = [\beta \ r]^{\mathrm{T}}$, these two equations are as follows:

$$\begin{bmatrix} \dot{\beta} \\ \dot{r} \end{bmatrix} = \begin{bmatrix} -\left(\dfrac{C_{\alpha f} + C_{\alpha r}}{mu_o}\right) & -\left(\dfrac{aC_{\alpha f} - bC_{\alpha r}}{mu_o^2}\right) - 1 \\ -\left(\dfrac{aC_{\alpha f} - bC_{\alpha r}}{I_z}\right) & -\left(\dfrac{C_{\alpha f}a^2 + C_{\alpha r}b^2}{I_z u_o}\right) \end{bmatrix} \begin{bmatrix} \beta \\ r \end{bmatrix} + \begin{bmatrix} \dfrac{C_{\alpha f}}{mu_o} \\ \dfrac{aC_{\alpha f}}{I_z} \end{bmatrix} \delta \tag{4.49}$$

When the relative motion of an automobile (relative to the road) is of interest, we must define two additional state variables: y, the lateral displacement, and ψ, the yaw angle of the automobile, relative to the road. Recall that $\tan \beta = \frac{v}{u_0}$ or $v \approx \beta u_0$, $\dot{y} = v + u_o\psi$ and $\dot{\psi} = r$, $\ddot{y} = \dot{v} + u_o r$, and with $\mathbf{x} = [y \ v \ \psi \ r]^{\mathrm{T}}$, the state-space equation is:

$$\frac{d}{dt}\begin{bmatrix} y \\ v \\ \psi \\ r \end{bmatrix} = \begin{bmatrix} 0 & 1 & u_o & 0 \\ 0 & -\left(\dfrac{C_{\alpha f} + C_{\alpha r}}{mu_o}\right) & 0 & \dfrac{-aC_{\alpha f} + bC_{\alpha r}}{mu_o} - u_o \\ 0 & 0 & 0 & 1 \\ 0 & \dfrac{-aC_{\alpha f} + bC_{\alpha r}}{I_z u_o} & 0 & -\left(\dfrac{C_{\alpha f}a^2 + C_{\alpha r}b^2}{I_z u_o}\right) \end{bmatrix} \begin{bmatrix} y \\ v \\ \psi \\ r \end{bmatrix} + \begin{bmatrix} 0 \\ \dfrac{C_{\alpha f}}{m} \\ 0 \\ \dfrac{aC_{\alpha f}}{I_z} \end{bmatrix} \delta$$

$$\tag{4.50}$$

When the road is curved and its orientation (i.e., yaw angle) is denoted as ψ_d (and its changing rate r_d), the vehicle-dynamic equations are then:

$$F_{yf} + F_{yr} = m(\ddot{y} + u_o r_d) \tag{4.51}$$

$$a F_{yf} - b F_{yr} = I_z \dot{r} \tag{4.52}$$

Substitute the following equations in Eqs. (4.51) and (4.52):

$$F_{yf} = C_{\alpha f}\alpha_f = C_{\alpha f}\left(\delta - \frac{v + ar}{u_o}\right) = C_{\alpha f}\left(\delta - \frac{\dot{y}}{u_o} - \frac{ar}{u_o} + (\psi - \psi_d)\right)$$

$$F_{yr} = C_{\alpha r}\alpha_r = C_{\alpha r}\left(\frac{br - v}{u_o}\right) = C_{\alpha r}\left(-\frac{\dot{y}}{u_o} + \frac{br}{u_o} + (\psi - \psi_d)\right)$$

to obtain:

$$m(\ddot{y} + u_o r_d) = -\left(\frac{C_{\alpha f} + C_{\alpha r}}{u_o}\right)\dot{y} + \left(\frac{-C_{\alpha f}a + C_{\alpha r}b}{u_o}\right)r + C_{\alpha f}\delta + (C_{\alpha f} + C_{\alpha r})(\psi - \psi_d) \tag{4.53}$$

$$I_z\dot{r} = -\left(\frac{aC_{\alpha f} - bC_{\alpha r}}{u_o}\right)\dot{y} - \left(\frac{C_{\alpha f}a^2 + C_{\alpha r}b^2}{u_o}\right)r + aC_{\alpha f}\delta + (aC_{\alpha f} - bC_{\alpha r})(\psi - \psi_d) \tag{4.54}$$

In state-space form, these can be written as Eq. (4.55):

$$\frac{d}{dt}\begin{bmatrix} y \\ \dot{y} \\ \psi - \psi_d \\ r \end{bmatrix} = \begin{bmatrix} 0 & 1 & 0 & 0 \\ 0 & -\left(\dfrac{C_{\alpha f} + C_{\alpha r}}{mu_o}\right) & \dfrac{C_{\alpha f} + C_{\alpha r}}{m} & \dfrac{-aC_{\alpha f} + bC_{\alpha r}}{mu_o} \\ 0 & 0 & 0 & 1 \\ 0 & \dfrac{-aC_{\alpha f} + bC_{\alpha r}}{I_z u_o} & \dfrac{aC_{\alpha f} - bC_{\alpha r}}{I_z} & -\left(\dfrac{C_{\alpha f}a^2 + C_{\alpha r}b^2}{I_z u_o}\right) \end{bmatrix}$$
$$\times \begin{bmatrix} y \\ \dot{y} \\ \psi - \psi_d \\ r \end{bmatrix} + \begin{bmatrix} 0 \\ \dfrac{C_{\alpha f}}{m} \\ 0 \\ \dfrac{aC_{\alpha f}}{I_z} \end{bmatrix}\delta + \begin{bmatrix} 0 \\ -u_o \\ -1 \\ 0 \end{bmatrix}r_d \tag{4.55}$$

A three-DOF vehicle has a nonrolling (i.e., unsprung) mass, m_{NR}, in the plane and a rolling (i.e., sprung) mass, m_R, which is constrained to rotate (i.e., roll) about an axis fixed in the nonrolling mass (Figure 4.12). In Figure 4.12, note that an xyz axis is fixed in the nonrolling mass such that the z axis is vertical and passes through the center of mass of the two mass systems and that the x axis is located in the vertical plane of symmetry and passes through the center of mass of the nonrolling mass. From Figure 4.14, that:

c is the distance along the x axis between the z axis and the CG of the rolling body.

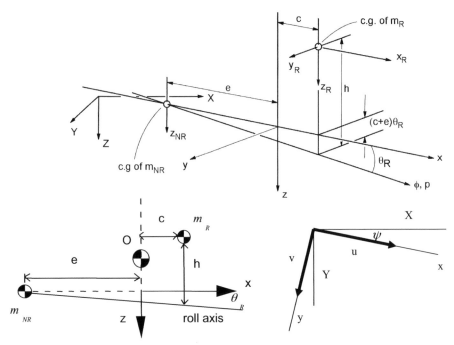

Figure 4.12. Schematic of the three-DOF model.

e is the distance along the x axis between the z axis and the CG of the nonrolling body.

θ_R is the angle downward from the horizontal of the axis (i.e., fixed in the nonrolling mass), about which the rolling mass is constrained to roll (i.e., the roll axis).

h is the vertical distance (in the z direction) between the CG of the rolling mass and the roll axis.

r is the yaw velocity, $\dot{\psi}$.

p is the roll velocity, $\dot{\phi}$.

u is the velocity in the x direction, $u = \dot{X} \cos \psi + \dot{Y} \sin \psi$.

v is the velocity in the y direction, $v = -\dot{X} \sin \psi + \dot{Y} \cos \psi$.

The derivation of the equations of motion is omitted here (see Appendix B).

In deriving the equations of motion, it is assumed that u represents a perturbation from a nominal forward velocity u_0, where u_0 is constant and large and u is small. The quantities u, v, r, p, and φ are all assumed to be small so their products can be neglected. Based on these assumptions and the definitions given previously, the linearized equations of motion for the three-DOF vehicle lateral dynamics are as follows:

$$
\begin{bmatrix}
mu_o & 0 & m_R h & 0 \\
0 & I_z & I_{xz} & 0 \\
m_R h u_o & I_{xz} & I_x & 0 \\
0 & 0 & 0 & 1
\end{bmatrix}
\begin{bmatrix}
\dot{\beta} \\
\dot{r} \\
\dot{p} \\
\dot{\phi}
\end{bmatrix}
+
\begin{bmatrix}
-Y_\beta & mu_o - Y_r & 0 & -Y_\phi \\
-N_\beta & -N_r & 0 & -N_\phi \\
0 & m_R h u_o & -L_p & -L_\phi \\
0 & 0 & -1 & 0
\end{bmatrix}
\begin{bmatrix}
\beta \\
r \\
p \\
\phi
\end{bmatrix}
=
\begin{bmatrix}
Y_\delta \\
N_\delta \\
0 \\
0
\end{bmatrix}
\delta
$$

$$(4.56)$$

where

$$m = m_{NR} + m_R$$

$$I_x = I_{xx|R} + m_R h^2 - 2\theta_R I_{xz|R} + \theta_R^2 I_{zz|R}$$

$$I_{xz} = m_R hc - I_{xz|R} + \theta_R I_{zz|R}$$

$$I_z = I_{zz|R} + I_{zz|NR} + m_R c^2 + m_{NR} e^2$$

The so-called stability derivatives used in Eq. (4.56) are defined as follows:

$$Y_\beta = -(C_{\alpha f} + C_{\alpha r}) \qquad Y_r = \frac{-aC_{\alpha f} + bC_{\alpha r}}{u_0}$$

$$Y_\phi = C_{\alpha r} \frac{\partial \delta_r}{\partial \phi} + C_{\gamma f} \frac{\partial \gamma_f}{\partial \phi} \qquad Y_\delta = C_{\alpha f}$$

$$N_\beta = -aC_{\alpha f} + bC_{\alpha r} \qquad N_r = -\left(\frac{a^2 C_{\alpha f} + b^2 C_{\alpha r}}{u_0} \right)$$

$$N_\phi = aC_{\gamma f} \frac{\partial \gamma_f}{\partial \phi} - bC_{\alpha r} \frac{\partial \delta_r}{\partial \phi} \qquad N_\delta = aC_{\alpha f}$$

$$L_p = -c_R \qquad L_\phi = m_R gh - k_R$$

Note that Eq. (4.56) is not in standard-state equation form but rather in the form $\mathbf{Ex} + \mathbf{Fx} = \mathbf{G}\delta$, where $\mathbf{x} = [\beta \ r \ p \ \phi]^T$. Furthermore, one of the combined inertia terms (I_z) in fact is state-dependent. In other words, Eq. (4.56) appears linear but in fact is nonlinear. Nevertheless, if we ignore the influence of roll angle on I_z (which is justified because the state-dependent term is of the second order), then Eq. (4.56) can be transformed into a state-space equation $\dot{\mathbf{x}} = \mathbf{Ax} + \mathbf{B}\delta$, where $\mathbf{A} = -\mathbf{E}^{-1}\mathbf{F}$ and $\mathbf{B} = \mathbf{E}^{-1}\mathbf{G}$. The state and input Matrices $\mathbf{A}$ and $\mathbf{B}$ can be obtained symbolically, but they also can be calculated numerically.

A steady-turning analysis can be performed using Eq. (4.56) by setting d/dt = 0. After some algebraic manipulations, we obtain a relationship similar to Eq. (4.32). However, this relationship accounts for the influence on understeer gradient of the roll DOF, ϕ, associated with the rolling (or sprung) mass (i.e., the third term in brackets on the right-hand side):

$$\delta = \frac{L}{R} + \left[\frac{mb}{C_{\alpha f} L} - \frac{ma}{C_{\alpha r} L} + \frac{m_R h}{L_\phi} \left(\frac{\partial \delta_r}{\partial \phi} - \frac{C_{\gamma f}}{C_{\alpha f}} \frac{\partial \gamma_f}{\partial \phi} \right) \right] \frac{u^2}{R} \qquad (4.57)$$

If the roll-related states (i.e., p and ϕ) are dropped, with $v = \beta u_0$, then Eq. (4.56) reduces to the two-DOF bicycle model shown in Eq. (4.49). Using the stability derivatives defined in the three-DOF model, the two-DOF model in Eq. (4.49) becomes:

$$\frac{d}{dt} \begin{bmatrix} v \\ r \end{bmatrix} = \begin{bmatrix} \dfrac{Y_\beta}{mu_0} & \dfrac{Y_r}{m} - u_0 \\ \dfrac{N_\beta}{I_z u_0} & \dfrac{N_r}{I_z} \end{bmatrix} \begin{bmatrix} v \\ r \end{bmatrix} + \begin{bmatrix} \dfrac{Y_\delta}{m} \\ \dfrac{N_\delta}{I_z} \end{bmatrix} \delta \qquad (4.58)$$

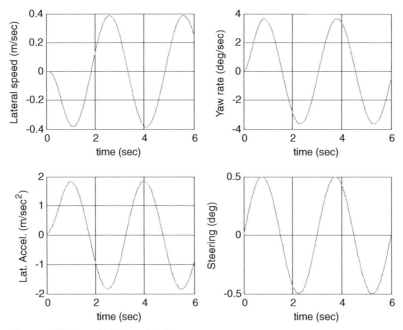

Figure 4.13. Two-DOF model time response.

EXAMPLE 4.6: SIMULATION OF AUTOMOBILE HANDLING. Although the equations presented here are linearized, they are sufficiently accurate under small steering inputs. In this example, the two-DOF model in Eq. (4.50) and the three-DOF model in Eq. (4.56) are both simulated and the results are shown in Figures 4.13 and 4.14. The lateral acceleration, defined as $a_y = u_0(\dot{\beta} + r) + \frac{m_R}{m}h\dot{p}$ for the

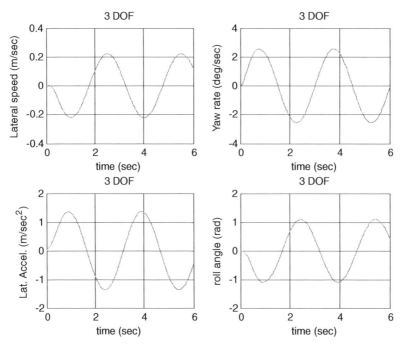

Figure 4.14. Three-DOF model response.

three-DOF model, is different than the two-DOF model (where $p = 0$). This, however, is not the main reason why the three-DOF response differs from that of the two-DOF. What do you think is the main reason?

```
% Ex4_6a.m
% 2DOF model
a   =  1.14;          % distance c.g. to front axle (m)
L   =  2.54;          % wheel base (m)
m   =  1500;          % mass (kg)
Iz  =  2420.0;        % yaw moment of inertia (kg-m^2)
Caf =  44000*2;       % cornering stiffness--front axle (N/rad)
Car =  47000*2;       % cornering stiffness-- rear axle (N/rad)
b=L-a; g=9.81;
Kus = m*b/(L*Caf) - m*a/(L*Car);        % (rad/(m/sec^2))
u_char = (L/Kus)^0.5;                   % understeer vehicle
u = u_char;

A=[-(Caf+Car)/(m*u),      (b*Car-a*Caf)/(m*u)-u
  (b*Car-a*Caf)/(Iz*u),   -(a^2*Caf+b^2*Car)/(Iz*u)];
B=[Caf/m; a*Caf/Iz];
C_lat = [1 0]; D_lat = 0;               % Lateral speed
C_yaw = [0 1]; D_yaw = 0;               % Yaw rate
C_acc=A(1,:) + u*[0,1];
D_acc = B(1);                           % Lateral acceleration
C = [C_lat; C_yaw; C_acc];
D = [D_lat; D_yaw; D_acc];
t=[0:0.01:6];
U=0.5*pi/180*sin(1/3*2*pi*t);           % 0.5 degree, 0.333Hz
                                        % sine steering
Y=lsim(A,B,C,D,U,t);                    % Note small lsim
subplot(221)
plot(t,Y(:,1),'r'); grid
xlabel('time (sec)')
ylabel('Lateral speed (m/sec)')

subplot(222)
plot(t,Y(:,2)*180/pi,'r'); grid
xlabel('time (sec)')
ylabel('Yaw rate (deg/sec)')

subplot(223)
plot(t,Y(:,3),'r'); grid
xlabel('time (sec)')
ylabel('Lat. Accel.(m/sec^2)')

subplot(224)
plot(t,U*180/pi,'r'); grid
```

```
xlabel('time (sec)')
ylabel('Steering (deg)')

% Ex4_6b.m
% 3DOF model
mR=1363.64; mNR=136.36;   m=(mR + mNR);   % Kg
IzzNR=220.0; IxxR=400.0; IxzR=75.0; IzzR=2200.0;      % Kg-m^2
c=0.14; e=1.4; h=0.35; b=1.4; a=1.14; % meters
L=a+b; g=9.81;
u=33.7256;              % vehicle speed
Theta_R=(5.0*pi/180);   % Theta_R = 5 degree
Caf=44000*2;            % cornering stiffness-
front axle (N/rad)
Car=47000*2;            % cornering stiffness-rear axle (N/rad)
Cgf=2000*2;             % camber thrust stiffness (N/rad)

dgfdf=0.8;              % degree incline change per degree roll
ddrdf=-0.095;           % degree rear steering per degree roll
kR=700*180/pi;          % N-m per radian of roll
cR=21.0*180/pi;         % N-m per rad/sec of roll rate

% Define the coefficients in the matrix equations:
Ix=IxxR + mR*(h^2) - 2*Theta_R*IxzR + (Theta_R^2) *IzzR;
Ixz=mR*h*c - IxzR + Theta_R*IzzR;
Iz=IzzR + IzzNR + mR*(c^2) + mNR*(e^2);

Yb = -(Caf+Car); Yr = (b*Car-a*Caf)/u;
Yf = (Car*ddrdf)+(Cgf*dgfdf); Yd = Caf;

Nb = b*Car - a*Caf; Nr = -(a^2*Caf+b^2*Car)/u;
Nf=a*Cgf*dgfdf - b*Car*ddrdf; Nd= a*Caf;
Lp= -cR; Lf = (mR*g*h-kR);
% Transform into state equation form:
E=[m*u 0 mR*h 0;
   0 Iz Ixz 0;
   mR*h*u Ixz Ix 0;
   0 0 0 1];
F=[-Yb (m*u-Yr) 0 -Yf;
   -Nb -Nr 0 -Nf;
   0 mR*h*u -Lp -Lf;
   0 0 -1 0];
G=[Yd;Nd;0;0];
A=-(inv(E)*F); B=inv(E)*G;

% Define the outputs as
% lat speed, lat accel, yaw rate and roll angle
C=[u 0 0 0
   u*(A(1,:)+[0 1 0 0])+(mR*h/m)*A(3,:)
```

```
    0  1  0  0
    0  0  0  1];
D=[0;u*B(1)+(mR*h/m)*B(3);0;0];

t=[0:0.01:6];
U=0.5*pi/180*sin(1/3*2*pi*t);   %0.5 degree, 0.333Hz
                                %sine steering

Y=lsim(A,B,C,D,U,t);
subplot, subplot(221)
plot(t,Y(:,1),'r'); title('3 DOF');
xlabel('time (sec)'); ylabel('Lateral speed (m/sec)'); grid;
subplot(222)
plot(t,Y(:,3)*180/pi,'r'); title('3 DOF');
xlabel('time (sec)'); ylabel('Yaw rate (deg/sec)'); grid;
subplot(223)
plot(t,Y(:,2),'r'); title('3 DOF');
xlabel('time (sec)'); ylabel('Lat. Accel.(m/sec^2)'); grid;
subplot(224)
plot(t,Y(:,4)*180/pi,'r'); title('3 DOF');
xlabel('time (sec)'); ylabel('roll angle (rad)'); grid;
```

4.4 Vertical Vehicle Motion

The vertical motion of a vehicle caused by irregular road surfaces and suspension characteristics is important in vehicle design. The human perception and tolerance of these vertical motions are critical factors in perceived "quality" of a vehicle. To understand these issues, we must be familiar with each of the following factors: (1) the excitation sources, (2) the vehicle dynamic response, and (3) the vehicle occupant ride perception. Characteristics of these factors are described in the following subsections.

Road Model

Although there are other sources of excitation (e.g., imperfections in the tire–wheel assembly and engine/transmission/driveline excitation), only road excitation is considered here. The road-surface profiles are stochastic in nature and can be represented by their statistical properties. A useful and compact representation is the power spectral density (PSD) of the road profile, which is found by taking the Fourier transform of the auto-correlation function of the measured road profile as a function of time. The Fourier transform essentially represents the measured time signal as a series of sinusoidal functions with varying amplitudes and phases. The PSD represents the power in the signal at a particular frequency and typically is plotted versus frequency, ω (in rad/s) or f (in Hertz = Hz = cycles/s). Note that $2\pi f = \omega$ and that for a vehicle moving at a constant longitudinal velocity, u_0, the distance traveled, x, and the time, t, are related by $x = u_0 t$. Thus, we can define a spatial frequency, $\omega' = \omega/u_0$ (in rad/m or rad/ft) or $f' = f/u_0$ (in cycles/m or cycles/ft), as shown in Figure 4.15.

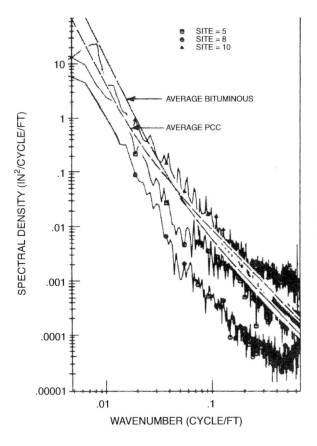

Figure 4.15. PSD of road elevation (Gillespie 1992).

Sometimes it is convenient, as discussed subsequently, to express the road input excitation as a velocity or acceleration input rather than a displacement (or elevation). Figure 4.16 shows the PSD curves for vertical displacement, velocity, and acceleration for a typical road, assuming a constant vehicle-forward speed of $u_0 = 50$ miles per hour (mph). Although the elevation PSD decreases significantly

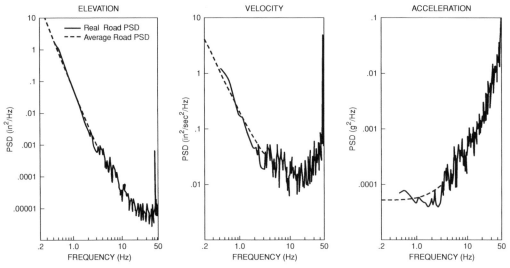

Figure 4.16. PSD of road elevation, velocity, and acceleration (Gillespie 1992).

Figure 4.17. Coloring filter for road-displacement profile.

with frequency, the velocity PSD remains essentially flat and the acceleration PSD increases significantly. A completely random signal (i.e., a "white-noise" signal) is characterized by a flat (or constant) PSD versus frequency plot. This fact can be used to advantage in simulating road profiles by treating a road profile as the output of a linear coloring filter $G(s = j\omega)$ (i.e., a dynamic system with transfer function G(s); Figure 4.17) with a random input. Thus, we can utilize a random-number generator available on a computer system as the basis for generating various road profiles.

The PSD of the road profile, $S_r(w)$, is given by:

$$S_r(\omega) = |G(\omega)|^2 (\sigma^2) \tag{4.59}$$

where σ^2 is the variance (i.e., square of the standard deviation) of a zero mean, normally distributed random input from a Gaussian (normal) random-number generator. Assuming a simple first-order filter, $G(s) = (\omega_0/(s + \omega_0))$, we obtain $|G(j\omega)|^2 = [(\omega_0/\omega)^2/((\omega_0/\omega)^2 + 1)]$ and:

$$S_r(f') = S_0 \left[\frac{1/(f'^2)}{1 + (f_0'/f')^2} \right] \tag{4.60}$$

where:

$S_r(f') =$ PSD of road displacement (elevation).
$S_0 = (\sigma f_0')^2$ is the roughness magnitude parameter
   (e.g., 1.25E-5 ft for a rough road and 1.25E-6 ft for a smooth road).
$f_0' = \omega_0/(2\pi u_0)$ is the cutoff spatial frequency
   (e.g., 0.05 cycle/ft for bituminous and 0.02 cycle/ft for Portland-cement roads).

It also is typical to represent the road-velocity profile – because its PSD is fairly flat (see Figure 4.16) – as a purely random (i.e., white-noise) input with $Sr(f') =$ constant $= (\sigma^2/2\pi)$.

EXAMPLE 4.7: SIMULATION OF ROAD PROFILE. A road-profile simulation can be carried out using Eq. (4.60) and the values $f_0' = 0.02$, $u = 80$ ft/sec, and $S_0 = 1.25 \times 10^{-5}$ ft. The simulation results are shown here and illustrate a fairly flat PSD for the random input variable, $w(t)$, and one that falls off with frequency (similar to Figure 4.15) for the simulated road elevation, $y(t)$. Unlike Figure 4.16, the frequency here is plotted on a linear rather than a logarithmic scale.

```
% Ex4_7.m
%
% Generate a zero-mean
% normally distributed sequence
% of 256 numbers with standard
% deviation = 1 as w
w=zeros(256,1);
Vel=80;      % ft/sec
w0 = 2.*pi*Vel*0.02;
% PCC road surface
```

```
S0 = 1.25e-5;      % rought road
sigma = (2*pi*Vel*sqrt(S0))/w0;
w = sigma*randn(size(w));
% Define the filter
T = 0.001; Fs = 1/T;
B = [1-exp(-w0*T)];
A = [1 -exp(-w0*T)];
% Obtained filter output y
% for the random input w
y=filter(B,A,w);
% Determine the power spectrum
% for y and w
P = spectrum(w,y,256);
specplot(P,Fs)
```

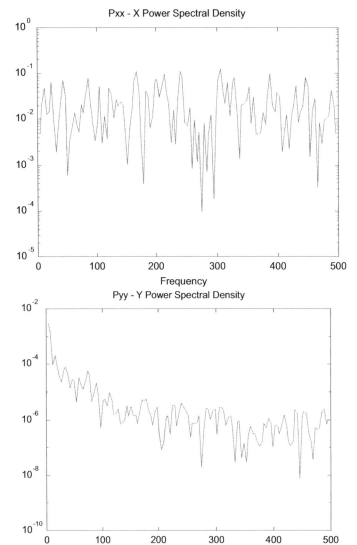

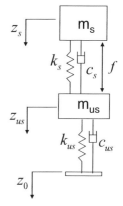

Figure 4.18. Standard quarter-car model.

Vertical-Vehicle-Motion Model

Ideally, the vehicle-ride dynamics consider the pitch and roll motions of the vehicle as well as the vertical (i.e., heave) motion. A half-car model accounts for both pitch and heave motions and leads to a four-DOF model. However, it can be shown that the coupling between the pitch-and-roll and the heave motions is not significant for typical passenger vehicles, and it is adequate to consider a so-called quarter-car model as illustrated in Figure 4.18. A two-DOF quarter-car model considers both the vehicle sprung mass and the unsprung mass associated with the wheel/tire/axle assembly. A one-DOF quarter-car model neglects the unsprung mass. Whereas both models are useful in suspension design and control, the two-DOF quarter-car model represents a good compromise between model simplicity and accuracy.

In the following analyses (which are used throughout this book), only the two-DOF quarter-car model of vertical-vehicle dynamics is considered. In the model, an active suspension is assumed to be in parallel with a passive suspension, k_s and c_s. The tire stiffness, k_{us}, and damping, c_{us}, also are modeled. Note that the tire damping, c_{us}, can sometimes be neglected. The dynamic equations that govern the motions of these two masses are as follows:

$$m_s \ddot{z}_s + c_s(\dot{z}_s - \dot{z}_{us}) + k_s(z_s - z_{us}) = -f \tag{4.61}$$

$$m_{us}\ddot{z}_{us} + c_s(\dot{z}_{us} - \dot{z}_s) + k_s(z_{us} - z_s) + c_{us}(\dot{z}_{us} - \dot{z}_0) + k_{us}(z_{us} - z_0) = f \tag{4.62}$$

The equations of motion in standard-state variable form are as follows:

$$\frac{d}{dt}\begin{bmatrix} z_{us} - z_0 \\ \dot{z}_{us} \\ z_s - z_{us} \\ \dot{z}_s \end{bmatrix} = \begin{bmatrix} 0 & 1 & 0 & 0 \\ -\dfrac{k_{us}}{m_{us}} & -\dfrac{(c_s + c_{us})}{m_{us}} & \dfrac{k_s}{m_{us}} & \dfrac{c_s}{m_{us}} \\ 0 & -1 & 0 & 1 \\ 0 & \dfrac{c_s}{m_s} & -\dfrac{k_s}{m_s} & -\dfrac{c_s}{m_s} \end{bmatrix} \begin{bmatrix} z_{us} - z_0 \\ \dot{z}_{us} \\ z_s - z_{us} \\ \dot{z}_s \end{bmatrix}$$

$$+ \begin{bmatrix} 0 \\ \dfrac{m_s}{m_{us}} \\ 0 \\ -1 \end{bmatrix} \frac{f}{m_s} + \begin{bmatrix} -1 \\ \dfrac{c_{us}}{m_{us}} \\ 0 \\ 0 \end{bmatrix} \dot{z}_0 \tag{4.63}$$

$$\Rightarrow \dot{\mathbf{x}} = \mathbf{A}\mathbf{x} + \mathbf{B}u + \mathbf{G}w$$

where $u(t)$ is the scalar control force (normalized by the sprung mass) and the state variables are the tire deflection $x_1 = z_{us} - z_0$, the unsprung-mass velocity $x_2 = \dot{z}_s$, the suspension stroke $x_3 = z_s - z_{us}$, and the sprung-mass velocity $x_4 = \dot{z}_s$. The coefficients of the state equation also can be presented in the following normalized form:

$$\mathbf{A} = \begin{bmatrix} 0 & 1 & 0 & 0 \\ -\omega_1^2 & -2(\rho\zeta_2\omega_2 + \zeta_1\omega_1) & \rho\omega_2^2 & 2\rho\zeta_2\omega_2 \\ 0 & -1 & 0 & 1 \\ 0 & 2\zeta_2\omega_2 & -\omega_2^2 & -2\zeta_2\omega_2 \end{bmatrix}; \quad \mathbf{B} = \begin{bmatrix} 0 \\ \rho \\ 0 \\ -1 \end{bmatrix}; \quad \mathbf{G} = \begin{bmatrix} -1 \\ 2\zeta_1\omega_1 \\ 0 \\ 0 \end{bmatrix}$$

(4.64)

where typical values of the parameters are $\rho = (m_s/m_{us}) = 10.0$, $\omega_1 = \sqrt{k_{us}/m_{us}} = 20\pi$ rad/s, $\omega_2 = \sqrt{k_s/m_s} = 2\pi$ rad/s, $\zeta_1 = c_{us}/(2m_{us}\omega_1) = 0.0$, and $\zeta_2 = c_s/(2m_s\omega_2) = 0.3$. The ground velocity input, $w(t) = z_0(t)$, is assumed to be zero mean and Gaussian with a variance of $2\pi Au_0$, where A is the ground-motion amplitude and u_0 is the vehicle-forward velocity; typical values are $A = 1.6 \times 10^{-5}$ ft and $u_0 = 80$ ft/s. If there is no active suspension-normalized control force, $u(t) = 0$, then the equations represent the vertical motion for a system with only a passive suspension.

EXAMPLE 4.8: SIMULATION OF VEHICLE-VERTICAL RESPONSE. Using the model in Eqs. (4.61)–(4.64) with the parameter values given immediately following those equations, we can simulate the response of the vehicle-vertical motion to a random ground-velocity input $w(t)$. The simulation results are shown here. The plots show tire deflection, suspension stroke, and vertical acceleration of the suspended mass versus time. A more useful output of the program is the frequency-response plots for the same variables (also shown here). The peaks at $\omega_2 = 2\pi$ rad/s (1 Hz) are associated with the suspension mode, whereas the peaks at $\omega_1 = 20\pi$ rad/s (10 Hz) are associated with the "wheel-hop" mode.

```
% Ex4_8.m
% x(1) = tire deflection (zus-z0)
% x(2) = velocity of unsprung mass (d(zus)/dt)
% x(3) = suspension stroke (zs-zus)
% x(4) = sprung mass speed (d(zs)/dt)
% cs = damping of passive suspension
% ks = stiff. of passive suspension
% ms = sprung mass
% mus = unsprung mass
% cus = tire damping coefficient
% kus = tire stiffness
% road displacement input = z0(t)
% road velocity input = d(z0)/dt
% Parameters:
w1 = 20*pi; % w1 = sqrt(kus/mus)
w2 = 2.0*pi; % w2 = sqrt(ks/ms)
z1 = 0.0; % z1 = cus/(2*mus*w1)
z2 = 0.3; % z2 = cs/(2*ms*w2)
```

```
rho = 10.; % rho = ms/mus
 % Open loop system equations:
 A = [0 1 0 0
      -w1^2 -2*(z2*w2*rho+z1*w1)     rho*w2^2     2*z2*w2*rho
      0 -1 0 1
      0 2*z2*w2 -w2^2 -2*z2*w2];
 B = [0 rho 0 -1]';
 G = [-1 2*z1*w1 0 0]';
 % Define outputs of interest:
C1=[1 0 0 0]; D1= 0.0;
% output=tire deflection
[num1, den1]=ss2tf(A,G,C1,D1,1);
C2=[0 0 1 0]; D2= 0.0;
% output=suspension stroke
[num2, den2]=ss2tf(A,G,C2,D2,1);
C3=[A(4,:)]; D3= 0.0;
% output=sprung mass accel.
[num3, den3]=ss2tf(A,G,C3,D3,1);
% Generate the white noise input w(t)
t=[0:0.001:1];
Amp=1.65E-5;
Vel=80; sigma=sqrt(2.*pi*Amp*Vel);
w=sigma*randn(size(t));
% Simulate the response of interest:
y1=lsim(num1,den1,w,t);
y2=lsim(num2,den2,w,t);
y3=lsim(num3,den3,w,t);
clf; subplot(321), plot(t,y1,'r');
title('Response to Road Velocity Input')
xlabel('Time, t(sec)'); ylabel('Tire Def')
subplot(322), plot(t,y2,'r');
title('Response to Road Velocity Input')
xlabel('Time, t(sec)'); ylabel('Susp Stroke')
subplot(323), plot(t,y3,'r');
title('Response to Road Velocity Input')
xlabel('Time, t(sec)'); ylabel('Sprung. mass accel')
% Obtain and plot the frequency response
freq=logspace(-1,2,100);
[mag1, phase1]=bode(num1,den1,freq);
subplot(324), loglog(freq,mag1,'r');
title('Frequency Response Magnitude');
xlabel('Frequency, rad/s');
ylabel('Tire def'); grid;
[mag2, phase2]=bode(num2,den2,freq);
subplot(325), loglog(freq,mag2,'r');
title('Frequency Response Magnitude');
xlabel('Frequency, rad/s');
```

```
ylabel('Susp stroke'); grid;
[mag3, phase3]=bode(num3,den3,freq);
subplot(326), loglog(freq,mag3,'r');
title('Frequency Response Magnitude');
xlabel('Frequency, rad/s');
ylabel('Sprung mass accel'); grid;
```

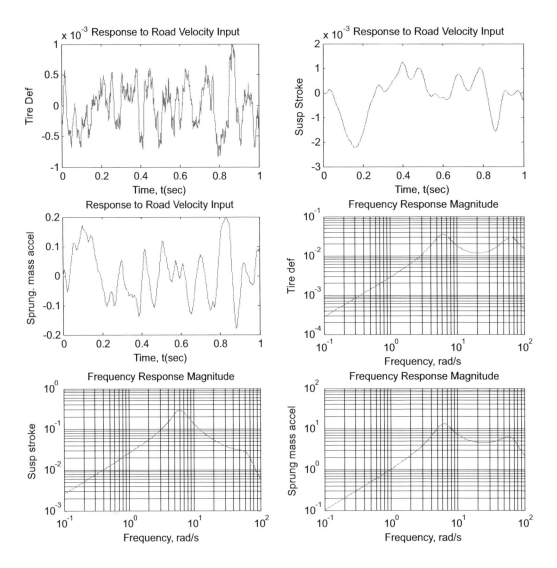

An interesting phenomenon of vehicle-suspension systems (both active and passive) is the existence of invariant points. Because the active suspension force acts between two masses, the overall dynamics of the suspension system is:

$$m_s \ddot{z}_s + m_u \ddot{z}_{us} + k_{us}(z_{us} - z_0) = 0 \qquad (4.65)$$

which is independent of the passive and active design and therefore is termed the *invariant equation*. In deriving Eq. (4.65), the tire damping, c_{us}, which is usually very

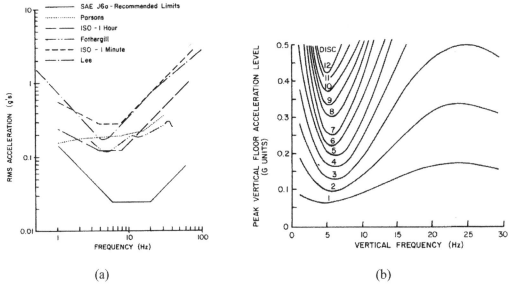

Figure 4.19. Constant comfort lines from various standards (Gillespie 1992).

small, is neglected. By defining three transfer functions for acceleration, rattle space, and tire deflection, as follows:

$$H_A(s) \equiv \frac{\ddot{z}_s(s)}{\dot{z}_0(s)} \quad H_{RS}(s) \equiv \frac{z_s(s) - z_{us}(s)}{\dot{z}_0(s)} \quad H_{TD}(s) \equiv \frac{z_{us}(s) - z_0(s)}{\dot{z}_0(s)} \quad (4.66)$$

the I/O relationship between the road excitation and the acceleration, rattle space (i.e., suspension stroke), and tire deflection can be studied. Hedrick and Butsuen (1988) show that because of this invariance property, a force (from a passive or active suspension) acting between the sprung and unsprung masses cannot independently affect all three transfer functions in Eq. (4.66). Specifically, they show that only one of these transfer functions can be independently specified (i.e., at certain frequencies, near the "wheel-hop" frequency, and in the frequency range of interest for suspension design). Levitt and Zorka (1991) also studied this problem, in the case in which the tire damping, C_{us}, is small but nonzero. They point out that this invariance property can be altered significantly, even for small tire damping, and they argue that tire damping should not be ignored in suspension design.

Ride Model

The perception of the ride by a vehicle occupant is the final criterion for how a suspension is judged. It is a subjective criterion based on the cumulative effects of many factors, including suspension design, seat design, and vibration response of the human body. Various researchers have attempted to quantify passenger perceptions of comfort by relating them to measurable quantities, such as vertical acceleration and "jerk" (i.e., the time rate of change of acceleration). Figure 4.19 shows lines of constant comfort as determined by various researchers on a plot of vertical acceleration versus excitation frequency. Because of different interpretations of comfort used in various studies, these results should be used only as a qualitative guide. The results all show, however, a minimum tolerance (i.e., maximum sensitivity) to vertical acceleration in the range of 4 to 8 Hz. This sensitivity is recognized as the result of vertical resonance of the abdominal cavity, which can lead to motion sickness. The ISO curves in Figure 4.19(a) show that the duration of the exposure also affects the maximum tolerable level of acceleration. The National Aeronautics and Space Administration (NASA) results in Figure 4.19(b) show that the shape of the curves depends on the acceleration level and tends to flatten out at low acceleration levels, and that discomfort is rather independent of frequency. Although the reduction of vibrations to improve ride is certainly the primary goal, some level of vibration may be desirable to provide "road feel," which is considered an essential element of feedback to the driver of a vehicle. In fact, suspension design – active or passive – typically involves several important tradeoffs: (1) improve passenger comfort, (2) reduce suspension stroke for packaging reasons, and (3) improve handling by reducing wheel hop and maintaining good tire–road contact.

EXAMPLE 4.9: TRADEOFFS IN SUSPENSION DESIGN. We assume that the suspension design problem can be formulated as a design-optimization problem, in which the goal is to select the suspension-stiffness parameter, k_s, to minimize the performance index (objective function):

$$J = \dot{x}^2_{4rms} + r_1 x^2_{1rms} + r_2 x^2_{3rms}$$

where

$\dot{x}_{4rms}$ is the root mean square (rms) sprung-mass acceleration and represents a measure of passenger-comfort goals.

x_{1rms} is the rms tire deflection and represents a measure of road-handling goals.

x_{3rms} is the rms suspension stroke and represents a measure of packaging goals.

r_1, r_2 are weights that quantify the tradeoffs desired and must be selected based on experience.

A plot of J versus $\omega_2 = \sqrt{k_s/m_s}$ is shown here and has a minimum near the nominal value of $\omega_2 = \pi$ rad/s, suggested previously.

The rms values of the variables of interest are calculated in MATLAB using special functions available for that purpose based on the model in Eqs. (4.61)–(4.64). For any system of the form $\dot{x} = Ax + Gw$, with a white-noise input vector w of covariance W, the steady-state solution for x can be obtained by solving the following algebraic matrix equation:

$$AX_{ss} + X_{ss}A^T = -GWG^T \qquad (4.67)$$

where X_{ss} is the steady-state covariance matrix for x and W is the covariance ("strength") of the random disturbance ($= 1$ here). Equation (4.67) is termed a Lyapunov equation and can be solved for X_{ss} given A, G, and W. In the following MATLAB example program, the steady-state covariance matrix (i.e., the solution of the previous Lyapunov equation) is obtained by calling a special function (lyap()).

```
% Ex4_9.m
%
% x(1) = tire deflection (zus-z0)
% x(2) = speed of unsprung mass
% x(3) = suspension stroke (zs-zus)
% x(4) = velocity of sprung mass
%
% cs = suspension damping
% ks = suspension stiffness
% ms = sprung mass
% mus = unsprung mass
% cus = damping coefficient of the tire
% kus = tire stiffness
% road displacement input = z0(t)
% road velocity input = d(z0)/dt
% Specify the parameter values here:
clear all;
w1 = 20*pi; % w1 = sqrt(kus/mus)
w2 = 2.0*pi; % w2 = sqrt(ks/ms)
z1 = 0.02; % z1 = cus/(2*ms*w1)
z2 = 0.3; % z2 = cs/(2*ms*w2)
rho = 10.; % rho = ms/mus
Amp=1.65E-5; Vel=80;
% Input velocity characteristics
for i=1:20,
 w(i) = i*(2.0*pi)/10.0;
 w2= w(i);
 % Open loop system equations:
 A = [0 1 0 0
 -w1^2 -2*(z2*w2*rho+z1*w1) rho*w2^2 2*z2*w2*rho
     0 -1 0 1
     0 2*z2*w2 -w2^2 -2*z2*w2];
 B = [0 rho 0 -1]';
 G = [-1 2*z1*w1 0 0]';
 %
 % calculate the rms response to
 % a unit variance white
 r1=5.0e4; r2=5.0E3;
 % Weights for typical (T) ride case
 Xss=lyap(A,G*G');
 x3_rms=sqrt(Xss(3,3));
 x1_rms=sqrt(Xss(1,1));
 x4dot_rms=sqrt(A(4,:)*Xss*A(4,:)'+ G(4)*G(4)');
 J(i)=(2.*pi*Amp*Vel)*(x4dot_rms^2+r1*(x1_rms^2)+r2*(x3_rms^2));
end;
clf;
plot(w/(2*pi),J,'r');
xlabel('w2 (Hz)'), ylabel('Cost function J'),grid;
```

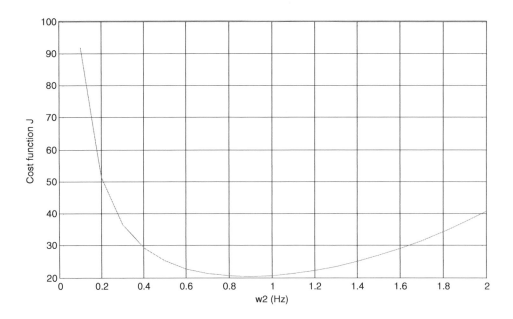

In this chapter, aspects of vehicle dynamics (i.e., those most useful for subsequent chapters on vehicle control) are reviewed. For example, vehicle longitudinal dynamics is useful when we consider the design of ABS, traction control, cruise control, and platooning. Vehicle lateral dynamics is useful for 4WS, vehicle stability control, active safety systems to prevent road-departure accidents, and automated steering for AHS. Similarly, vehicle vertical dynamics is important for active suspension design. Clearly, there are many other interesting topics in vehicle dynamics (e.g., rollover, nonconventional vehicles, articulated vehicles, and sloshing of liquid in tankers) that are not reviewed here. Interested readers may refer to additional information in the references provided.

PROBLEMS

1. For a passenger car with the following parameters (see Figure 4.5, m: mass; f: rolling resistance coefficient; g: gravitational acceleration):

$$m = 1,000 \text{ kg} \quad h = 0.6 \text{ m} \quad a = 1.0 \text{ m} \quad b = 1.5 \text{ m} \quad f = 0.02 \quad g = 10 \text{ m/sec}^2$$

Assume for this vehicle that it is FWD, the radius of the front tires is 0.3 m, the road is flat, the surface is dry, and therefore no excessive tire slip will occur.
Please answer the following:

    (a) When a driving torque of 600N-m is applied to the front axle, calculate the vehicle startup longitudinal acceleration (acceleration at $t = 0+$).

    (b) When a vehicle is driving from standstill under the same conditions described previously but now on an upslope of 5 percent (for simplicity, assume $\sin \theta = 0.05$ and $\cos \theta = 1$), calculate the normal load on the front and rear axles at initial startup ($t = 0+$).

2. Reconsider Example 4.3. If a constant braking force of $F_b = 1{,}000.0$ N is applied at time $t = 0$ seconds, estimate the stopping time and stopping distance. Explain your assumptions and calculations.

3. Reconsider the vehicle-braking simulation in Example 4.3 (see also the MATLAB program and the parameter values given there). Compare the stopping distances and times that can be achieved by two slightly different vehicles. The braking force is given by:

$$F_b = (T_b/r_w) + \left((I_w/r_w^2)\right)(du/dt) \quad \text{if } F_b < mF_z$$
$$F_b = mF_z \text{ if } F_b \geq mF_z$$

The first vehicle is equipped with an ABS such that it can maintain $\mu \approx \mu_p$, where $\mu_p = 0.7$. The second vehicle does not have ABS and may experience wheel lockup shortly after braking, such that $m = 0.5$. Use the values $I_w = 0.003$ kg-m^2, $r_w = 0.3$ m, and $F_z = mg = (2{,}000 \text{ kg})(9.8 \text{ m/s}^2) = 19{,}600$ N. Consider the brake-torque level $T_b = 7{,}000$ N-m.

4. For a vehicle with ideal tires (i.e., F_y/F_z is proportional to the tire-slip angle):

    (a) If we redesign the powertrain location so that the vehicle weight does not change but the CG moves forward (i.e., new $a = 90$ percent of the old a, where a is the distance from the CG to the front axle; the length l is not changed), will the vehicle become more understeer? More oversteer? By how much?

    (b) If we compare an empty car to a car with a full load in the trunk (i.e., m increased by 40 percent, a increased by 10 percent), will the vehicle become more understeer? More oversteer? By how much?

5. Recall that the yaw-rate gain is the gain of the transfer function from steering angle to vehicle yaw rate at steady-state, and that flat tires have lower cornering stiffness than normally inflated tires. Provide equations and/or calculations as a basis for discussing whether the following statements are true or false:

    (a) "Flat rear tires make a vehicle more understeer."
    (b) "Flat front tires reduce the yaw-rate gain."
    (c) "If all vehicle parameters (including tire cornering stiffness) are fixed and only the vehicle CG is moved backward, this vehicle becomes more understeer."

6. In this problem, the performance of a RWD versus a FWD vehicle is considered. Consider the longitudinal motion of an automobile traveling on a slope where the angle of the slope is small (i.e., $\theta \approx 0$).

    (a) Let the rolling resistance $R_x = 0$. Show that:

$$F_{zf} = \frac{b}{L}W - \frac{h}{L}F_x$$
$$F_{zr} = \frac{a}{L}W + \frac{h}{L}F_x$$

    where F_{zf} and F_{zr} are the axle loads on the front and rear wheels and F_x is the total traction force.

(b) Find the maximum tractive force possible for the FWD and RWD vehicles assuming that the coefficient of static friction is μ.

(c) If $a = l/4$ (i.e., the center of mass is toward the front of the vehicle because of the large engine and transmission components mass), $h = 2l/7$, and assuming $\mu = 0.7$, compare the tractive forces of a FWD versus RWD automobile.

7. Using the two-DOF linearized handling model in Eq. (4.49), simulate the response of a vehicle to a one-degree change in the steering angle. Plot the results for the lateral acceleration and the yaw rate. The vehicle model to be used in the simulation is the following:

$$\begin{bmatrix} (mu_0 D + Y_\beta) & (mu_0 + Y_r) \\ N_\beta & (I_z D + N_r) \end{bmatrix} \begin{Bmatrix} \beta \\ r \end{Bmatrix} = \begin{Bmatrix} Y_\delta \\ N_\delta \end{Bmatrix} \delta_f$$

where D denotes the derivative operator d/dt and

$$Y_b = C_{af} + C_{ar} \qquad Y_r = \frac{C_{af}a}{u_0} - \frac{C_{ar}b}{u_0} \qquad\qquad Y_d = C_{af}$$

$$N_b = aC_{af} - bC_{ar} \qquad N_r = (a^2/u_0)C_{af} + (b^2/u_0)C_{ar} \qquad N_d = aC_{af}$$

The parameter values to be used in the simulation are as follows:

$$l = 2.54 \text{ m} \qquad\qquad a = 1.14 \text{ m} \qquad\qquad b = l - a = 1.40 \text{ m}$$
$$g = 9.81 \text{ m/s}^2 \qquad\qquad u_0 = 20.0 \text{ m/s} \qquad\qquad m = 1{,}000 \text{ kg}$$
$$I_z = 1{,}200.0 \text{ kg-m}^2 \qquad C_{af} = 2{,}400.0 \text{ N/deg} \qquad C_{ar} = 2{,}056.0 \text{ N/deg}$$

Is this vehicle understeer or oversteer? What is the understeer coefficient? What is the critical or characteristic speed?

8. Consider a one-DOF suspension model (i.e., neglect the unsprung mass). Refer to the following figure in which z_0 is the vertical ground motion input and z_s is the deflection of the vehicle mass. The passive-suspension stiffness $k_l z + k_{nl} z^3$ is non-linear, c is the passive-suspension damping, and z is the extension of the suspension spring around its nominal position.

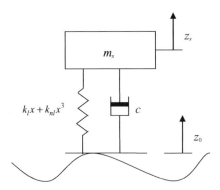

(a) Write the equation of motion for the sprung mass (it is helpful to draw the free-body diagram of the sprung mass).

(b) Use the Taylor series and linearize the equation about equilibrium (i.e., $z_s = 0$ and $z_0 = 0$). Neglect any effects due to gravity. Clearly show the steps used to linearize the system.

(c) Write the transfer function from road input, Δz_0, to displacement of the sprung mass, Δz_s.

(d) Let $m = 1$ kg, $k_l = 2$ N/m, $k_{nl} = 1$ N/m^3, and $c = 2$N-s/m. If the road input is $z_0 = \sin(8\pi t)$, what will the magnitude of z_s be at steady-state?

9. Refer to the suspension design in Example 4.9. Plot the performance index J versus ω_2 as in Example 4.9 but show on the same plot the curves for each of the following values of the mass ratio, ρ: $\rho = 8$, $\rho = 10$, and $\rho = 12$. How sensitive are the value of J and the location of the minimum point to the value of ρ?

10. When a vehicle drives on a "washboard" road surface, it is generally known that a smoother ride is possible if we drive either very slowly or very fast. Explain this phenomenon.

REFERENCES

Asgari, J., and D. Hrovat, 1991, "Bond Graph Models of Vehicle 2D Ride and Handling Dynamics," in S. A. Velinsky, R. H. Fries, I. Haque, and D. Wang (eds.), *Advanced Automotive Technologies–1991*, ASME DE–Vol. 40, New York, pp. 391–406.

Bakker, E., L. Nyborg, and H. B. Pacejka, 1987, "Tire Modeling for Use in Vehicle Dynamics Studies," SAE Paper 870421.

Bakker, E., H. B. Pacejka, and L. Linder, 1989, "A New Tyre Model with an Application in Vehicle Dynamics Studies," SAE Paper 89007.

Bosch, R., 2009, *Automotive Handbook*, Wiley & Sons, New York.

Chen, H. H., A. A. Alexandridis, and R. M. Chalasani, 1991, "Longitudinal Dynamics and Performance Assessment of All-Wheel-Drive Vehicles," in S. A. Velinsky, R. H. Fries, I. Haque, and D. Wang (eds.), *Advanced Automotive Technologies–1991*, ASME DE–Vol. 40, New York, pp. 483–99.

Clark, S. D., 1971, *Mechanics of Pneumatic Tires*. National Bureau of Standards Monograph 122, November 1971.

Dixon, J. C., 1991, *Tyres, Suspension and Handling*, Cambridge University Press.

Ellis, J. R., 1966, *Vehicle Dynamics*, London.

Gillespie, T., 1992, *Fundamentals of Vehicle Dynamics*, Society of Automotive Engineers.

Godthelp, J., G. J. Blaauw, and A. R. A. van der Horst, 1982, "Instrumented Car and Driving Simulation: Measurements of Vehicle Dynamics," Institute for Perception TNO, Report IZF 1982–37, The Netherlands.

Hedrick, J. K. and T. Butsuen, 1988, "Invariant Properties of Automotive Suspension," *Advanced Suspensions, Proc. of the Institute of Mechanical Engineers*, Paper No. C423/88, October 1988.

Heydinger, G. J., P. A. Grygier, and S. Lee, 1993, "Pulse Testing Techniques Applied to Vehicle Handling Dynamics," Proceedings of the 1993 SAE International Congress and Exposition, SAE Paper 930828.

Kiencke, U., and L. Nielsen, 2005, *Automotive Control Systems for Engine, Driveline and Vehicle*, Springer Publishing Co.

Levitt, J. A. and N. G. Zorka, 1991, "The Influence of Tire Damping in Quarter Car Active Suspension Models," *ASME Journal of Dynamic Systems, Measurement and Control*, Vol. 113, March 1991, pp. 134–7.

Lin, C. F., A. G. Ulsoy, and D. J. LeBlanc, 2000, "Vehicle Dynamics and External Disturbance Estimation for Vehicle Path Prediction," *IEEE Transactions on Control System Technology*, Vol. 8, No. 3, May 2000, pp. 508–18.

MacAdam, C. C., 1989a, "Static Turning Analysis of Vehicles Subject to Externally Applied Forces – A Moment Arm Ratio Formulation," *Vehicle System Dynamics*, Vol. 18, No. 6, December 1989.

MacAdam, C. C., 1989, "The Interaction of Aerodynamic Properties and Steering System Characteristics on Passenger Car Handling," *Proceedings of the 11th IAVSD Symposium of the Dynamics of Vehicles on Roads and Tracks*, ed. R. Anderson, Kingston, Canada, Swets & Zeitlinger B.V. – Lisse, 1989.

MacAdam, C. C., et al., 1990, "Crosswind Sensitivity of Passenger Cars and the Influence of Chassis and Aerodynamic Properties on Driver Preferences," *Vehicle System Dynamics*, Vol. 19, No. 4, 1990, pp. 201–36.

Pacejka, H. B., 2006, *Tyre and Vehicle Dynamics*, Elsevier.

Rajamani, R., 2006, *Vehicle Dynamics and Control*, Springer Publishing Co.

SAE, 1976, *Vehicle Dynamics Terminology: SAE J670e*, Society of Automotive Engineers.

SAE, 1989, *Vehicle Dynamics Related to Braking and Steering*, Society of Automotive Engineers.

Sayers, M., 1991, *Introduction to AUTOSIM*, University of Michigan, Ann Arbor.

Segel, L., 1966, "On the Lateral Stability and Control of the Automobile as Influenced by the Dynamics of the Steering System," *ASME Journal of Engineering for Industry*, 66, August 1966.

Segel, L., 1975, "The Tire as a Vehicle Component," in B. Paul, K. Ullman, and H. Richardson (eds.), *Mechanics of Transportation Suspension Systems*, ASME, New York, AMD–Vol. 15.

Segel, L., 1990, *Vehicle Dynamics*, Course Notes for ME558, University of Michigan, Ann Arbor.

Segel, L., and C. C. MacAdam, 1987, "The Influence of the Steering System on the Directional Response to Steering," *Proceedings of the 10th IAVSD Symposium of the Dynamics of Vehicles on Roads and Tracks*, Prague, Czech Republic, 1987.

Segel, L., and R. G. Mortimer, 1970, "Driver Braking Performance as a Function of Pedal-Force and Pedal-Displacement Levels," *SAE Transactions*, SAE Paper 700364.

Venhovens, P. J. T., 1993, *Optimal Control of Vehicle Suspensions*, Doctoral Dissertation, Delft University, The Netherlands.

Wong, J. Y., 2008, *Theory of Ground Vehicles*, Wiley & Sons, New York.

5 Human Factors and Driver Modeling

It often is necessary to consider the human role (i.e., drivers and passengers) in the design of automotive systems. For example, this is evident in the discussion of passenger comfort as a key criterion for suspension design in Chapter 4. *Human factors*, also known as *human engineering* or *human-factors engineering*, consist of the application of behavioral and biological sciences to the design of machines and human–machine systems (Sheridan 2002). The term *ergonomics* is used as a synonym for human factors; however, it often is associated with narrower aspects that address anthropometry, biomechanics, and body kinematics as applied to the design of seating and workspaces. The terms *cognitive engineering* and *cognitive ergonomics* also are used to describe the sensory and cognitive aspects of human interactions with designed systems. All major automotive companies, as well as many government agencies (e.g., U.S. Department of Transportation, U.S. Department of Defense, NASA, and Federal Aviation Administration), have research and engineering groups that address human factors. This chapter is a brief introduction to human factors, especially as they apply to automotive control system design. The introduction is followed by a discussion of driver models, especially for vehicle steering.

5.1 Human Factors in Vehicle Automation

Humans (i.e., drivers and passengers) clearly interact with automotive control systems in many ways. Commercial success of a new control technology for vehicles may depend on not only the effectiveness of that technology but also acceptance by customers. Established automotive technologies (e.g., automatic transmissions and cruise control) are widely used in the United States but are less widely adopted in Europe. Navigation systems are more successful in Japan than in the United States. Customer acceptance often depends on many difficult-to-quantify factors. New automotive control technologies in which human factors must be considered carefully include electric vehicles, hybrid electric vehicles, traction control, ABS, intelligent (or adaptive) cruise control, cruise control, airbags, active suspensions, and navigation systems. For example, ABS design must consider not only prevention of vehicle skidding on various road surfaces during braking but also whether to provide any feedback to the driver (e.g., pedal vibration) about ABS activation. ABS is most effective when a driver applies the brakes continuously rather than pumping the brake pedal. Thus, the ABS design must consider how the combined human–ABS system will perform and interact. This is a difficult task because driver behavior is quite variable among different drivers as well as over time. For example, some drivers prefer to have brake-pedal-vibration feedback when the ABS is activated, whereas others are startled and actually stop the braking action in the presence of the vibrations. Drivers also adapt to a particular technology and their behavior changes with experience. Thus, a major issue that must be considered is driver adaptation; that is, the presence of a new technology may cause drivers to change their behavior. For example, once drivers adapt to the improved braking capabilities of an ABS, they may drive more aggressively.

Human-factor issues in design also can be complicated by legal considerations. For example, airbags are effective in reducing deaths due to car crashes among the general public. However, they actually can increase the risk for drivers (or passengers) of small physical stature, such as children. Thus, any provisions must be designed so that they will be used by those who benefit while also enabling those at higher risk of injury to deactivate them. Alternatively, smart airbags that deploy differently must be developed. Another example of the influence of legal considerations is intelligent (or autonomous) cruise-control systems that control vehicle headway as well as speed. To avoid potential litigation, some of these systems do not control the brakes, despite the potential for better performance.

As vehicles become more automated and ITS also plan to implement more intelligence in the highway infrastructure, it is necessary to assess how helpful these new technologies are to drivers (Barfield and Dingus 1998). For example, driver distraction caused by using a cellphone (although it can be an important asset in a roadside emergency) is currently a controversial topic, which has led to bans in some localities. Making the situation more complicated is the fact that whereas some research finds that cellphone use increases accidents, other studies have not found any noticeable effects compared to other distractions. Vehicle navigation systems (based on GPS technology) also have been introduced and must be carefully designed to minimize driver distraction (Eby and Kostyniuk 1999).

With platoons of vehicles under automated lateral and longitudinal control, AHS were demonstrated on a large scale on a special segment of highway in San Diego, California. However, despite great technological success in the demonstrations, AHS is viewed with skepticism. This is partially due to concerns regarding interactions between humans and the automated system – for example, transfer of control at entrances and exits and authority/responsibility between controller and driver, especially when failures occur.

Consequently, the design of vehicle-control systems must include consideration of the human role. Although this is a difficult task, human-factor experiments can be conducted using vehicle experiments on test tracks or roadways, as well as driving simulators that include actual drivers in the system. Human-factor experiments must be designed carefully to include the factors of interest and to exclude those that are not a focus of a study. For example, possible factors to consider in the design of experiments may include the driving scenario and the driver's experience, alertness, health, physical characteristics, age, and gender. Due to significant variations in driver behavior (both over time and among drivers), the results of human-factor tests typically must be evaluated statistically. It may be necessary to randomize the

Figure 5.1. Single-loop driver-vehicle control-system block diagram.

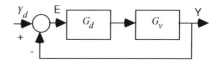

order of tests (e.g., when comparing two candidate system designs) to eliminate any bias (Charlton and O'Brien 2002; Fuller and Santos 2002).

In the following section, the *crossover-model principle*, which refers to an experimentally observed characteristic of all human–machine systems, is described. Driver adaptation and risk-homeostasis theory (Wilde 1994) then are introduced. Finally, the use of driving simulators for human-factor engineering is described briefly.

Crossover-Model Principle

This principle is based on an experimental observation from many tests performed on a variety of human–machine systems. It also has been found empirically to be applicable to driver steering and it is in that context that it is discussed here. Thus, any good driver model is expected to yield results consistent with this principle. To explain the principle, refer to the block diagram in Figure 5.1, which is a SISO model of the driver–vehicle system. In experiments in which human operators are asked to reduce the error, e, to zero (e.g., in computer simulations or in lane-following driving tasks), it was observed that regardless of the characteristics of the vehicle-transfer function, $G_v(s)$, human operators adjust their characteristics (i.e., $G_d(s)$) such that the loop-transfer function, $G_d(s)G_v(s)$, has an invariant property in the vicinity of the crossover frequency, ω_c.

This is illustrated in Figure 5.2, which shows the magnitude of the loop-transfer function, $|G_d(j\omega)G_v(j\omega)|$, versus frequency, ω. The observation, termed the *crossover model*, is that this plot exhibits a slope of -20 dB/decade in the vicinity of the crossover frequency, ω_c, which is the frequency at which $|G_d(j\omega)G_v(j\omega)|$ crosses the 0 dB line (i.e., where the open-loop gain equals 1). This implies that around the frequency point, $\omega = \omega_c$, the loop-transfer function, $G_d(s)G_v(s)$, can be approximated by $\frac{\omega_c e^{-sT}}{s}$. The crossover frequency and, consequently, the closed-loop-system bandwidth decrease with the difficulty of the control task (i.e., the specific $G_v(s)$). However, the basic crossover-model principle applies regardless of the $G_v(s)$ characteristics.

The crossover principle applies for frequencies near the crossover frequency but not at much higher or lower frequencies. Nevertheless, for single-loop systems, this principle essentially leads to correct closed-loop characteristics because the shape of the open-loop transfer function that is far from the crossover frequency, ω_c, usually has little effect on closed-loop dynamics. In other words, the crossover model can be

Figure 5.2. Frequency-response characteristic of the crossover model.

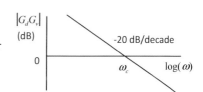

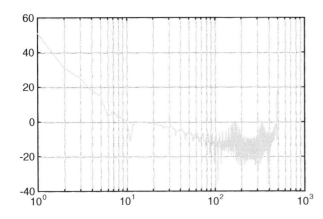

Figure 5.3. FFT of loop-transfer function in Example 5.3.

used with good accuracy to generate the closed-loop transfer function of the system. The phase margin ($\frac{\pi}{2} - \omega_c T$) of the closed-loop system also can be estimated from the approximation but usually should be treated with caution.

EXAMPLE 5.1: ILLUSTRATION OF THE CROSSOVER-MODEL PRINCIPLE. We consider here a vehicle transfer function, $G_v(s)$, and a driver transfer function, $G_d(s)$, which are described in detail in Example 5.3. The MATLAB program in that example also calculates the magnitude of the loop-transfer function (i.e., $|G_d(j\omega)G_v(j\omega)|$). The magnitude of the loop-transfer function can be calculated directly from the transfer functions using the MATLAB "bode" command. In this example, the calculation is performed from the simulation data in Example 5.3 using the "fft" (i.e., Fast Fourier Transform) function available in MATLAB. The results, illustrated in Figure 5.3, show that the crossover frequency is $\omega_c \approx 3\pi$ rad/s (1.5 Hz) and that the loop-transfer function exhibits a slope of approximately −20 dB/decade for $\omega \approx \omega_c$. This is in accordance with the crossover-model principle outlined herein. Thus, the simulated driver in Example 5.3 gives results consistent with the crossover model.

Risk-Homeostasis Theory

It is perhaps not surprising that as vehicles and highways are improved to increase driver and passenger safety, drivers will push the limits of a new technology. This hypothesis is termed *risk homeostasis* (Hoyes et al. 1996; MacGregor and Slovic 1989; Summala 1996) and is a controversial theory. It assumes that a driver behaves in a manner as to maintain a certain (acceptable) level of risk, the so-called target risk. Clearly, this level of risk varies from person to person; for example, as road curvature varies, drivers increase or reduce their vehicle-longitudinal velocity to maintain an acceptable level of risk. The adaptation of driver behavior to maintain a certain level of risk can be an important factor in vehicle design as well as many active safety systems for vehicles.

For example, ABS are designed to reduce not only stopping distances but also the occurrence of vehicle-lateral instability during braking while steering, especially in icy conditions. ABS have been proven in many vehicle tests to be quite effective. When ABS was first introduced commercially as an option, the insurance premiums of owners of ABS-equipped vehicles were reduced with the expectation that the incidence of accidents would decrease. However, an extensive study subsequently conducted by the insurance industry showed no statistically significant reduction in accidents of vehicles equipped with ABS. Although there was much debate about the reasons for this surprising finding, a popular explanation is based on driver adaptation and risk-homeostasis theory. Another possible explanation is related to the fact that human drivers need time to adapt to the changed vehicle behavior. From statistics, the reduction in rear-end crashes is offset by the surge in the number of vehicle rollovers – likely due to the fact that human drivers accustomed to locked wheels oversteer to correct vehicle motion.

It has been shown that modifying vehicle design to improve handling in a steering maneuver does not have the desired result due to driver adaptation (Mitschke 1993). The steering performance remains the same and the driver adapts to the vehicle characteristics so as to reduce the effort associated with the steering task. Similar results also were reported in studies of an active safety system to reduce the incidence of single-vehicle roadway-departure (SVRD) accidents (Chen and Ulsoy 2002), as discussed in Example 2 herein.

Driving Simulators

Driving simulators come in various types and are described in this section, including a human driver and a simulated vehicle and driving scenario. They enable the safe and inexpensive testing of new automotive control technologies as well as testing of prototype systems during the development process. Because human factors often are critical, driving-simulator studies can be an important tool in the development of automotive control technologies.

In simulator experiments to test braking reactions and following distances of drivers under various conditions, it was found that following distance correlates surprisingly poorly with the speed of the traffic stream – to the point that most drivers could not stop in time to avoid a collision if the driver of the vehicle ahead were to apply the brakes suddenly (Chen et al. 1995). Due to safety concerns, such studies are difficult to perform in actual driving experiments. Similarly, driving-simulator studies were used to develop and evaluate active safety systems to reduce SVRD accidents (Chen and Ulsoy 2002). The studies showed the effectiveness of these systems but also provided early identification of issues that must be considered in their design (e.g., driver adaptation).

The simplest driving simulators run on personal computers with joystick (or mouse) controls and they are similar to flight simulators available commercially. More advanced versions may have improved driver controls, such as foot pedals for braking and acceleration, as well as a steering wheel, which also may have passive or active force feedback to provide a realistic feel in handling. A key element of such a simulator is vehicle dynamics, which is simulated based on models such as those introduced in Chapter 4. Vehicle dynamics must be sufficiently high-fidelity to provide a realistic driving experience and typically includes nonlinearities (e.g., tire models) as well as wind disturbances, ground-disturbance input, and so on. Simulators also include a graphical display of what a driver sees during driving. These displays can range from very simple to very complex, and they have an important role

Figure 5.4. Sample screen display of the driving simulator.

in a driver's perception of the driving task. Some displays include realistic depictions of actual roadway-driving tasks of up to several hours. Basic driving simulators are on a fixed base; however, full-scale driving simulators with a motion base also are available (e.g., the National Advanced Driving Simulator at the University of Iowa and large-scale driving simulators at companies including Deere, Ford, Mercedes and Toyota). These devices have hydraulically actuated platforms that provide the feel of acceleration and deceleration to simulate a realistic driving experience.

EXAMPLE 5.2: DRIVER ADAPTATION IN AN ACTIVE SAFETY SYSTEM FOR SVRD ACCI-DENTS. Referring to the block diagram in Figure 5.1, the proposed active safety system can be viewed as a controller, $G_s(s)$, placed between the driver steering-command output and the actual steering input applied to the vehicle. Consequently, the product of $G_s(s)$ and $G_v(s)$ can be viewed as an equivalent vehicle-steering dynamics model. Essentially, the $G_s(s)$ is designed to provide more consistent lane-keeping performance for the closed-loop driver–vehicle system, thereby reducing the occurrence of SVRD accidents. The details of the design of $G_s(s)$ are not discussed here but readers are referred to Chen and Ulsoy (2001 and 2002).

A personal-computer–based driving simulator was used to evaluate the effectiveness of this controller in reducing lane-departure accidents (Chen and Ulsoy 2002). It has a simple graphical display (Figure 5.4) with fairly sophisticated vehicle-dynamics models. The driving simulator was used for straight-road lane-keeping tasks, subject to lateral wind disturbances.

Two types of experiments were conducted to evaluate the controllers using the simulator with and without the designed steering-assist controller, as follows:

Long Test. These experiments are designed to address the effect of fatigue. Therefore, the six subject drivers are required to become inattentive or drowsy during the experiments. Although systematic measures of driver fatigue have

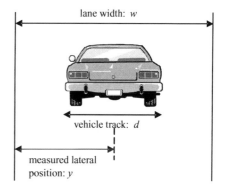

Figure 5.5. Vehicle location relative to roadway.

been reported in human-factors research, an accurate and quantitative measure of driver fatigue usually requires complex instrumentation. Instead, the subject drivers are asked to follow guidelines to achieve a certain level of fatigue (e.g., arise at 7:30 a.m. or earlier, experiments scheduled just before the "circadian dip of alertness" in the early afternoon, and no coffee). Consequently, the six subject drivers become drowsy and performance degradation is observed in the experiments. The driving task is at least 60 minutes long of straight-road driving with a wind disturbance. The human driver functions as a regulator to maintain the vehicle at the center of the lane. The data from the long driving experiments are divided into 1-minute segments to observe how the driving behavior varies with time.

Short Test. These experiments consist of short driving with artificially large lateral-position error. To generate consistent lateral-position error, the ten subject drivers are asked to steer the simulated vehicle to the left until the center of the vehicle reaches the centerline of the road (i.e., the left edge of the lane). They then steer the simulated vehicle back to the right until the vehicle center reaches the right edge of the lane. After that, the ten drivers are asked to steer the vehicle back to the center of the lane as rapidly as possible. The total experimental time for one test run is approximately 15 seconds. The same test is repeated forty times for each driver (both with and without the controller) to generate a large dataset for statistical analysis.

The lane-keeping performance, with and without the controller, was then assessed using the following metrics. First, we consider two time-domain metrics described as follows.

Percentage of Road Departure (PRD). Figure 5.5 shows the relative location of the vehicle and the roadway. Denote the lateral-position error as $y(i)$, $i = 1, 2, \ldots N$, where N is the number of datapoints for each data segment to be analyzed. *Road departure* is defined as when $y \leq \frac{d}{2}$ or $y \geq w - \frac{d}{2}$. These two cases represent situations in which the vehicle edges exceed the road edges. PRD is defined as the amount of time in which road departure occurs divided by the overall simulation time; that is:

$$PRD = \frac{1}{N} \sum_{i=1}^{N} f(y(i)), \quad \text{where} \begin{cases} f(y(i)) = 1, & \text{if } y(i) \leq \frac{d}{2} \text{ or } y(i) \geq w - \frac{d}{2} \\ f(y(i)) = 0, & \text{if } \frac{d}{2} < y(i) < w - \frac{d}{2} \end{cases}$$

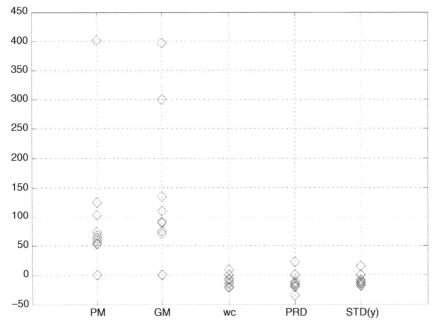

Figure 5.6. Percentage change in lane-keeping metrics for a case with controller as compared to a case without controller in short driving tests on the driving simulator.

For the long driving experiments, the overall simulation time is 60 seconds because each segment of data is 1-minute long. The data show a higher PRD observed in the later part of the long driving task, which indicates more fatigued drivers. For the short driving experiments, the overall simulation time is 15 seconds. The objective of this test is to verify that the PRD, or the variation in PRD, can be reduced when the controller is implemented.

Standard Deviation of Lateral Position (STD(y)). STD(y) is defined as $\sqrt{\frac{1}{N-1}\sum_{i=1}^{N}(y(i)-\frac{1}{N}\sum_{i=1}^{N}y(i))^2}$. The standard deviation (or variance) of lateral-position error is used in the literature as a performance index for vehicle lane-keeping. With the controller, the goal is to show that STD(y) can be maintained at a low level or that it will not vary with time (for the long driving experiments) as much as in the case without the controller.

Also, frequency domain metrics (i.e., gain margin, phase margin, and crossover frequency for the loop-transfer function, $G_d G_s G_v$) can be used. Using the driving-simulator data, the driver model can be computed. The driver models then can be combined with the vehicle and controller models to obtain the loop-transfer-function frequency response. The stability margins and crossover frequency then are computed and used to evaluate the benefits of the controllers. Improvement in lane-keeping performance is indicated by increased stability (i.e., gain and phase) margins. The crossover frequency can be viewed as a measure of driver effort, with lower values indicating less effort to maintain a given level of lane-keeping performance.

The results obtained are interesting (Chen and Ulsoy 2002b): In the short driving tests, all of the metrics show statistically significant improvement in lane-keeping performance, based on a student t-test. Figure 5.6 shows the percentage change for the system with the controller compared to the system

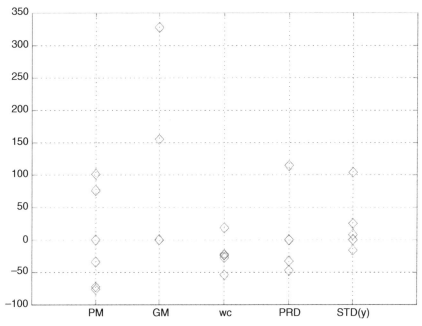

Figure 5.7. Percentage change in lane-keeping metrics for a case with controller as compared to a case without controller in the long driving tests on the driving simulator.

without the controller in the various metrics for the ten drivers in the short tests. In most cases, the metrics are improved (i.e., gain and phase margins are increased and PRD and STD(y) are reduced). Note also that the crossover frequency is reduced, indicating less effort on the part of a driver to achieve this level of performance.

Although these results are encouraging, the long driving tests provide a different conclusion. Figure 5.7 shows no clear (i.e., statistically significant) change in the lane-keeping metrics (i.e., gain and phase margins, PRD, and STD(y)). In some cases, the metrics improve; in others, they remain the same or worsen. There is only a statistically significant reduction in crossover frequency, indicating a reduced effort by a driver to achieve the same (statistically speaking) lane-keeping performance. The likely explanation is based on risk-homeostasis theory. That is, drivers adapt their behavior – by using a more relaxed driving style – to maintain the same level of risk in terms of lane-keeping performance. In the short driving tests, there is not enough time for this adaptation to occur; therefore, it is observed only in the long tests. Although not shown here, these conclusions are confirmed by computing driver models from the various test data, as described in (Chen and Ulsoy 2002).

5.2 Driver Modeling

The driving function can be decomposed into three parts: *navigation*, *path planning*, and *control*. Navigation addresses the overall selection of a route. Path planning is concerned with the recognition, decision, and selection of a path responding to traffic situations (e.g., overtaking). The control process actuates the steering, acceleration, and brake systems to follow the selected path determined by the two higher levels. In this section, several driver-control models are presented.

In Chapter 4, vehicle dynamics is considered without consideration of the driver's role. It is obvious that the behavior of a vehicle as a closed-loop system (i.e., with the driver closing the loop and serving simultaneously as the sensor, the controller, and the actuator) is different than the behavior of a free vehicle. This is because the driver is an active element of the dynamic system. For example, by steering, the driver clearly has an important role in vehicle-lateral motion. The driver also affects vehicle-longitudinal motions by controlling the tractive and braking forces through the accelerator and brake pedals. For example, a good driver braking on an icy surface often "pumps" the brakes to maintain lateral stability – much like the ABS that is discussed in Chapter 13. Although drivers cannot affect directly the vehicle-vertical motion, they are an important passive element in the suspension-design problem. Drivers also must be considered a passive element in many other vehicle-control-system design problems (e.g., 4WS and transmission control). In this chapter, the driver's role as an active element in a closed-loop system is considered, particularly in terms of vehicle steering.

To date, there have been numerous studies about developing "driver models" for steering behavior. These studies are useful in the following ways:

- *Safety Issues.* How will the combined (i.e., driver and vehicle) system behave in certain emergency situations and maneuvers (e.g., collision avoidance and rollover)? Problems such as truck rollover and jackknifing, vehicle rollover (e.g., minivans and sport utility vehicles [SUVs]), and emergency handling have long been significant issues in vehicle design.
- *Vehicle Handling.* How should a vehicle be designed so that the combined system will perform well and the driver will not be easily fatigued but rather find the vehicle pleasurable to drive? Studies in this area show that vehicles that allow drivers to perform well are not necessarily those that are the most pleasurable to drive.

Figure 5.8 illustrates a driver-model block diagram and delineates three types of actions that a driver can perform while driving (Weir and McRuer 1968). This block diagram shows that a driver may have several types of control action that can be used, including the following:

- *Precognitive behavior*, in which a driver has some internally generated maneuver commands. These usually are rote maneuvers, such as pulling into a driveway.
- *Pursuit behavior*, which corresponds to a feed-forward or preview action based on knowledge of an upcoming road path. This is discussed in more detail herein.
- *Compensatory behavior*, which reacts to sensed errors. This is the transfer-function approach and is a subsequent discussion herein.

Various driver-steering models have been proposed in the past thirty years (Hess and Modjtahedzadeh 1990; MacAdam 1989; Weir and McRuer 1968). We can classify these into two major categories: (1) transfer-function models, and (2) preview/predictive models. The transfer-function models mainly describe the compensatory behavior of a driver, whereas preview/predictive models focus on a driver's pursuit behavior. Each type of driver models is briefly described in the following sections.

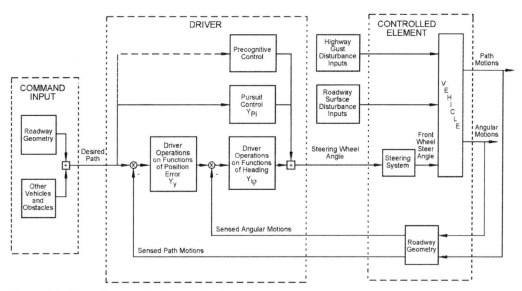

Figure 5.8. Block diagram for human driving task.

Transfer-Function Models

Consider a simple closed-loop system (see Figure 5.1) for lateral-motion control of a vehicle through steering. The vehicle dynamics is represented by the transfer function, $G_v(s)$, and the driver (i.e., controller) has the transfer function, $G_d(s)$. Here, it is assumed implicitly that the driver acts on feedback of the vehicle-lateral position, y. This assumption is a basic problem with the transfer-function approach; in fact, the driver of a vehicle acts on many sensed feedback signals (e.g., lateral position and velocity, yaw and angle and rate, roll angle and rate, and pitch angle and rate). Thus, there are multiple feedback loops and, typically, different gains and transfer functions must be associated with each. Such a system becomes complicated and it is difficult to determine experimentally the control actions associated with each feedback loop. However, the simple configuration shown in Figure 5.1 is suitable for the experimental determination of driver control laws and gains. Another advantage of this simple approach is that it is generalized; no assumptions must be made about the driving strategy used by a driver. Possible approaches to determining the driver transfer function, $G_d(s)$, are as follows: (1) Assume that $G_v(s)$ is completely known from a vehicle-dynamics model and that the form of $G_d(s)$ is known; the parameters of $G_d(s)$ then can be determined based on measured values of the error, e, and the steer angle, δ. (2) Alternatively, assume that the form of the combined product $G_d(s)G_v(s)$ is known; then determine $G_d(s)$ from measurements of e and y. In the first case, the experimental data required are e and δ; in the second case, they are e and y.

Hess and Modjtahedzadeh (1990) provided a transfer-function model of driver steering, illustrated in Figure 5.9. The driver is represented by a low-frequency compensation block (i.e., a PD controller, as in Example 5.3) and a high-frequency (i.e., within one decade of the crossover frequency) compensation (or structural model) block, which includes the driver delay, τ_0. The block labeled G_{NM} is a simple second-order model of the neuromuscular system of a driver's arms. A more detailed

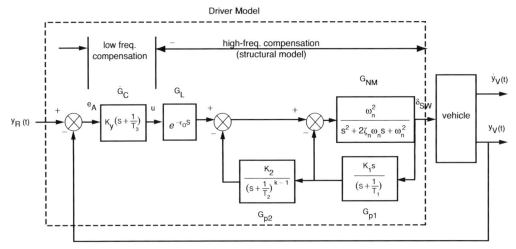

Figure 5.9. Transfer-function driver model (Hess and Modjtahedzadeh 1990).

description of the driver model, as well as typical values of the model parameters, is in Hess and Modjtahedzadeh (1990). A simpler version of this model with only low-frequency compensation and delay is provided in Example 5.3.

EXAMPLE 5.3: DRIVER HANDLING SIMULATION. A vehicle-transfer function $G_v(s)$ with front-wheel-steering angle δ as the input and lateral displacement (y) as the output signal in the straight-road following case can be obtained from this state-space model:

$$\frac{d}{dt}\begin{bmatrix} y \\ v \\ \psi \\ r \end{bmatrix} = \begin{bmatrix} 0 & 1 & u_o & 0 \\ 0 & -\dfrac{C_{\alpha f}+C_{\alpha r}}{mu_o} & 0 & \dfrac{bC_{\alpha r}-aC_{\alpha f}}{mu_o}-u_o \\ 0 & 0 & 0 & 1 \\ 0 & \dfrac{bC_{\alpha r}-aC_{\alpha f}}{I_z u_o} & 0 & -\dfrac{a^2C_{\alpha f}+b^2C_{\alpha r}}{I_z u_o} \end{bmatrix} \begin{bmatrix} y \\ v \\ \psi \\ r \end{bmatrix} + \begin{bmatrix} 0 \\ \dfrac{C_{\alpha f}}{m} \\ 0 \\ \dfrac{aC_{\alpha f}}{I_z} \end{bmatrix} \delta$$

Consider a simple PD controller with delay as the driver model:

$$G_d(s) = (\delta(s)/e(s)) = (K_d + K_{dd}s)e^{-sT} \tag{5.1}$$

If the delay T is sufficiently small, then one of the following Padé approximations can be used:

$$e^{-sT} \approx \frac{(Ts)^2 - 6(Ts) + 12}{(Ts)^2 + 6(Ts) + 12}; \quad \text{or} \quad e^{-sT} \approx \frac{1 - 0.5Ts}{1 + 0.5Ts}$$

In the following discussion, the second (i.e., first-order) approximation for the time delay is used. The vehicle parameters are assumed to be as follows:

$m = 1{,}500$ kg	$I_z = 2{,}420$ kg-m^2	$g = 9.81$ m/s^2
$l = 2.54$ m	$a = 1.14$ m	$b = l - a = 1.4$ m
$u_0 = 20$ m/s	$C_{af} = 2{,}050$ N/deg	$C_{ar} = 1{,}675$ N/deg

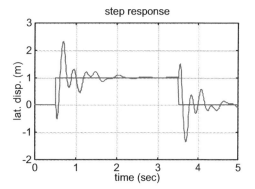

Figure 5.10. Response of the transfer-function model driver–vehicle system.

If we consider only proportional control with neural delay (i.e., $K_{dd} = 0$), then the system is found to be unstable (which can be verified easily by modifying the example program presented herein). However, if we add phase lead (e.g., use PD instead of P control such that both K_d and K_{dd} are nonzero), then it is possible to find gains that stabilize the system even with a significant time delay.

The step response of a closed-loop system with a simple PD control and a 0.1-second delay is shown in Figure 5.10. Apparently, the response is oscillatory and although it controls the lateral motion of the vehicle, it does not predict accurately the human-driver steering response. In part, this is due to neglecting the high-frequency compensation block in Figure 5.9. However, it also is known that human drivers actually use feedback signals in addition to lateral displacement to stabilize a vehicle, including path angle, path rate, heading angle, and heading rate.

```
% Ex5_3.m
a=1.14; l=2.54; b=l-a;
g=9.81; u0=20.0; m=1500; Iz=2420.0;
Caf=2050.0*57.2958; Car=1675*57.2958;
% Define the coefficients
Yb=-(Caf+Car); Yr=(Car*b/u0)-(Caf*a/u0); Yd= Caf; Nb=b*Car-
a*Caf; Nd= a*Caf; Nr=-(a^2/u0)*Caf - (b^2/u0)*Car;
A=[0 1 u0 0;
  0 Yb/m/u0 0 Yr/m-u0;
  0 0 0 1;
  0 Nb/Iz/u0 0 Nr/Iz];
B=[0;Yd/m;0;Nd/Iz];
C=[1 0 0 0]; D=0;
[num,Gvden]=ss2tf(A,B,C,D,1);
Gvnum=num(3:5);
Kd = 0.02; Kdd = 0.33;
T = 0.1;
Gdnum=conv([Kdd Kd],[-T/2 1]);
Gdden=conv([1e-6 1], [T/2 1]);
Gcnum=conv(Gdnum,Gvnum);
Gcden=conv(Gdden,Gvden) + [0 0 Gcnum];
t=[0:0.01:5];
```

```
yd=[zeros(1,51), ones(1,300), zeros(1,150)];
y=lsim(Gcnum,Gcden,yd,t);
e=lsim(Gcden-[0 0 Gcnum],Gcden,yd,t);
e = yd'-y;
plot(t,yd,'b',t,y,'r'); grid
title('step response')
xlabel('time (sec)')
ylabel('lat. disp. (m)'), pause
w=logspace(0, 2, 100);
bode(conv(Gvnum, Gdnum), conv(Gvden, Gdden),w); pause
%Calculate mean, standard deviation, and rms for e(t):
em=mean(e);
es=std(e);
erms=sqrt(mean(e.^2));
% Calculation of open-loop bode plot from simulation data:
Y=fft(y); E=fft(e);
H=Y./E; magH=20*log10(abs(H)); phaseH=angle(H)*(180/pi);
semilogx(magH); grid; pause;
semilogx(phaseH); grid;
```

Preview/Predictive Models

Another group of models uses a combined pursuit/compensatory action and mimics
human preview and predictive behavior by incorporating a forward model; these
are known as the preview/predictive models. In these models, human drivers scan
through a future desired road path within a finite future distance when performing
a driving task. This behavior is captured in the preview/predictive block. The terms
preview and *predictive*, respectively, refer to a driver's ability to see the future
desired path and to predict future vehicle response. Driver models that use preview
information over a horizon often generate approximate inverse control actions,
resulting in superior control quality when compared with transfer-function models.
This is especially true for high-lateral-acceleration path-following tasks (e.g., a sharp
curve or a double-lane-change maneuver). A well-known preview/predictive model
was proposed by MacAdam (1980) in which the driver was assumed to behave like
a preview optimal controller with delay. Because this model is widely verified and
it is closely related to the model proposed herein, the details are presented in the
following discussion.

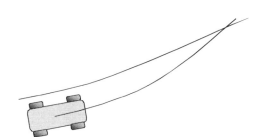

Figure 5.11. Preview/predicted paths.

MacAdam's model is based on the optimal preview control framework for SISO linear systems. Given the state equation (e.g., bicycle model) of a vehicle:

$$\dot{x} = Ax + Bu$$
$$y = Cx \tag{5.2}$$

the control (i.e, steering) signal minimizing a quadratic cost function is solved. The cost function proposed by MacAdam has the following form:

$$u_{opt}(t) = \min_{u} \left\{ \int_{t}^{t+T_p} \{[y_d(\eta) - y(\eta)]^2 \delta(\eta - t)\} d\eta \right\} \tag{5.3}$$

where $y_d(t)$ is the desired lateral displacement, $y(t)$ is the actual lateral displacement, and $\delta(t)$ is the weighting function over the preview window. In general, $u(t)$ could vary within the preview window $[t, t + T_p]$. However, the solution of this problem involves solving a partial-differential equation and may be unnecessary for approximating human behavior. A simpler problem can be formulated by assuming $u(t + \tau) = u_{opt}(t)$, $\forall \tau \in [0, T_p]$, which can be solved more easily. The output of the linear dynamics in Eq. (5.2) is decomposed into zero-input response and zero-state response:

$$y(t + \tau) = Ce^{A\tau}x(t) + C \left(\int_{0}^{\tau} e^{A\eta} d\eta \right) Bu(t) \equiv F(\tau)x(t) + G(\tau)u(t) \tag{5.4}$$

The optimal solution for Eq. (5.3) then can be obtained by substituting Eq. (5.4) in Eq. (5.3) and setting the partial derivative of the cost function J with respect to u as zero:

$$u_{opt}(t) = \frac{\int_{t}^{t+T_p} \{[y_d(\eta) - F(\eta - t)x(t)] G(\eta)\delta(\eta - t)\} d\eta}{\int_{t}^{t+T_p} G(\eta)^2 \delta(\eta - t) d\eta} \tag{5.5}$$

The optimal solution $u_{opt}(t)$ shown here can be viewed as a "proportional feedback" controller operating on the error between the desired output and predicted zero I/O over the preview window $[t, t + T_p]$ rather than for a single point in time. This "previewed proportional control" was found to result in a control law that well approximates average human-driver behavior and has been implemented in commercial software packages (i.e., CARSIM and TRUCKSIM). Both this approach and the transfer-function approach discussed previously may need to be made adaptive to account for variations in driver behavior. Experimental studies confirm that drivers indeed exhibit adaptive behavior in steering tasks, as discussed previously in Example 5.3.

EXAMPLE 5.4: PREVIEW DRIVER MODEL. A simple example of the preview-driver steering model is presented here. The preview model is not the one proposed by MacAdam referenced previously; rather, it is a simple linear model that can be simulated easily using MATLAB. Assume that the desired lateral position is known at time $t + T_p$, where t is the current time and T_p is the preview time. In

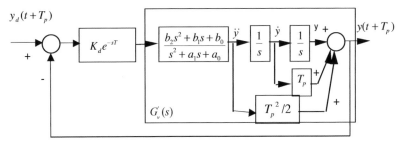

Figure 5.12. Block diagram of the preview/predictive driver–vehicle system.

other words, the value $y_d(t + T_p)$ is a known reference input. The actual path at time $t + T_p$ is estimated using the simple model:

$$y(t + T_p) = y(t) + (T_p)\dot{y}(t) + (T_p^2/2)\ddot{y}(t) \tag{5.6}$$

Then, the control action can be defined – for example, if we use a proportional gain plus delay driver model with preview – as follows (with some abuse of notation):

$$\delta(t) = K_d e^{-sT}(y_d(t + T_p) - y(t + T_p)) \tag{5.7}$$

The vehicle-transfer function $G_v(s)$ was found to be of the form:

$$G_v(s) = \frac{Y(s)}{\delta(s)} = \frac{b_2 s^2 + b_1 s + b_o}{s^2(s^2 + a_1 s + a_o)} \tag{5.8}$$

Therefore, the overall closed-loop system can be presented in the block-diagram form as shown in Figure 5.12.

Using block-diagram algebra leads to an equivalent vehicle-dynamics model of the following form:

$$G_v' = \frac{((T_p^2/2)s^2 + T_p s + 1)(b_2 s^2 + b_1 s + b_0)}{s^2(s^2 + a_1 s + a_0)} \tag{5.9}$$

This system can be simulated using a MATLAB program similar to the one shown previously in Example 5.3; only the numerator of the vehicle-dynamics transfer function is modified. Notice that the preview/predictive action adds derivative and double-derivative terms, thereby stabilizing the closed-loop system. The simulation results, with $K_d = 0.05$ and $T = 0.1$, are presented in Figure 5.13.

```
% Ex5.4.m
a=1.14; l=2.54; b=l-a;
g=9.81; u0=20.0; m=1500; Iz=2420.0;
Caf=2050.0*57.2958; Car=1675*57.2958;
% Define the coefficients
Yb=-(Caf+Car); Yr=(Car*b/u0)-(Caf*a/u0); Yd= Caf;
   Nb=b*Car-a*Caf; Nd= a*Caf; Nr=-(a^2/u0)*Caf - (b^2/u0)*Car;
A=[0 1 u0 0;
   0 Yb/m/u0 0 Yr/m-u0;
   0 0 0 1;
   0 Nb/Iz/u0 0 Nr/Iz];
```

```
B=[0;Yd/m;0;Nd/Iz];
C=[1 0 0 0]; D=0;
[num,Gvden]=ss2tf(A,B,C,D,1);
Gvnum=num(3:5);
Kd = 0.05; T = 0.1; Tp=0.5;
Gvnum_pv=conv(Gvnum, [Tp^2/2 Tp 1]);
Gdnum=Kd*[-T/2 1];
Gdden=[T/2 1];
Gcnum=conv(Gdnum,Gvnum_pv);
Gcden=conv(Gdden,Gvden) + Gcnum;
t=[0:0.01:5];
yd=[zeros(1,51), ones(1,300), zeros(1,150)];
y_p=lsim(Gcnum,Gcden,yd,t);
e=yd-y_p';
% calculate the true lateral displacement
steer=lsim(Gdnum,Gdden,e,t);
y=lsim(Gvnum, Gvden, steer,t);
plot(t,yd, '-g', t,y_p, 'r', t-Tp, y,'-.b'); grid
xlabel('time (sec)')
title('Lat. disp. (m)')
legend('Yd', 'Y_p', 'Y')
```

Longitudinal-Driver Models

The state of knowledge on driver models for the longitudinal control of vehicles is poor compared with lateral control. This does not mean that there are few longitudinal driver models; in fact, there are numerous publications and a short review is provided herein. However, the accuracy of the longitudinal models is not as good as for the lateral models for several reasons. First, the lane-keeping task is well delineated. A driver simply must stay inside a tightly defined 12-foot lane, which means

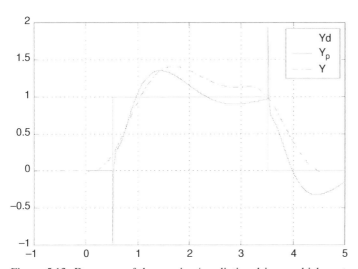

Figure 5.13. Response of the preview/predictive driver–vehicle system.

that the maximum error on either side of the vehicle is 3 feet for passenger cars and less than 2 feet for larger trucks! For comparison, in theory, a human driver has an infinite margin for maintaining a large range between the controlled and the lead vehicles. Second, vehicle longitudinal dynamics varies more among vehicles. Third, many previous longitudinal-driver models were aimed at representing traffic-flow behavior (i.e., macroscopic) rather than accurate vehicle-following behavior. The research results from a macroscopic viewpoint, however, could provide an average longitudinal model.

Since the early 1950s, many longitudinal-driving models have been proposed, forming the basis for microscopic traffic-simulation studies. Pipes (1953) suggested a linear "follow-the-leader" model, which assumes that the driver aims to accelerate the vehicle in proportion to the speed difference between the lead and the following vehicles. The proportional constant is termed the *sensitivity* of the driver model, and this desired acceleration is realized after a neuromuscular delay. In mathematical form, this model can be described as follows:

$$\dot{u}_{n+1}(t + T) = \lambda(u_{n+1} - u_n) \tag{5.10}$$

This is one example of the stimulus–response model, in which the stimulus is the relative speed and the response is the acceleration with time delay T. An apparent drawback of this model is that the desired speed and spacing between vehicles were not used explicitly. In other words, this equation is not a satisfactory vehicle-following model if what we are interested in is the vehicle-level response. Chandler (1958) identified the parameters (i.e., sensitivity and delay) of the Pipes model based on measured vehicle-following data. Gazis et al. (1961) extended the Pipes's model by assuming that the sensitivity of the follow-the-leader model is proportional to the mth power of velocity over the lth power of range error. In other words:

$$\ddot{x}_n(t + \tau) = \frac{c\dot{x}_n(t + \tau)^l}{[x_{n-1}(t) - x_n(t)]^m}[\dot{x}_{n-1}(t) - \dot{x}_n(t)] \tag{5.11}$$

Newell (1961) proposed a different model based on the assumption that a human driver has a desired speed and a natural tendency to converge exponentially to this desired speed. Tyler (1964) formulated the human driver as a linear optimal controller; the cost function being optimized is a quadratic function of range error and range-rate error. Later, Burnham et al. (1974) modified Tyler's model to include human reaction time and vehicle nonlinearities. Gipps (1981) proposed a switching-vehicle-speed model based on two mutually exclusive considerations: (1) to keep a safe distance from the lead vehicle; and (2) to converge to the desired free-flow speed. The minimum of the two speeds is used:

$$v_n(t + \tau) = \min \begin{cases} v_n(t) + 2.5a_n\tau\left(1 - \frac{v_n(t)}{V_n}\right)\sqrt{0.025 + \frac{v_n(t)}{V_n}} \\ \\ b_n\tau + \sqrt{b_n^2\tau^2 - b_n\left[2[x_{n-1}(t) - s_{n-1} - x_n(t) - v_n(t)\tau] - \frac{v_{n-1}(t)^2}{\hat{b}}\right]} \end{cases} \tag{5.12}$$

To predict vehicle behavior under both free and congested traffic with a single equation, Bando (1995) devised the "optimal velocity model" that assumes a special basis function to describe human behavior. She also identified these parameters using Japanese highway-traffic data and obtained good results.

Other longitudinal models proposed in the literature include the look-ahead model (which uses the status of the three closest lead vehicles to compute acceleration command), the time-to-collision model, and the state-transition model (in which a human driver is assumed to switch among strategies according to several driving states).

PROBLEMS

1. Use the two-DOF vehicle-lateral-dynamics model given here. Simulate the response of the same system in terms of lateral acceleration and lateral displacement using a closed-loop vehicle-handling model that includes a simple (i.e., proportional plus delay) driver model of the following form:

$$G_d(s) = (\delta(s)/e(s)) = K_d e^{-sT}$$

where $K_d = 1$ and $T = 0.015$ sec. Instead of the pure delay, use the approximation:

$$e^{-sT} \approx \frac{(Ts)^2 - 6(Ts) + 12}{(Ts)^2 + 6(Ts) + 12}; \quad or \quad e^{-sT} \approx \frac{1 - 0.5Ts}{1 + 0.5Ts}$$

Is the response stable? If not, find a satisfactory pair (K_d, T) for stable response. Also, provide the simulation result for this set of (K_d, T) parameter values and comment on your results.

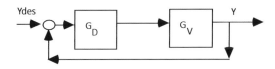

Numerical Values of Vehicle Parameters

Using the notation for parameters as listed in Chapter 4, a vehicle is defined by the following parameter values:

$l = 2.54$ m	$a = 1.14$ m	$b = l - a = 1.40$ m
$g = 9.81$ m/s^2	$u_0 = 30.0$ m/s	$m = 1,000$ kg
$I_z = 1,200.0$ kg-m^2	$C_{af} = -2,400.0$ N/deg	$C_{ar} = -2,056.0$ N/deg
$\Theta = 0.0$ rad	$f = 0.02$	$\rho = 1.202$ kg/m^2
$C_d = 0.4$	$A = 1$ m^2	57.2958 deg $= 1$ rad

$$K = \left(\frac{-mb}{lC_{af}} - \frac{-ma}{lC_{ar}} \right) \qquad u_{char} = \sqrt{l/K} = 113.2 \text{ m/s}$$
$$= 1.98 \times 10^{-4} \text{ rad/(m/s}^2)$$

DOF Vehicle Handling Model

$$\begin{bmatrix} (mu_0 D - Y_\beta) & (mu_0 - Y_r) \\ -N_\beta & (I_z D - N_r) \end{bmatrix} \begin{Bmatrix} \beta \\ r \end{Bmatrix} = \begin{bmatrix} Y_{\delta f} & Y_{\delta r} \\ N_{\delta f} & N_{\delta r} \end{bmatrix} \begin{Bmatrix} \delta_f \\ \delta_r \end{Bmatrix}$$

where:

$Y_\beta = C_{\alpha f} + C_{\alpha r} = -2.5531 \times 10^5$ N/rad

$Y_r = \dfrac{C_{\alpha f} a}{u_0} - \dfrac{C_{\alpha r} b}{u_0} = 407.9461$ N·S/rad

$Y_{\delta f} = -C_{\alpha f} = 1.3751 \times 10^5$ N/rad

$Y_{\delta r} = -C_{\alpha r} = 1.1780 \times 10^5$ N/rad

$N_\beta = a C_{\alpha f} - b C_{\alpha r} = 8.1589 \times 10^3$ N·m/rad

$N_r = (a^2/u_0) C_{\alpha f} + (b^2/u_0) C_{\alpha r}$
$\quad = -2.0480 \times 10^4$ N·m·s/rad

$N_{\delta f} = -a C_{\alpha f} = 1.5676 \times 10^5$ N·m/rad

$N_{\delta f} = b C_{\alpha r} = -1.6492 \times 10^5$ N·m/rad

2. From a two-DOF vehicle lateral-dynamics model, the following transfer function is obtained:

$$G'v(s) = (y(s)/\delta(s)) = \frac{1380s^2 + 2400s + 34287}{s^2(s^2 + 29.832s + 224.523)}$$

A preview-based driver model can be developed as follows: (1) the driver model is assumed to be $G_d(s) = K_d\, e^{-sT}$; (2) the error, e, is calculated as $e(t + T_p) = r(t + T_p) - y(t + T_p)$; and (3) it is assumed that the desired path at some time in the future, $r(t + T_p)$, is known and that the actual position at that same time, $y(t + T_p)$, is estimated from the simple model $y(t + T_p) = y(t) + T_p\,\dot{y}(t) + (T_{p2}/2)\,\ddot{y}(t)$, where T_p is preview time.

(a) Draw a more detailed version of the block diagram given here so that it also shows the lateral velocity, $v_y(t) = \dot{y}(t)$, and acceleration, $ay(t) = \ddot{y}(t)$, as well as the preview action. (*Hint:* This will help to obtain the answer to [b].)

(b) Refer to the block diagram here and your block diagram from (a). Determine what $G_v(s)$ must be for this system with preview. (*Hint:* $G'v(s)$ is given but $G_v(s)$ must be an equivalent vehicle-dynamics transfer function that incorporates the preview action.)

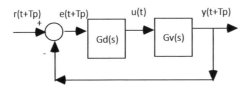

3. From a two-DOF vehicle-lateral-dynamics model, the following transfer function is obtained:

$$G_v(s) = \frac{y(s)}{\delta(s)} = \frac{b_0 s^2 + b_1 s + b_2}{s^2(s^2 + a_1 s + a_2)}$$

From experiments and using a Padéapproximation to the delay, a driver steering model was found to be:

$$G_d(s) = \frac{\delta(s)}{e(s)} \approx K_d \frac{1 - 0.5Ts}{1 + 0.5Ts}$$

The control input, $u(t)$, in the following figure corresponds to the steering input, $\delta(t)$.

(a) Obtain the closed-loop transfer function $G_c(s) = \frac{y(s)}{r(s)}$.

(b) What will the steady-state response (y) to a unit step change in the reference (i.e., desired) lateral displacement (r) be for the combined driver and vehicle in the block diagram?

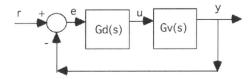

REFERENCES

Barfield, W., and T. A. Dingus, 1998, *Human Factors in Intelligent Transportation Systems*, Lawrence Erlbaum Associates.

Bando, M., et al., 1995, "Phenomenological study of dynamical model of traffic flow," *J. Phys. I France*, vol. 5, pp. 1389–99.

Bekey, G. A., 1962, "The Human Operator as a Sampled-Data System," *IRE Transactions on Human Factors in Electronics*, HFE-3, September.

Brown, I. D., A. H. Tickner, and D. C. V. Simmonds, 1969, "Interference between Concurrent Tasks of Driving and Telephoning," *Journal of Applied Psychology*, 53(5), 419–24.

Burnham, G. O., J. Seo, and G. A. Bekey, 1974, "Identification of Human Driver Models in Car Following," *IEEE Transactions on Automatic Control*, Vol. AC-19, pp. 911–15.

Chandler, R. E., Herman, R. and E. W. Montroll, 1958, "Traffic Dynamics: Studies in Car Following," *Operations Research*, Vol. 6, No. 2, Mar.–Apr., 1958.

Charlton, S. G., and T. G. O'Brien (eds.), 2002, *Handbook of Human Factors Testing and Evaluation*, second edition, Lawrence Erlbaum Associates, Mahwah, NJ.

Chen, S., T. B. Sheridan, H. Kusunoki, and N. Komoda, 1995, "Car-Following Measurements, Simulations, and a Proposed Procedure for Evaluating Safety," *Proceedings of the IFAC Symposium on Analysis, Design and Evaluation of Man-Machine Systems*, Pergamon Press, pp. 603–8.

Chen, L. K., and A. G. Ulsoy, 2001, "Identification of Driver Steering Model, and Model Uncertainty, from Driving Simulator Data," *ASME Journal of Dynamic Systems Measurement Control*, Vol. 123, No. 4, pp. 623–29, December.

Chen, L. K., and A. G. Ulsoy, 2002, "Design of a Vehicle Steering Assist Controller Using Driver Model Uncertainty," *International Journal of Vehicle Autonomous Systems*, Vol. 1, No. 1, 2002, pp. 111–32.

Chen, L. K. and A. G. Ulsoy, 2006, "Experimental Evaluation of a Vehicle Steering Assist Controller Using A Driving Simulator," *Vehicle System Dynamics*, Vol. 44, No. 3, March, pp. 223–45.

Davis, J. R., and C. M. Schmandt, 1989, "The Back Seat Driver: Real Time Spoken Driving Instructions," pp. 146–50, in D. H. M. Reekie, E. R. Case, and J. Tsai (eds), *First Vehicle Navigation and Information Systems Conference (VNIS'89)*, New York: Institute of Electrical and Electronics Engineers.

Dingus, T. A., J. F. Antin, M. C. Hulse, and W. W. Wierwille, 1988, "Human Factors Issues Associated with In-Car Navigation System Usage," *Proceedings of the Human Factors Society-32nd Annual Meeting*, pp. 1448–50.

Dudek, C. L., R. D. Hutchinson, W. R. Stockton, R. J. Koppa, S. H. Richards, and T. M. Mast, 1978, *Human Factors Requirements for Real-Time Motorist Information Displays, Volume I – Design Guide* (Technical Report (FHWA-RD-78-5), Washington, DC, U.S. Department of Transportation, Federal Highway Administration.

Eby, D. W., and L. P. Kostyniuk, 1999, "An On-the-Road Comparison of In-Vehicle Navigation Assistance Systems," *Human Factors*, Vol. 41, pp. 295–311.

Esterberg, M. A., E. D. Sussman, and R. A. Walter, 1986, *Automotive Displays and Controls – Existing Technology and Future Trends* (Report PM-45-U-NHT-86-11), Washington, DC, U.S. Department of Transportation, Federal Highway Administration, National Highway Traffic Safety Administration.

Fuller, R., and J. A. Santos (eds.), 2002, *Human Factors for Highway Engineers*, first edition, Pergamon Press.

Gazis, D. C., R. Herman, and R. W. Rothery, 1961, "Nonlinear Follow-the-Leader Models of Traffic Flow," *Operations Research*, Vol. 9, pp. 545–66.

Gipps, P. G., 1981, "A Behavioral Car-Following Model for Computer Simulation," *Transportation Research-B*, Vol. 15B, pp. 105–11.

Godthelp, H., and W. Kappler, 1988, "Effects of Vehicle Handling Characteristics on Driving Strategy," *Human Factors*, Vol. 30, No. 2, pp. 219–29.

Godthelp, H., P. Milgram, and G. J. Blaauw, 1984, "The Development of a Time-Related Measure to Describe Driving Strategy," *Human Factors*, Vol. 26, No. 3, pp. 257–68.

Green, P., 1990, *Vehicle Control and Human Factors*, Notes for IVHS Short Course, University of Michigan.

Hess, R. A., and A. Modjtahedzadeh, 1990, "A Control Theoretic Model of Driver Steering Behavior," *IEEE Control Systems Magazine*, August, pp. 3–8.

Hosman, R. 1985, "Laboratory and Moving-Base Simulator Experiments on Speed and Accuracy of Visual and Whole-Body Motion Perception," *Proceedings of the IFAC Man-Machine Systems*, Varese, Italy.

Hoyes, T. W., N. A. Stanton, and R. G. Taylor, 1996, "Risk Homeostasis Theory: A Study of Intrinsic Compensation," *Safety Science*, Vol. 22, No. 1–3, pp. 77–86.

Hulse, M. C., T. A. Dingus, T. Fischer, and W. W. Wierwille, 1989, "The Influence of Roadway Parameters on Driver Perception of Attentional Demand," *Advances in Industrial Ergonomics and Safety I*, edited by Anil Mital, Taylor and Francis, pp. 451–6.

International Standards Organization, 1977, *Road Vehicles – Passenger Cars – Location of Hand Controls, Indicators and Tell-Tales* (International Standard 4040), Geneva, Switzerland ISO.

Kleinman, D. L., S. Baron, and W. H. Levison, 1970, "An Optimal Control Model of Human Response, Part I: Theory and Validation," *Automatica*, Vol. 6, pp. 357–69.

Kondo, M., and A. Ajimine, 1968, "Driver's Sight Point and Dynamics of the Driver-Vehicle System Related to It," SAE Automotive Engineering Congress, Detroit, MI, SAE Paper 680104.

Li, Y. T., L. R. Young, and J. L. Meiry, 1965, "Adaptive Functions of Man in Vehicle Control Systems," in *Proceedings of the International Federation of Automatic Control Symposium*, Paddington, England.

MacAdam, C. C., 1980, "An Optimal Preview Control for Linear Systems," *ASME Journal of Dynamic Systems, Measurement, and Control*, Vol. 102, September, pp. 188–90.

MacAdam, C. C., 1981, "Application of an Optimal Preview Control for Simulation of Closed-Loop Automobile Driving," *IEEE Transactions on Systems, Man, and Cybernetics*, SMC-11, June, pp. 393–9.

MacAdam, C. C., 1983, "Frequency Domain Methods for Analyzing the Closed-Loop Directional Stability and Maneuverability of Driver/Vehicle Systems," *Proceedings of the Conference on Modern Vehicle Design Analysis*, edited by M. Dorgham, London.

MacAdam, C. C., 1985, "Computer Model Predictions of the Directional Response and Stability of Driver Vehicle Systems During Anti-Skid Braking," *Proceedings of the IMech E Conference on Antilock Braking Systems for Road Vehicles*, edited by P. Newcomb, London.

MacAdam, C. C., 1988, "Development of Driver/Vehicle Steering Interaction Models for Dynamic Analysis," Final Technical Report, UMTRI-88-53, TACOM Contract DAAE07-85-C-R069.

MacAdam, C. C., 1989, "Mathematical Modeling of Driver Steering Control at UMTRI – An Overview," *UMTRI Research Review*, July–August, pp. 1–13.

MacAdam, C. C., 2003, "Understanding and Modeling the Human Driver," *Vehicle System Dynamics*, Vol. 40, Nos. 1–3, pp. 101–34.

MacAdam, C. C., Sayers, M. W., Pointer, J. D. and M. Gleason , 1990, "Crosswind Sensitivity of Passenger Cars and the Influence of Chassis and Aerodynamic Properties on Driver Preferences," *Vehicle System Dynamics*, Vol. 19, No. 4.

Macgregor, D. G. and P. Slovic, 1989, "Perception of Risk in Automotive Systems," *Human Factors: The Journal of the Human Factors and Ergonomics Society*, August 1989, Vol. 31, No. 4 377–89.

McLean, D., T. P. Newcomb, and R. T. Spurr, 1976, "Simulation of Driver Behavior During Braking," *I. Mech. E. Conference on Braking of Road Vehicles*, IME Paper C41/176.

McLean, J. R., and E. R. Hoffman, 1971, "Analysis of Drivers' Control Movements," *Human Factors*, Vol. 13, May, pp. 407–18.

McLean, J. R., and E. R. Hoffmann, 1972, "The Effects of Lane Width on Driver Steering Control and Performance," *Proceedings of the Sixth Australian Road Research Board Conference*, Vol. 6, No. 3, 418–40.

McLean, J. R., and E. R. Hoffmann, 1973, "The Effects of Restricted Preview on Driver Steering Control and Performance," *Human Factors*, Vol. 15, No. 6, pp. 421–30.

McRuer, D. T., Hofmann, L. G., Jex, H.R., Moore, G.P., Phatak, A.V., Weir, D.H. and Wolkovitch, J., 1968, "New Approaches to Human-Pilot/Vehicle Dynamic Analysis," Technical Report AFFDL-TR-67-150, WPAFB.

McRuer, D. T., R. W. Allen, and R. H. Klein, 1977, "New Results in Driver Steering Control Models," *Human Factors*, Vol. 19, No. 4, pp. 381–97.

McRuer, D. T., and R. Klein, 1975a, "Comparison of Human Driver Dynamics in Simulators with Complex and Simple Visual Displays and in an Automobile on the Road," *11th Annual Conference on Manual Control*.

McRuer, D. T., and R. Klein, 1975b, "Effects of Automobile Steering Characteristics on Driver Vehicle System Dynamics in Regulation Tasks," *11th Annual Conference on Manual Control*, pp. 408–39.

Miller, D. C. and J. I. Elkind, 1967, "The Adaptive Response of the Human Controller to Sudden Changes in Controlled Process Dynamics," *IEEE Transactions on Human Factors in Electronics*, HFE-8, 3.

Mitschke, M., 1993, "Driver-Vehicle Lateral Dynamics under Regular Driving Conditions," *Vehicle System Dynamics*, Vol. 22, pp. 483–92.

Modjtahedzadeh, A., and R. A. Hess, 1991, "A Model of Driver Steering Dynamics for Use in Assessing Vehicle Handling Qualities," in S. A. Velinsky, R. H. Fries, I. Haque, and D. Wang, D. (eds.), *Advanced Automotive Technologies-1991*, ASME DE-Vol. 40, New York, pp. 41–56.

Mortimer, R. G., and C. M. Jorgeson, 1972, "Eye Fixations of Drivers as Affected by Highway and Traffic Characteristics and Moderate Doses of Alcohol," *Proceedings of the 16th Annual Meeting, Human Factors Society*, pp. 86–92.

Nagai, M., and M. Mitschke, 1985, "Adaptive Behavior of Driver-Car Systems in Critical Situations: Analysis by Adaptive Model," *JSAE Review*, December, pp. 82–9.

Newcomb, T. P., 1981, "Driver Behavior During Braking," SAE/I. Mech. E. Exchange Lecture. SAE Paper 810832.

Newell, G. F., 1961, "Nonlinear Effects in the Dynamics of Car Following," *Operations Research*, Vol. 9, pp. 209–29.

Pew, R. W., and S. Baron, 1982, "Perspectives on Human Performance Modeling," *Proceedings of IFAC Man-Machine Systems*, Baden-Baden, Federal Republic of Germany.

Phatak, A. V., and G. A. Bekey, 1969, "Model of the Adaptive Behavior," *IEEE Transactions on Man-Machine Systems*, MMS-10, September, pp. 72–80.

Pilutti, T., U. Raschke, and Y. Koren, 1990, "Computerized Defensive Driving Rules for Highway Maneuvers," *Proceedings of the American Control Conference*, San Diego, CA, May, pp. 809–11.

Pilutti, T., and A. G. Ulsoy, 1999, "Identification of Driver State for Lane-Keeping Tasks," *IEEE Transactions on Systems, Man and Cybernetics*, Vol. 29, No. 5, September 1999, pp. 486–502.

Pipes, L. A., 1953, "An Operational Analysis of Traffic Dynamics," *Journal of Applied Physics*, Vol. 24, No. 3, pp. 274–81, March 1953.

Rashevsky, N., 1968, "Mathematical Biology of Automobile Driving," *Bulletin of Mathematical Biophysics*, 30, p. 153.

Sanders, M. S., and E. J. McCormick, 1987, *Human Factors in Engineering and Design* (6th edition), New York: McGraw-Hill Publishers.

Sheridan, T. B., 1964, "Control Models of Creatures Which Look Ahead," *Proceedings of the 5th National Symposium on Human Factors in Electronics*.

Sheridan, T. B., 1966, "Three Models of Preview Control," *IEEE Transactions on Human Factors in Electronics*, HFE-7, June.

Sheridan, T. B., 1985, "Forty-Five Years of Man-Machine Systems: History and Trends," *Proceedings of IFAC Man-Machine Systems*, Varese, Italy.

Sheridan, T. B., 2002. *Humans and Automation: System Design and Research Issues*. John Wiley & Sons, Inc., New York, NY, USA.

Snyder, H. L., and R. W. Monty, 1985, "Methodology and Results for Driver Evaluation of Electronic Automotive Displays," *Ergonomics International '85*, 514–17.

Summala, H., 1996, "Accident Risk and Driver Behaviour," *Safety Science*, Vol. 22, Nos. 1–3, pp. 103–17.

Tyler, J. S., 1964, "The characteristics of model following systems as synthesizes by optimal control," *IEEE Trans. Automatic Control*, Vol. AC-9, pp. 485–98.

Weir, D. H., and A. V. Phatak, 1968, "Model of Human Response to Step Transitions in Controlled Element Dynamics," *Technical Report NASA* CR-671.

Weir, D., R. J. DiMarco, and D. T. McRuer, 1977, "Evaluation and Correlation of Driver/Vehicle Data, Vol. II," Final Technical Report, NHTSA, DOT-HS-803-246.

Weir, D. H., and D. T. McRuer, 1968, "A Theory for Driver Steering Control of Motor Vehicles," *Highway Research Record*, Vol. 247, pp. 7–28.

Wierwille, W. W., J. F. Antin, T. A. Dingus, and M. C. Hulse, 1988, "Visual Attentional Demand of an In-Car Navigation Display," in A. G. Gale et al. (eds.), *Vision in Vehicles II*, Amsterdam: Elsevier.

Wierwille, W. W., G. A. Gagne, and J. R. Knight, 1967, "An Experimental Study of Human Operator Models and Closed-Loop Analysis Methods for High-Speed Automobile Driving," *IEEE Transactions on Human Factors in Electronics*, HFE-8, September.

Wilde, G. J. S., 1994, "Risk Homeostasis Theory and Its Promise for Improved Safety," in R. M. Trimpop and G. J. S. Wilde (eds.), *Challenges to Accident Prevention: The Issue of Risk Compensation Processes*, Groningen, The Netherlands: Styx Publications.

Wolf, J. D., and M. F. Barrett, 1978, *Driver Vehicle Effectiveness Model Volume II*: Appendices, Report DOT-HS-804-338, Washington, DC, U.S. Department of Transportation.

Young, L. R. and L. Stark, 1965, "Biological Control Systems – A Critical Review and Evaluation," *NASA Contractor Report*, NASA CR-190.

PART II

POWERTRAIN CONTROL SYSTEMS

6 Air–Fuel Ratio Control

Mathematical models such as those described in Chapter 3 can be used as the basis for engine control-system design (Kiencke and Nielsen 2005). This chapter discusses control of the air–fuel mixture ratio that enters the cylinders (i.e., the combustion chamber) (Grizzle et al. 1990, 1991; Kiencke 1988; Kuraoka et al. 1990; Moklegaard et al. 2001; Ohyama 1994; Sweet 1981).

6.1 Lambda Control

Air–fuel ratio control sometimes is referred to as *lambda control* because the ratio is defined as follows:

$$\lambda = \frac{\dot{m}_a/\dot{m}_f}{\lambda_{ST}} \qquad (6.1)$$

where λ is the normalized air–fuel ratio, $\dot{m}_a$ is the mass airflow into the engine, $\dot{m}_f$ is the metered-fuel mass flow rate, and λ_{ST} is the air–fuel mass ratio for a stoichiometric mixture (i.e., the air–fuel mass ratio is equal to 14.64). The driver regulates the amount of air, $\dot{m}_a$, entering the cylinders, and the correct quantity of fuel, $\dot{m}_f$, must be added by the controller. Undesirable emission gases (e.g., NO_x, HC, and CO) are produced by the combustion process (see Figure 1.4). A catalytic converter reduces the emissions but operates efficiently only if the value of λ is near 1 (see Figure 1.5). For efficient operation of the converter, the average lambda offset must be maintained within 0.1 percent. During engine transients, short excursions of up to 3 percent may be permitted. Thus, the control specifications are quite demanding.

An EGO sensor provides the information needed for feedback control (Figure 6.1). The sensor is highly nonlinear but produces a voltage proportional to the amount of oxygen in the exhaust gas. In turn, this indicates whether the air–fuel mixture is *rich* (i.e., <14.64) or *lean* (i.e., >14.64). The sensor response time to a step change in lambda is about 300 ms or less. In the air–fuel-ratio control problem, the engine can be modeled simply by a pure delay plus a first-order lag:

$$G_p(s) = \frac{K_M e^{-sT_L}}{\tau_M s + 1} \qquad (6.2)$$

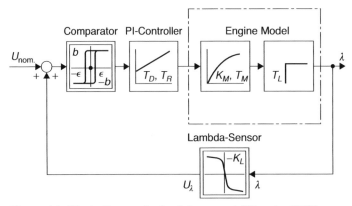

Figure 6.1. Block diagram for lambda control (Kiencke 1988).

where the delay, T_L, varies from about 100 to 1,000 ms and the lag, τ_M, varies from about 50 to 500 ms.

A PI controller typically is used in this application. However, the nonlinearity of the EGO sensor is a problem, and there is considerable interest in the development of an inexpensive linear sensor for use with lambda control. The variation of the model parameters also has created considerable interest in adaptive lambda controllers that can estimate the parameters during operation (Jones et al. 1995). Also the measurement made with the EGO sensor is an average lambda value for the engine; research to control the lambda value in each cylinder also is of interest (Kiencke 1988).

Figure 6.1 shows a block diagram of a lambda controller. The controller includes a nonlinear (i.e., hysteresis) element, which addresses the nonlinearity of the EGO sensor and the pure delay T_L.

6.2 PI Control of a First-Order System with Delay

In this section, we first review the control of a first-order system using PI control and then the control of a first-order system with delay using the Smith Predictor. Example 1 discusses the application of these methods to air–fuel ratio control.

PI Controller Design for a Process Described by a First-Order Transfer Function

Given $G_p(s) = K/(\tau s + 1)$ and $G_c(s) = K_P + K_I/s$, the closed–loop-system transfer function is $Y(s)/U_c(s) = G_c\,G_p/(1 + G_c\,G_p)$. The closed-loop system therefore has the following characteristic equation:

$$s^2 + ((1 + KK_P)/\tau)s + (KK_I/\tau) = 0 \tag{6.3}$$

Comparing this to a standard second-order system, we can identify:

$$2\zeta\omega_n = (1 + KK_P)/\tau; \quad \text{and} \quad \omega_n^2 = (KK_I/\tau) \tag{6.4}$$

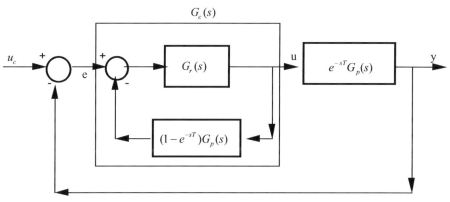

Figure 6.2. Smith Predictor for control of a system with delay.

Because $\omega_n^2 > 0$, for stability we must have $\zeta > 0$. Various other performance measures can be expressed in terms of the natural frequency, ω_n, and the damping ratio, ζ. The 1 percent settling time is $t_s = (4.6)/(\zeta\,\omega_n)$. The maximum percentage overshoot (for $0 \le \zeta \le 1$) is $M_p = \exp(-\pi\zeta/\sqrt{1 - \zeta^2}) \approx 1 - (\zeta/0.6)$ for $0 \le \zeta \le 0.6$. For example, by specifying the desirable value of M_p, we can determine ζ; then, from ζ and the specified value of t_s, we can determine the value of ω_n. Finally, from the ζ and ω_n values, we can determine the PI controller gains, K_P and K_I. This second-order closed-loop system also includes a zero; therefore, the formulas given herein for settling time and overshoot are only approximations (Franklin et al. 2010).

Controlling a System with Delay (the Smith Predictor)

Consider the control of a process with pure delay:

$$Y(s)/U(s) = e^{-sT}G_p(s) \tag{6.5}$$

The Smith Predictor approach is to design a controller, $G_r(s)$, for the process without delay, $G_p(s)$. This can be accomplished, for example, as described in the previous section. Then, we implement the feedback controller $G_c(s)$ for the process with the delay, where:

$$U(s)/E(s) = G_c(s) = \frac{G_r(s)}{1 + G_p(s)G_r(s) - G_p(s)G_r(s)e^{-sT}} \tag{6.6}$$

Then, the closed-loop system becomes:

$$Y(s)/U_c(s) = \frac{G_c G_p e^{-sT}}{1 + G_c G_p e^{-sT}} = e^{-sT}\frac{G_r G_p}{1 + G_r G_p} \tag{6.7}$$

The closed-loop system has the response designed without the delay plus the process delay and is illustrated in the block diagram in Figure 6.2.

EXAMPLE 6.1: SMITH PREDICTOR PI CONTROL OF AIR–FUEL RATIO. In this example, it is assumed that a linear EGO sensor is available to measure the normalized air–fuel ratio, λ, and that the process is well modeled by a first-order lag with delay, as in Eq. (6.2). The desired value of the air–fuel ratio is $\lambda_d = 1.0$ and

$e = \lambda_d - \lambda$. Furthermore, it is assumed that the controller is given by a PI controller where:

$$U(s) = \left(K_P + \frac{K_I}{s} \right) E(s) \tag{6.8}$$

With this controller, the closed-loop transfer function becomes:

$$\frac{\lambda(s)}{\lambda_d(s)} = \frac{K_M(K_P s + K_I)}{\tau_M s^2 + (1 + K_M K_P)s + (K_M K_I)} \tag{6.9}$$

The characteristic equation is found by setting the denominator of this transfer function to zero:

$$s^2 + ((1 + K_M K_P)/\tau_M)s + (K_M K_I/\tau_M) = 0 \tag{6.10}$$

Using the analogy to a second-order system, as in Eq. (6.4), we can obtain the controller gains:

$$\begin{aligned} K_I &= \frac{\omega_n^2 \tau_M}{K_M} \\ K_P &= \frac{2\zeta \omega_n \tau_M - 1}{K_M} \end{aligned} \tag{6.11}$$

Now, to account for the process delay, T_L, we can use a Smith Predictor design based on this controller (Franklin et al. 2010). Thus, the controller is given by:

$$G_c(s) = \frac{G_r(s)}{1 + G_p(s)G_r(s) - G_p(s)G_r(s)e^{-sT_L}} \tag{6.12}$$

where $G_p(s)$ is the process model in Eq. (6.2) but without the delay:

$$G_p(s) = \frac{K_M}{\tau_M s + 1} \tag{6.13}$$

and

$$G_r(s) = \left(K_P + \frac{K_I}{s} \right) \tag{6.14}$$

The response of the closed-loop system then is determined by the selected values of the design parameters, ζ and ω_n, and also includes the same delay, T_L, as the open-loop process. The Smith Predictor is only one approach to the control of systems with delay, and it is effective when the delay time T_L and the process model $G_p(s)$ are well known. Other methods for the control of systems with delays are available and may be more effective in the presence of model uncertainties (Yi et al. 2010).

PROBLEMS

1. For the air–fuel-ratio control problem, draw a block diagram (similar to Figure 6.2) that shows the process model G_p, the controlled variable y, the control input u, and the reference input r. Also list the necessary sensors and actuators.

2. Air–fuel ratio control can be designed as in Example 6.1 (see Figure 6.2). Assume that a linear EGO sensor was developed so that the sensor nonlinearity can be neglected. Use the values $T_L = 1.0$ s, $\tau_M = 0.5$ s, and $K_M = 1$.

    (a) First, consider the process without the delay to design a PI controller $G_r(s) = K_P + K_I/s$. This can be accomplished by trial and error or by any standard control-design procedure. Closed-loop specifications are the settling time $t_s = 0.46$ seconds and the percentage overshoot $M_p = 0$.

    (b) Next, use the Smith Predictor approach to derive the controller $G_c(s)$ for the process with delay.

    (c) Use Simulink to simulate and compare the closed-loop systems designed in (a) and (b), respectively.

REFERENCES

Franklin, G. F., J. D. Powell, and A. Emami-Naeni, 2010, *Feedback Control of Dynamic Systems*, Pearson Publishing.

Grizzle, J. W., J. A. Cook, and K. L. Dobbins, 1990, "Individual Cylinder Air–Fuel Ratio Control with a Single EGO Sensor," *Proceedings of the American Control Conference*, San Diego, CA, May 1990, pp. 2881–6.

Grizzle, J. W., K. Dobbins, and J. Cook, 1991, "Individual Cylinder Air–Fuel Ratio Control with a Single EGO Sensor," *IEEE Transactions on Vehicular Technology*, Vol. 40, February 1991, pp. 280–6.

Jones, V. K., B. A. Ault, G. F. Franklin, and J. D. Powell, 1995, "Identification and Air–Fuel Ratio Control of a Spark Ignition Engine," *IEEE Transactions on Control Systems Technology*, Vol. 3, No.1, March 1995.

Kiencke, U., 1988, "A View of Automotive Control Systems," *Control Systems Magazine, IEEE*, Vol. 8, August 1988, pp. 11–19.

Kiencke, U. and L. Nielsen, 2005, *Automotive Control Systems*, Springer-Verlag.

Kuraoka, H., N. Ohka, M. Ohba, S. Hosoe, and F. Zhang, 1990, "Application of H-Infinity Design to Automotive Fuel Control," *IEEE Control Systems Magazine*, April 1990, pp. 102–6.

Moklegaard L., M. Maria Druzhinina, and A. G. Stefanopoulou, 2001, "Brake Valve Timing and Fuel Injection: A Unified Engine Torque Actuator for Heavy-Duty Vehicles," *Journal of Vehicle System Dynamics*, Vol. 36, Nos. 2–3, September 2001.

Ohyama, Y., 1994, "Air–Fuel Ratio Characteristics of Engine Drivetrain Control System Using Electronically Controlled Throttle Valve and Air Flow Meter," *Proceedings of AVEC' 94*, pp. 497–502.

Sweet, L. M., 1981, "Control Systems for Automotive Vehicle Fuel Economy: A Literature Review," *ASME Journal of Dynamic Systems, Measurement, and Control*, Vol. 103, September 1981, pp. 173–80.

Yi, S., P. W. Nelson, and A.G. Ulsoy, 2010, *Time Delay Systems: Analysis and Control Using the Lambert W Function*, World Scientific Publishing Company, Inc., Hackensack, NJ.

7 Control of Spark Timing

The focus of this chapter is the control of spark timing. As discussed in Chapter 3, the spark is ignited in advance of TDC during the compression stroke. The exact timing can influence performance, fuel economy, emissions, and knock. As discussed in Chapter 1, advancing the spark timing can improve performance and reduce fuel consumption. However, advanced spark timing also can lead to engine knocking and potential engine damage. Spark-timing control can be used, for example, in idle-speed control (see Chapter 8) with throttle control. In this chapter, we focus on the occurrence of engine knock and the control of knock by adjustment of spark timing.

7.1 Knock Control

Knock occurs when an unburned part of the air–fuel mixture within the combustion chamber explodes prematurely. This is called *knocking* because it generates resonating gas-pressure oscillations, which are heard as a knocking sound. Knocking can lead to serious engine damage (Heywood 1989). Historically, a low-compression ratio or conservative spark timing was used to ensure that knocking did not occur; however, this approach sacrifices performance and fuel economy. Knock control can be used when a feedback sensor becomes available, which adjusts the spark timing based on a measured variable that indicates knock. Suitable measurements include the cylinder pressure (e.g., the 5- to 15-kHz region was found to be a good knock indicator), engine-block vibrations, light emission within the combustion chamber, and ion current through the gas mixture.

When the occurrence of knock is detected by one of these means, the ignition timing is retarded by the amount $\Delta\alpha_R$; when knocking does not occur, the spark timing is advanced. With this scheme, the ignition angle is constantly adjusted about a value, which is close to the knock limit. The control scheme here is simple but the measurement and detection of knock are challenging, as is the selection of an appropriate $\Delta\alpha_R$. Figure 7.1 illustrates the knock-control strategy described herein. When the occurrence of knock is detected – as indicated by spikes in the top half of Figure 7.1 – the ignition angle is retarded by the amount $\Delta\alpha_R$. When no knock occurs, the ignition angle is allowed to increase with time at a rate of $d\alpha_A/dt$. It is obvious

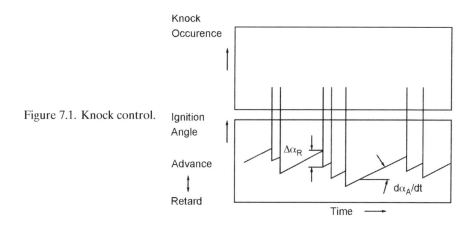

Figure 7.1. Knock control.

that the selection of these two quantities is critical to the controller performance, as is the development of an accurate knock-detection system.

PROBLEM

1. Write a MATLAB program to implement the simple knock-control scheme described in this chapter. In the program, allow $\Delta\alpha_R$ and $d\alpha_A/dt$ to be user selectable. Assume that a function *knockdetector* is available and returns a value of 1 when knock is detected and a value of 0 otherwise. Test your results with a simulated knock sequence qualitatively similar to the one shown in Figure 7.1, and use this sequence to select effective values for the user-selectable control parameters.

REFERENCES

Entenmann, R., S. Unland, O. Torno, and W. Haeming, 1997, "Method for the Adaptive Knock Control of an Internal Combustion Engine," U.S. Patent 5,645,034.

Heywood, J. B., 1989, *Internal Combustion Engine Fundamentals*, McGraw-Hill, New York.

Morita, S., W. Fukui, and S. Wada, 1997, "Knock Control System for an Internal Combustion Engine," U.S. Patent 5,694,900.

Sawamoto, K., Y. Kawamura, T. Kita, and K. Matsushita, 1987, "Individual Cylinder Knock Control of Detecting Cylinder Pressure," Passenger Car Meeting and Exposition, Dearborn, MI.

Schmillen, K. P., 1991, "Different Methods of Knock Detection and Knock Control," SAE International Congress and Exposition, Dearborn, MI, February 1991.

8 Idle-Speed Control

One of the most important and basic engine-control functions is idle-speed control. It requires consideration of the complete engine dynamics (as described in Chapter 3 about engine modeling) and has been a focus of various researchers to improve the performance of current and future engine designs (Grizzle et al. 2001; Wang et al. 2001). This chapter discusses engine idle-speed control.

The engine speed at idle is maintained at a desired value despite changes in engine loads (e.g., due to accessories such as an air-conditioning compressor). The controlled variable is engine idle speed and it is measured as discussed previously. The variables manipulated by the controller include the throttle angle and the spark advance. An optimal control approach provides an effective framework for the study of engine idle-speed control (Hrovat and Powers 1988). In this approach, we consider a vector, $\mathbf{x}$, of state variables and a vector, $\mathbf{u}$, of control variables. The problem then becomes finding the optimal controls, $\mathbf{u}^*(t)$, which minimize an objective function $J(\mathbf{x}, \mathbf{u})$ subject to constraint equations $g(\mathbf{x}, \mathbf{u}) = 0$. In general, this is a complex problem because J and g are difficult to determine and the problem, once formulated, is difficult to solve.

However, engineering judgment can be used to reduce the problem to a manageable but sufficiently accurate form. Assuming that during idle the EGR is not in operation, the control inputs are considered to be $\mathbf{u}$ = [throttle rate, spark rate, fuel-flow rate]. Rates of the actual control variables are used in this formulation to separate the effects of actuator dynamics. The engine model described in Chapter 3, or a more complex version of it with about twenty state variables, typically is used. The model relating these variables must be calibrated with empirical data. This formulation leads to a simulation model that is used to study the dynamics of the system (the results are shown in Figure 8.1) and then to obtain a reduced-order model with five state variables (i.e., perturbations in engine speed, manifold pressure, throttle angle, throttle-motor rate, and spark advance) and two control inputs (i.e., perturbations in the throttle-rate command and the spark-advance command). The reduced-order model is linearized about a particular nominal operating condition and solved as a linear quadratic (LQ) optimal-control problem. In the LQ problem, linear-state equations describe the process model and the objective function, $J(\mathbf{x}, \mathbf{u})$, to be minimized, is quadratic in $\mathbf{x}$ and $\mathbf{u}$. The solution to

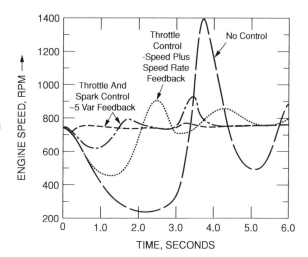

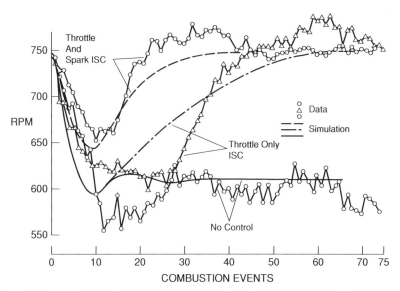

Figure 8.1. Idle-speed-control simulation results (Hrovat and Powers 1988).

LQ problems can be obtained readily using standard numerical algorithms in computer-aided control-system design packages such as MATLAB.

Figure 8.2 illustrates the experimental results for idle speed due to a sudden change in engine load at time $t = 0$ as obtained using a throttle-only idle-speed controller and a throttle and spark idle-speed controller. The figure compares these results to those obtained from the simulation studies using the twenty-state simulation model. The idle-speed response with no control also is shown for comparison purposes. Clearly, both controllers provide improvement in the idle speed over the no-control case, and the combined throttle and spark idle-speed control strategy is most effective. The inclusion of spark advance improves the response because the spark command acts more rapidly than the throttle command, which has actuator and manifold lags.

Figure 8.2. Idle-speed-control experimental and simulation results (Hrovat and Powers 1988).

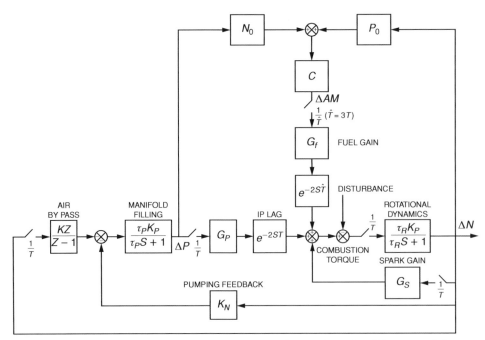

Figure 8.3. Linearized six-cylinder engine idle-speed-control model (Hrovat and Powers 1988).

EXAMPLE 8.1: SIMPLIFIED IDLE-SPEED CONTROL. The idle-speed control problem, with adjustment of only throttle and not spark advance, is discussed in Chapters 1 and 3. Here, a combined throttle- and spark-control system is described based on the linearized engine model presented in Chapter 3, thereby obtaining the block diagram shown in Figure 8.3. It is a modified version of the block diagram shown in Figure 3.6 and it uses Eqs. (3.3) and (3.4) to estimate the mass airflow quantity $\dot{M}$. The control uses a pure integral action (with gain K in Figure 8.3) of airflow through a closed throttle-bypass valve to ensure steady-state accuracy and a proportional control (with gain G_s) of the spark timing to enhance the speed of response. The system in Figure 3 is given as a discrete-time system because implementation of the controller is digital. Simulation results are shown in Figure 8.4 for the engine parameters given in Table 3.1 and for the systems with (i.e., closed-loop) and without (i.e., open-loop) idle-speed control. Clearly, the closed-loop control reduces the engine-speed variation and eliminates the steady-state error (i.e., offset) in engine speed due to constant torque-disturbance input.

PROBLEMS

1. *Closed-loop engine dynamic simulation.* Simulate the closed-loop idle-speed control system given in Example 8.1 (see Figure 8.3) to a unit-step torque-disturbance input. In performing the simulation, neglect the IP delays and the effects of sampling; replace the controller ($K_z/(z-1)$) by the integral (I) controller (K_I/s). Notice from Figure 8.3 that the system includes fuel control based on estimation of the mass flow

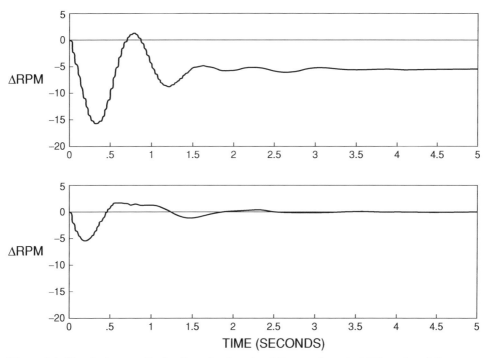

Figure 8.4. Simulation results for linearized engine idle-speed control (Hrovat and Powers 1988).

rate as in Eq. (3.4). Thus, in this controller, $\Delta\theta = -(K_I/s)\,\Delta N$, $\Delta\delta = G_s\,\Delta N$ and, from Eq. (3.4), $\Delta F_d = cN_0\,\Delta P + cP_0\,\Delta N$.

(a) Use the model parameter values $\tau_P = 0.21$, $K_P = 0.776$, $G_P = 13.37$, $\tau_R = 3.98$, $K_R = 67.2$, $K_N = 0.08$, $G_f = 36.6$, $K_\theta = 20$, and $G_d = 10$. Fix the controller parameter values $G_s = -0.005$, $c = 0.000125$, $N_0 = 600$, and $P_0 = 12$. Try several values of the controller integral action gain, K_I, until a good response qualitatively similar to that shown in Figure 8.4 is achieved.

(b) Use $G_s = 0$ (i.e., no coordination with SI control); then select a K_I that results in a reasonable (i.e., stable) response.

(*Hint:* Using the relationships given for the controller, neglecting the IP delays and the effects of sampling, and using Eqs. (3.1)–(3.8), obtain the following closed-loop system equations:

$$\dot{\mathbf{x}} = \mathbf{A}\mathbf{x} + \mathbf{b}_v v, \quad \text{and} \quad \mathbf{y} = \mathbf{C}\mathbf{x}$$

where $v = \Delta T_d$ and

$$\mathbf{x} = \begin{Bmatrix} \Delta\theta \\ m_a \\ \Delta P \\ \Delta N \end{Bmatrix} \quad \mathbf{A} = \begin{bmatrix} 0 & 0 & 0 & -K_I \\ K_\theta & 0 & 0 & 0 \\ K_P K_\theta & 0 & -1/\tau_P & -K_P K_N \\ 0 & 0 & K_R(G_P + G_f cN_0) & (K_R G_\delta G_s - 1/\tau_R + K_R G_f cP_0) \end{bmatrix}$$

$$\mathbf{b}_v = \begin{Bmatrix} 0 \\ 0 \\ 0 \\ -K_R \end{Bmatrix}$$

and $\mathbf{C} = [0\ 0\ 0\ 1]$.)

2. Repeat the closed-loop engine simulations in Problem 1 but with the following alternative control strategies:

(a) Try a PI control where $\Delta\theta = -(K_I/s)\,\Delta N - K\,\Delta N$ with $G_s = 0$ and determine whether satisfactory results can be obtained.

(b) Try feed forward plus PI control $\Delta\theta = -(K_I/s)\,\Delta N - K\Delta N + K_{FF}\,\Delta T_d$ (i.e., assume the disturbance ΔT_d is measured or can be estimated). Again, fine-tune the gains until a good response is obtained.

(c) Use maximum control design DOF (Spark + PI + feedforward) and fine-tune all of the gains until the results are satisfactory.

(*Hint:* The $\mathbf{A}$ and $\mathbf{b_v}$ matrices in Problem 1 must be modified for the PI control algorithm and the feed-forward control actions.)

REFERENCES

Grizzle, J. W., J. Buckland, and J. Sun, 2001, "Idle Speed Control of a Direct Injection Spark Ignition Stratified Charge Engine," *International Journal of Robust and Nonlinear Control*, Vol. 11, pp. 1043–71.

Hrovat, D., and J. Sun, 1997, "Models and Control Methodologies for IC Engine Idle Speed Control Design," *Control Engineering Practice*, Vol. 5, No. 8, August, pp. 1093–100.

Hrovat, D., and W. Powers, 1988, "Computer Control Systems for Automotive Power Trains," *IEEE Control Systems Magazine*, Vol. 8, pp. 3–10.

Wang, Y., A. Stefanopoulou, and R. Smith, 2001, "Inherent Limitations and Control Design for Camless Engine Idle Speed Dynamics," *International Journal of Robust and Nonlinear Control*, Vol. 11, pp. 1023–42.

Ye, Z., 2007, "Modeling, Identification, Design, and Implementation of Nonlinear Automotive Idle Speed Control Systems – An Overview," *IEEE Transactions on Systems, Man, and Cybernetics, Part C: Applications and Reviews*, Vol. 37, No. 6, November, pp. 1137–51.

9 Transmission Control

Automotive transmissions are a key element in the powertrain that connects the power source to the drive wheels. To improve fuel economy, reduce emissions, and enhance drivability, many new technologies have been introduced in automotive transmissions in recent years (Sun & Hebbale 2005). Electronic control of automatic and continuously variable transmissions (CVTs) is considered briefly in this chapter, as well as the related topic of clutch control for AWD vehicles.

9.1 Electronic Transmission Control

Electronically controlled transmissions (ECT) are used to improve fuel economy, performance, drivability, and shift quality. It even may be possible – because of the flexibility provided by microcomputer software – to allow for "adaptive" shift schedules, which can be tailored for improved fuel economy, performance, or comfort. Because of the flexibility offered in ECT, the industry is seeing an accelerated trend away from traditional mechanical/hydraulic transmission control and a move toward the integrated ECT/engine/traction/speed control functions. In the following discussion, the major benefits of ECT are illustrated:

(1) *Precise Lockup Control*. Historically, vehicles with an automatic transmission usually are about 10 percent less efficient than those with a manual transmission. The efficiency loss arises mainly from the slip of the torque converter. Torque-converter lockup has been used widely to improve fuel economy. With electronic control, it is possible to coordinate shift point, lockup schedule, and accurate timing of lockup release to reduce shock at gear shifting. Consequently, in recent years, the fuel economy of an ECT is only slightly worse than for a manual transmission.

(2) *Better Shift Quality*. For the same reasons, clutch pressures can be controlled for improved shift quality. Coordinating with the engine-control unit (i.e., reducing engine torque during gear shifting) can improve substantially shift quality and reduce shock load on shift elements.

(3) *Flexible Driving*. Different gear-shift patterns (e.g., power and economy) can be selected by a driver to better adapt to driving conditions. In general, if the information is available, the shift schedule can be programmed to consider vehicle

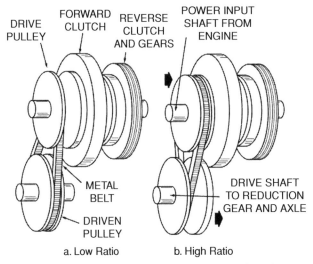

Figure 9.1. Continuously variable transmission (CVT).

status (e.g., warm-up and load), driving conditions (e.g., local or highway), and even driver status (e.g., passive or aggressive).

(4) *Weight Reduction.* Both size and weight can be reduced because of reduced complexity. The number of component parts also was found to be greatly reduced, which is translated into higher reliability and lower cost.

(5) *Integrated Vehicle Control.* Engine, cruise, and traction control functions can be integrated for superior performance.

(6) *Foolproof Design.* Improper operation by a driver (e.g., down shifting or shifting into reverse gear at high vehicle speed) can be detected to avoid severe damage to the transmission.

ECTs are divided into two major categories: discrete ECTs and CVTs. The CVT is an area of ongoing research activity, and the basic principle of operation is illustrated in Figure 9.1. Mechanical belt/chain–type with pulley-driven CVTs are among the most advanced with still (relatively) small-scale production. A modern CVT system of this type usually consists of a steel belt that runs between two variable-width pulleys. In the new push-belt design, the belt is pushed toward the output pulley, which can transmit torques that exceed the capacity of conventional pull-type belts. The distance between pulley cones can be varied to change the gear ratio between shafts, which effectively generates an infinite number of "gears." A CVT usually is less efficient than a standard discrete automatic transmission because of the losses in the belt–pulley system. However, it enables the engine to operate in the optimized speed range and therefore promises increased overall drivetrain efficiency. Initial testing on a midsized engine (i.e., 3.3L) reports a 10 percent improvement in fuel economy (Kluger and Fussner 1997). Although CVTs are not widely available on passenger vehicles in the United States, they are installed on numerous compact models in Europe and Japan, with an installation base of more than 1 million. We can expect to see wider CVT applications in the next decade.

The control of CVTs is actually somewhat simpler than that of the discrete ECTs. In the following discussion, discrete ECTs are the main consideration. There are many functions that an ECT may perform – for example, torque-converter lockup, lube/clutch cooling, shift scheduling, control of slip in the torque-converter bypass clutch, strategies for smooth transition from and to neutral idle, speed-ratio control for CVTs, line-pressure control, belt-load control in CVTs, driving from a stop for ECTs without torque converters, and shift-execution control (Hrovat and Powers 1988). In the following example, both the shift scheduling and the shift execution for a discrete ECT are discussed. The example is for the prototype four-speed transaxle ECT shown in Figure 9.2(a) (Hrovat and Powers 1988).

The transmission consists of a torque converter, reverse, low and high clutches, and a single shift actuator (a so-called dog actuator). The low clutch is constantly engaged in the first gear and the high clutch is constantly engaged in the fourth gear. The dog actuator is used to engage second gear when displaced to the left or third gear when displaced to the right. Power is transmitted to the front wheels through the differential and CV-joint–equipped axles. The dog actuator essentially creates a rigid link or engagement, thereby tolerating very small speed differentials at the time of engagement. Considered here is a power-on upshift from first to second gear. As illustrated in Figure 9.2(b), the shift consists of three phases: torque, inertia, and level-holding. During the torque phase, the engine torque is transferred from the low to the high clutch (see pressure traces in Figure 9.2[b]). When the low clutch is unloaded, the high clutch controls the turbine speed to the new synchronous level by following the speed-ratio ramp shown in Figure 9.2(b). This constitutes the inertia phase of the shift. Subsequently, the turbine speed is held at the second-gear synchronous level. This level-holding phase facilitates the dog-actuator engagement, which is the most critical phase of the shift because of the precise speed-ratio-control requirement. When the dog actuator has been engaged, the shift is completed by releasing the high clutch.

Closed-loop speed-ratio control is an important function in light of the stringent requirement for the inertia and level-holding phases. As a first step, this requires the development of a process model for the powertrain. The model required includes models for the engine, the torque converter, and an automatic clutch with an electrohydraulic control valve. A simplified version of the model results in a fifth-order linear model suitable for controller design. A block diagram for the system is shown in Figure 9.3(a), the digital implementation of the controller is shown in Figure 9.3(b), and the validation results are shown in Figure 9.4.

These results show both good ramp-following during the inertia phase and precise level-holding during the level-holding phase. Note that the initial level-holding during engagement of the dog actuator is also quite good. The effect of making the shift faster also is shown in Figure 9.4. Faster shifts yield improved fuel economy, whereas slower shifts lead to improved comfort.

9.2 Clutch Control for AWD

A clutch controls the energy flow from the engine to the transmission (Figure 9.5). Power can be transmitted from the engine to the output shaft by engaging and disengaging the clutch. A conventional clutch is known as a *compressed clutch* and consists of two basic components: the clutch cover and the disc. The clutch cover is an outer shell that contains the friction plate and the drive block. The friction plate is

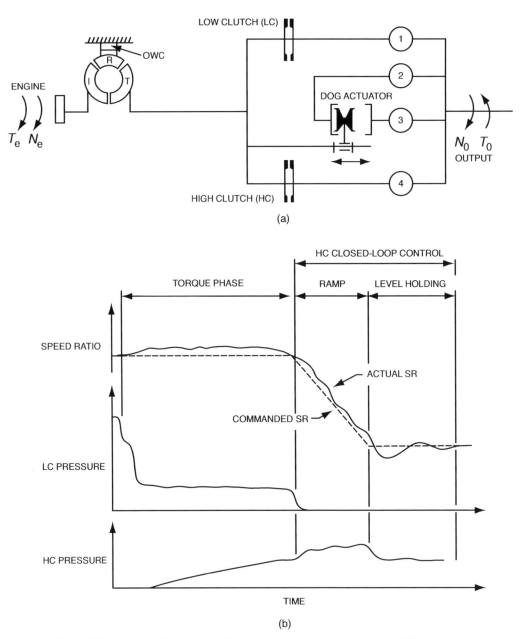

Figure 9.2. (a) Example schematic for electronic transmission control. (b) One-two power-on upshift execution phases (Hrovat and Powers 1988).

a cast piece that provides a pivot point for the diaphragm as well as a friction surface for the steel plate and mounting surface for the drive block. The drive block, driven by a hydraulic system or a magnetic-coil system, translates and provides pressure between the steel plate and the friction plate. The clutch housing protects the clutch components and also works as a heat sink for the coolant, which helps with heat dissipation during operation.

Clutch systems for various vehicles are different in functionality and capacity. In this section, a clutch system for an automatic AWD vehicle is studied. Automatic

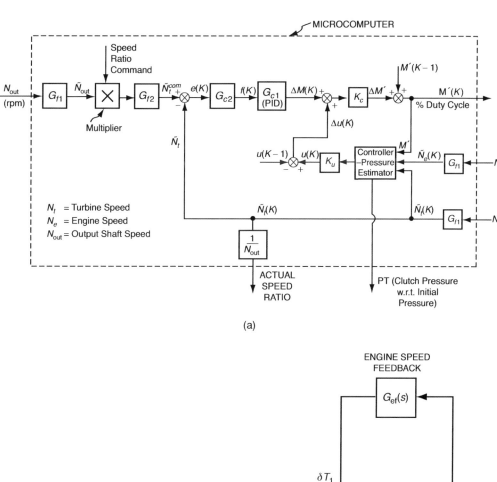

Figure 9.3. (a) Digital shift controller block diagram. (b) Shift control system block diagram (Hrovat and Powers 1988).

AWD (Hallowell 2005), which also is called "active AWD" or "smart AWD," is essentially a complex 2WD system. The torque from the engine is not transferred to all four wheels all of the time but rather on demand. The primary shaft (either front or rear shaft) is always engaged with the engine shaft, whereas the secondary shaft is engaged through an AWD clutch system when necessary, thus switching the powertrain system from 2WD to 4WD mode. For example, when slippage occurs

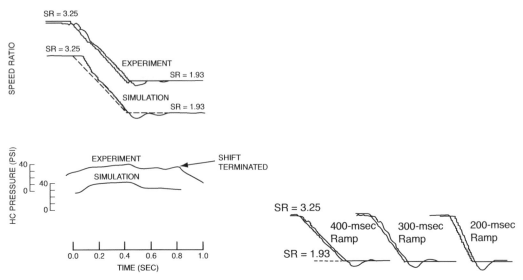

Figure 9.4. Simulation and experimental results for one-two power-on upshift (Hrovat and Powers 1988).

on the primary wheels, the clutch system engages and powers the secondary shaft to enable the vehicle to move smoothly Other situations, such as turning, climbing, and accelerating, also require traction on the secondary wheels for improved vehicle dynamics. The automatic AWD is implemented widely in many types of vehicles (e.g., Honda CRV, Toyota RAV4, Land Rover Freelander, Isuzu Trooper, and Jeep Grand Cherokee).

The clutch in automatic AWD systems provides more sophisticated functionalities compared with those in 2WD systems (Figure 9.6). The control of the clutch

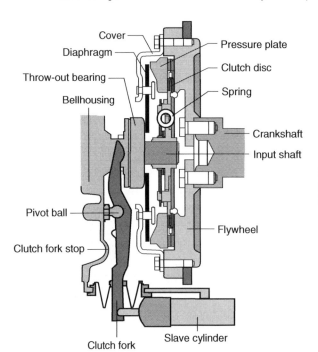

Figure 9.5. Structure of a clutch system. (*Source:* http://img528.imageshack.us/ img528/575/25838775lk1.jpg.)

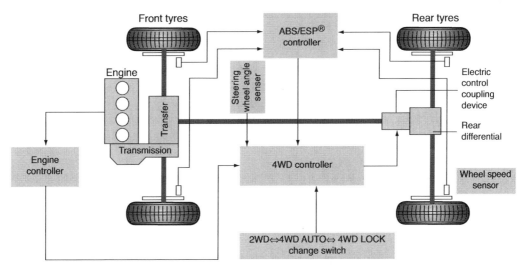

Figure 9.6. An automatic AWD system. (*Source:* http://www.autopressnews.com/2006/m03/
suzuki/sx4_iawd_system.jpg.)

considers not only the impact and friction mechanism during engagement but also
the delivery of a certain amount of torque to the secondary shaft to achieve a proper
torque distribution all of the time. The clutch-control system in automatic AWD systems can be divided further into two parts. The first part, based on vehicle dynamics
and current vehicle status (e.g., longitudinal velocity, yaw rate, and lateral acceleration), calculates the desired torque distribution and the desired output torque on
the secondary shaft. The second part calculates the control effort for the actuator
to compress the plate and transfer the requested torque under the current state of
the clutch. In this book, we focus on the design of the second part of the control for
a particular type of AWD clutch (Figure 9.7). This wet-friction type of clutch does
not have an active pump system for cooling; instead, two thirds of the clutch plate is
immersed in the oil, in which the heat from the plate is dissipated. A coil system is
used as an actuator and generates electromagnetic forces to provide a compressive
force between plates.

   The objective of the clutch control is to achieve fast and accurate tracking of the
reference torque signal under different operating conditions and to protect clutch
components from overheating and failure. However, there are several challenges
in designing an effective control for the clutch system. The dynamics of the system
changes with different input and operating conditions – especially the temperature
states of the components – and exhibits a highly nonlinear behavior. During engagement, the friction between the steel plate and the friction plate generates a significant
amount of heat and may raise the temperature of the components to 180 degrees C
in typical operations. Such a large increase in temperature can change significantly
many mechanical properties of the components – such as the friction coefficient
between the plate and the coil resistance of the actuator – and result in a noticeable
fluctuation in the output torque. Furthermore, even without temperature variation,
the mapping between the input (i.e., coil voltage) and the output torque is nonlinear.
Thus, designing a control that can achieve consistent performance under different
conditions is challenging for such a nonlinear system.

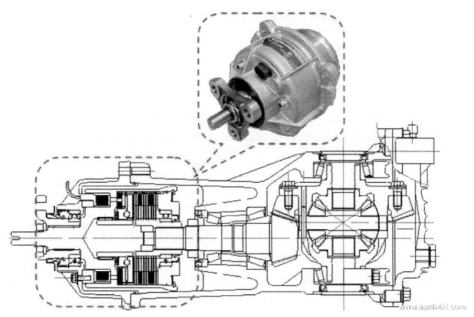

Figure 9.7. An AWD clutch system. (*Source:* http://www.awdwiki.com/images/hyundai-tuscon-electromagnetic-clutch.jpg.)

Existing designs for the clutch-control system typically are an open-loop feed-forward proportional control with an open-loop observer (Figure 9.8). The controller estimates the error between the reference torque and the output torque using an open-loop observer and then computes the control for driving the plant to yield the desired output. There are several major drawbacks to the current designs. First, due to the lack of feedback, it is difficult to achieve satisfactory performance with an open-loop observer because uncertainty in the system leads to poor state estimates. Second, the improper setting of initial conditions affects system performance for a long period because of the slow dynamics of the thermal system. Third, the pro-portional control cannot achieve zero steady-state error. Fourth, a single controller cannot offer consistent performance for different operating conditions. Fifth, there is no analytical basis for selecting the controller gain, which is fine-tuned by trial and error. Finally, the stability of the system, the effects of model uncertainty, and the effects of the potential time delay in the controller have not been investigated.

In Duan et al. (2011), the design of a more effective control for the clutch system via the piecewise affine (PWA) system framework is presented. Based on that theoretical framework, a piecewise control is proposed to achieve consistent performance under different operating conditions. The controller gains are selected analytically, and the stability of the closed-loop system in the presence of uncertainty and controller delay is guaranteed by analyzing the global stability of the system.

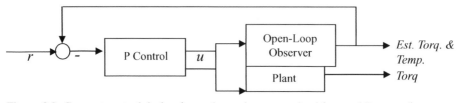

Figure 9.8. Current control design (open-loop observer + feed forward P control).

PWA systems are defined by dividing the state space into a finite number of polyhedral regions in which the dynamics within the region is described via a local linear model. The switching between these local models may depend on both inputs and states or only on states. This structure provides a flexible and traceable framework to model a large class of nonlinear systems as well as a suitable platform for rigorous analysis and design. In addition to the flexibility for modeling, the PWA system framework provides a good platform for controller analysis and design. The application of the PWA system framework to the modeling and control of a nonlinear automotive AWD clutch system is presented in Duan et al. (2011). The nonlinear clutch system is formulated into the PWA system framework first; subsequently, a switched control system is designed to ensure the closed-loop stability in the presence of uncertainties and delays through Lyapunov stability analysis. The results show significant improvements in performance over current designs and demonstrate the tradeoff among performance, robustness to modeling uncertainty, and time delays.

Model Development

Because temperature fluctuation introduces significant disturbances to the clutch system and may lead to premature degradation of the components, estimation of temperature states is necessary for temperature compensation and failure prevention. A model is developed to estimate the temperature states of the clutch components. Three first-order differential equations are obtained for the thermal system:

$$\dot{\mathbf{x}}_{\mathbf{T}}(t) = \mathbf{A}_{\mathbf{T}}\mathbf{x}_{\mathbf{T}}(t) + \mathbf{B}_{\mathbf{T}}\mathbf{u}_{\mathbf{w}}(t)$$

$$y_T(t) = \mathbf{C}_{\mathbf{T}}\mathbf{x}_{\mathbf{T}}(t) \tag{9.1}$$

where:

$$\mathbf{A}_{\mathbf{T}} = \begin{bmatrix} -\dfrac{R_{po}+R_{pa}+R_{pc}}{C_p} & \dfrac{R_{po}}{C_p} & \dfrac{R_{pc}}{C_p} \\[3mm] \dfrac{R_{po}}{C_o} & -\dfrac{R_{po}+R_{oa}+R_{oc}}{C_o} & \dfrac{R_{oc}}{C_o} \\[3mm] \dfrac{R_{pc}}{C_c} & \dfrac{R_{oc}}{C_c} & -\dfrac{R_{oc}+R_{pc}+R_{ca}}{C_c} \end{bmatrix}, \mathbf{B}_{\mathbf{T}} = \begin{bmatrix} \dfrac{1}{C_p} & \dfrac{R_{pa}}{C_p} \\[3mm] 0 & \dfrac{R_{oa}}{C_s} \\[3mm] 0 & \dfrac{R_{ca}}{C_c} \end{bmatrix},$$

$$\mathbf{C}_{\mathbf{T}} = [0 \quad 0 \quad 1]; \quad \mathbf{x}_{\mathbf{T}}(t) = \begin{bmatrix} T_p(t) \\ T_o(t) \\ T_c(t) \end{bmatrix}$$ is the temperature-state vector; $\quad \mathbf{u}_{\mathbf{w}}(t) =$

$\begin{bmatrix} P_{mech}(t) \\ T_a(t) \end{bmatrix} = \begin{bmatrix} y_o(t)|\omega_o(t) - \omega_i(t)| \\ T_a(t) \end{bmatrix}$ is the disturbance-input vector; and $y_T(t) =$ $T_c(t)$ is the temperature-system output. Also, T_a is the ambient temperature; y_o is the output torque; ω_i is the input shaft speed; $\omega_o \in \mathfrak{R}^1$ is the output shaft speed; T_p, T_o, T_c are the temperature states of the clutch plate, oil, and coil, respectively; C_p, C_o, C_c are the constant scalars representing corresponding thermal capacities; and R_{po}, R_{pa}, R_{pc}, R_{oa}, R_{oc}, R_{ca} are the constant thermal resistances among the three components. The mechanical power P_{mech}, generated by friction, is estimated as the product between absolute slip speed and output torque. The inputs T_a, ω_i, and ω_o are measured in the current design. The coefficients for thermal capacitance and thermal resistance are identified using experimental data. This temperature model is used in the design of a temperature-disturbance observer and for feed-forward control.

The resistor–inductor circuit system generates magnetic compressive force to engage the clutch. To identify the mapping between the control input (i.e., coil voltage) and the output torque, a first-order differential equation is used first to describe the relationship between the coil voltage and the coil current:

$$\frac{d}{dt}i_c(t) = A_{qc}(T_c)i_c(t) + B_q u(t) \tag{9.2}$$

where:

$$A_{qc}(T_c) = A_q + \Delta A_q(T_c) = -\frac{\Omega_{c0}}{L_c} - \frac{\Omega_{c0}\alpha(T_c(t) - T_{c0})}{L_c} \tag{9.3}$$

$B_q = \frac{1}{L_c}$, $u(t) = V_c(t)$ is the coil-voltage input, $i_c(t)$ is the coil current, L_c is the coil inductance, α is the temperature coefficient of the coil resistance, Ω_{c0} is the nominal coil resistance, and T_{c0} is the nominal coil temperature.

Then, a model is developed to estimate the output torque, which is determined by both the coil current and the plate temperature:

$$y_o(t) = \mu(T_p)C_q(i_c) \tag{9.4}$$

where $y_o(t)$ is the output torque; $\mu(T_p)$ is a scalar function and represents the effects of plate temperature, T_p, on the μ factor (i.e., plate-friction coefficient); and $C_q(i_c)$ is the nominal output torque under nominal plate temperature T_{p0}. Assuming $\mu(T_p)$ and $C_q(i_c)$ are independent and can be approximated by the polynomial functions in Eqs. (9.5) and (9.6), respectively:

$$\mu(T_p) = 1 + \sum_{j=1}^{m} l_j(T_p(t) - T_{p0})^j \tag{9.5}$$

$$C_q(i_c) = \begin{cases} \sum_{k=1}^{n} p_k i_c(t)^k, & \text{if } i_c > 0 \\ 0, & \text{if } i_c \leq 0 \end{cases} \tag{9.6}$$

where T_{p0} is the nominal plate temperature and l_j, p_k are the coefficients that later are identified empirically. Note that when a negative current is applied, the clutch will disengage and no output torque will be transferred (i.e., $C_q(i_c)$ saturates at zero when $i_c \leq 0$). Thus, the temperature effects introduce nonlinearities in Eqs. (9.1) and (9.4).

Control Design and Stability Analysis

The proposed controller, as shown in Figure 9.9, contains three parts: (1) an observer that estimates the unmeasured temperature states; (2) a feed-forward control that

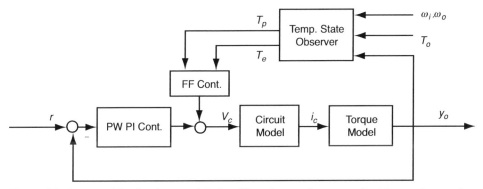

Figure 9.9. Proposed feedback control design (disturbance observer + feed forward control + piecewise PI feedback control).

compensates for the disturbances (i.e., temperature variation); and (3) a piecewise PI control that realizes torque-feedback control:

$$u(t) = u_{ff}(t) + u_{fb}(t) + r(t) \qquad (9.7)$$

Three steps are required to develop the controller. First, an observer is designed based on the temperature model in Eq. (9.1) to estimate the temperature disturbance. Second, a feed-forward controller is designed to remove the temperature effects in Eqs. (9.5) and (9.6), rendering the dynamics of the nominal plant system (i.e., the plant and feed-forward system) independent of temperature states. Third, approximating the nonlinearity in Eq. (9.4) via a piecewise linear function leads to a PWA model for the nominal plant system. Based on the PWA model, a piecewise PI-feedback control is developed. Compared to the original controller in Figure 9.8, three additional sensors (i.e., coil-current sensor, output-torque sensor, and coil-temperature sensor) are proposed to be implemented to realize this new design. The coil-temperature sensor provides temperature measurements and helps the temperature estimates of the observer to converge to the actual states. The torque sensor and current sensor provide signals to realize the piecewise PI feedback control. These sensors are discussed in detail in Duan et al. (2011).

Experiments

To calibrate the parameters of the clutch model, experimental tests are conducted. The clutch is mounted on a test bed where different operating conditions are simulated. The outside shell of the clutch is fixed on the body of the test bed and the input and output shafts are connected to two motors, the speed of which can be changed arbitrarily. Applying a voltage in the clutch coil generates a magnetic force and engages the clutch. Torque sensors and angular-velocity sensors are placed on the motors for monitoring the I/O speed and torque. Several thermistors are placed on the clutch for measuring the temperature of the plate, oil, and coil. The measurements in these calibration experiments are T_p, T_o, T_c, T_a, ω_i, ω_o, i_c, V_c, and Q.

A test profile, shown in Figure 9.10, is used for calibrating the temperature model. It consists of a series of input and slip speeds, with a sequence of constant-power torque steps at each combination. Coil current is applied periodically to generate frictional energy in the system. The temperature of the components increases when

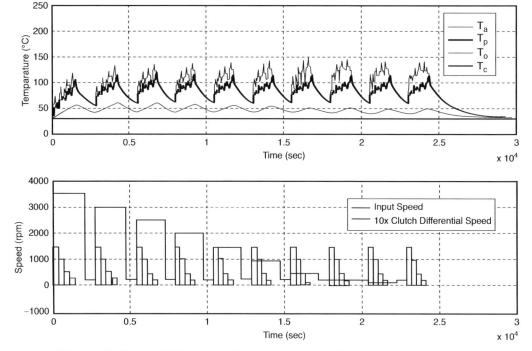

Figure 9.10. Test profile for clutch thermal system modeling.

there is an energy input and decreases when no energy is supplied. Therefore, the temperature curves take on a zigzag pattern. Input speeds range from 125 to 3,500 RPM. At each input speed, the slip speed starts at 150 RPM and is reduced by steps. The sampling rates for all the data are 10 Hz.

The parameters of the circuit model can be obtained from the tests. To calibrate the torque model, two experiments with a similar setup are conducted to identify the functions $\mu(T_p)$ and $C_q(i_c)$, respectively. A constant current input is used in the first experiment to investigate the relationship between the plate temperature and the μ factor. In the second experiment, different coil currents are applied sequentially with sufficient cooling time in between to characterize the relationship between the coil current and the output torque under nominal plate temperature.

Open-Loop System Results

Using the experimental data collected and the system identification toolbox from MATLAB, the model parameters are as follows:

$$\mathbf{A_T} = \begin{bmatrix} -0.05 & 1.02 & 0.395 \\ 0.004 & -0.026 & 0.096 \\ 0.004 & 0.025 & -0.049 \end{bmatrix}, \quad \mathbf{B_T} = \begin{bmatrix} 0.074 & 0.33 \\ 0 & 0.004 \\ 0 & 0.01 \end{bmatrix}, \quad \mathbf{C_T} = \begin{bmatrix} 0 & 0 & 1 \end{bmatrix}$$

$$\Omega_{c0} = 2.00 \text{ Ohm}, L_c = 0.121 \text{ H}, \alpha = 0.004°\text{C}^{-1}.$$

$$\mu(T_p) = 1 - 0.0014(T_p - 40) - 6.8 \times 10^{-6}(T_p - 40)^2$$

$$y_0(i_c) = \begin{cases} -51.2585i_c + 218.6445i_c^2 - 56.9184i_c^3 + 4.4563i_c^4, & \text{if} \quad i_c > 0 \\ 0, & \text{if} \quad i_c \leq 0 \end{cases}$$

Table 9.1. *Partition over space of coil current*

Region	1	2	3	4
i_c	<0 A	0~0.8 A	0.8~3.3 A	> 3.3 A
$q_1^i i_c + q_2^i$	0	$103 i_c$	$250 i_c - 117$	$101 i_c + 372$

The polynomial model for μ is calibrated using the experimental data shown in Figure 9.11. Based on the nonlinearity of y_o, a PWA function with the partition in Table 9.1 is used to approximate y_o, as shown in Figure 9.12.

For these types of clutch systems, the error in the plate-temperature estimates is typically within 15°C. Thus, the bounds for the uncertainty in the model can be estimated; the step response of the open-loop system under the temperature disturbance in Figure 9.13 is shown in Figure 9.14. Note that the variation of the temperature states results in significant fluctuation and a decrease in the output torque for the open-loop system.

Closed-Loop System

We now compare the existing controllers, as in Figure 9.8 with $K_P = 0.1$ here to the proposed controller in Duan et al. (2011). The gains of the proposed controller are listed in Table 9.2 and vary for each region listed in Table 9.1.

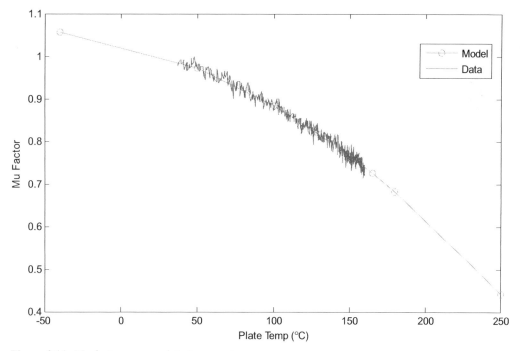

Figure 9.11. Mu factor versus plate temperature.

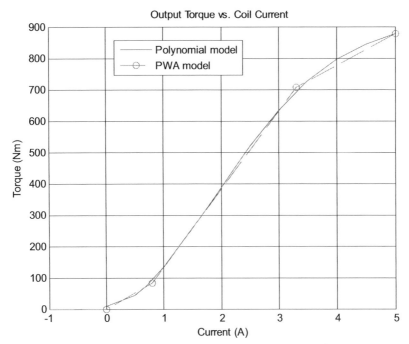

Figure 9.12. Nonlinear torque model versus PWA approximation.

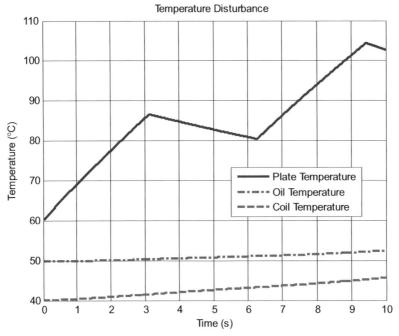

Figure 9.13. Simulated temperature fluctuation.

Table 9.2. *Controller gains for each region*

Region	1	2	3	4
K_P^i	0.03	0.0296	0.012	0.0301
K_I^i	0	1.0597	0.426	1.0781
K_Z^i	40	0	0	0

As expected, both controlled systems perform better than the open-loop system in Figure 9.14. The comparison between the proposed design and the current design (see Figures 9.8 and 9.9, respectively) is provided in Figure 9.15, in which the performance of the proposed design is much better than the current design, which has a significant steady-state error.

Stability Analysis

The stability of the closed-loop system also was examined using Lyapunov stability methods to ensure system performance (Duan et al. 2011). The stability analysis considers uncertainty caused by approximation of the nonlinear system as a PWA system and the effects of time delays in the controller. The tradeoff among delay, uncertainty, and performance (i.e., location of the closed-loop eigenvalues) is provided in Figure 9.16. The bottom-left area is the feasible design region with guaranteed sta-

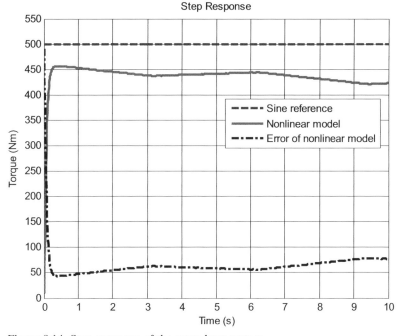

Figure 9.14. Step response of the open-loop system.

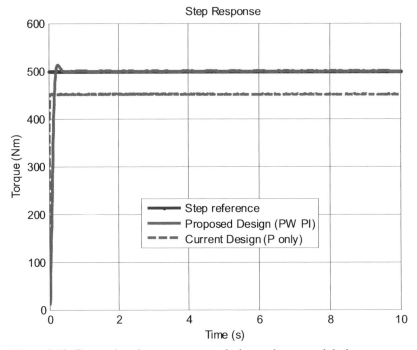

Figure 9.15. Comparison between current design and proposed design.

bility. As shown in the figure, the increase in delay reduces the maximum tolerable uncertainty for a particular control design. A larger control gain improves the speed of the system but renders it less robust against uncertainty and delay. A map such as this can assist in the selection of controller parameters to obtain a sufficiently robust system with satisfactory performance in the presence of delay.

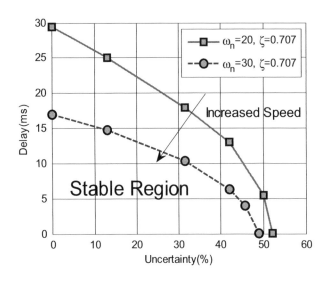

Figure 9.16. Map of feasible design region showing the tradeoff between speed of response, uncertainty, and delay.

PROBLEMS

1. Patent databases include a number of patents awarded for automotive ECTs. Search for and select one and summarize (i.e., one paragraph) the innovation being claimed in that patent.

2. Figure 9.16 shows that as the closed-loop system bandwidth (ω_n) increases, delays and uncertainty are more likely to destabilize the system. If the delay is 15 ms and the uncertainty is 25 percent, what is the system bandwidth at the guaranteed stability limit?

REFERENCES

Duan, S., J. Ni, and A. G. Ulsoy, 2011, "Modeling and Control of an Automotive All-Wheel Drive Clutch as a Piecewise Affine System," *ASME Journal of Dynamic Systems, Measures, and Controls* (in press).

Hallowell, S. J., 2005, *Torque Distribution Systems and Methods for Wheeled Vehicles*, U.S. Patent 6, 909–959.

Hrovat, D., and W. F. Powers, 1988, "Computer Control Systems for Automotive Power Trains," *IEEE Control Systems Magazine*, August 1988, pp. 3–10.

Kobayashi, N., T. Tokura, K. Shiiba, T. Fukumasu, T. Asami, and A. Yoshimura, 2008, "Development of a New Sport Shift Control System for Toyota's Automatic Transmission," *SAE World Congress and Exhibition*, Dearborn, MI, April 2008.

Kluger, M., and D. Fussner, 1997, "An Overview of Current CVT Mechanisms, Forces and Efficiencies," *SAE Technical Paper* 970688.

Sun, Z., and K. Hebbale, 2005, "Challenges and Opportunities in Automotive Transmission Control," *Proceedings of the American Control Conference*, Portland, OR, June 2005.

Zheng, Q., K. Srinivasan, and G. Rizzoni, 1998, "Dynamic Modeling and Characterization of Transmission Response for Controller Design," SAE Technical Paper 981094.

10 Control of Hybrid Vehicles

Hybrid vehicles, especially hybrid electric vehicles (HEVs), demonstrate significant potential in reducing fuel consumption and exhaust emissions while maintaining driving performance. Hybrid vehicles are equipped with more than one power source, and at least one should be reversible. The reversible power source serves as an energy storage device, whereas the other power source is either the primary power source or a "range extender." Most hybrid vehicles use a battery as the energy buffer, in which case they are known as HEVs. They can be classified as series, split, and parallel hybrids (Figure 10.1). The performance potential of the different configurations and the associated control problems are quite different.

The lower fuel consumption of a HEV typically is the result of several design features: (1) right sizing of the prime mover (i.e., internal combustion engine [ICE]); (2) load-leveling and engine shutdown to avoid inefficient engine operation; (3) regenerative braking; and (4) enhanced CVT function in certain configurations (i.e., series and split). Due to the multiple power sources and the complex configuration and operation modes associated with them, the control strategy for a hybrid vehicle is more complicated than for an engine-only vehicle. This chapter focuses on the design of control algorithms for series, parallel, and split hybrid vehicles. However, we first discuss the layout and pros and cons of the three configurations.

10.1 Series, Parallel, and Split Hybrid Configurations

For series hybrids, the mechanical power from the internal-combustion engine is converted immediately to electrical power by the generator, which is properly sized to match the engine. The electrical power from the generator then drives the motor (or motors), which can be augmented (positively or negatively) by the battery power. Because there is no mechanical connection between the engine and the wheels, there is total flexibility in engine operations (i.e., the generator provides a virtual CVT function for the engine) and the drivetrain becomes simple. The electric-power distribution provides greater packaging flexibility compared to parallel or split hybrids – the latter two configurations both have direct mechanical connections between the engine and the wheels. In addition, because the propulsion power is distributed in electric form, it is possible to implement in-hub motors, which provide AWD and enhanced vehicle-stability control (VSC) functions (see Chapter 14).

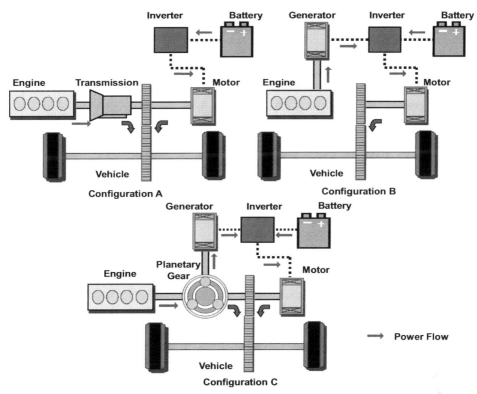

Figure 10.1. Hybrid vehicle configurations: A. Parallel; B. Series; C. Power-Split (Parallel/Series).

In a typical VSC execution, braking is applied to a wheel or wheels on one side of the vehicle to create a yaw moment. With in-hub motors, it is possible to apply traction power to the other side, doubling the yaw-moment capacity. The power capacity of the electric motor (or motors) must be large because the engine does not drive the wheel directly. The most significant drawback of the series-hybrid design is the low efficiency inherent from the multiple energy conversions. Each energy conversion (i.e., from mechanical to/from electrical or electrical to/from chemical) is associated with a loss of between 5 and 20 percent. The system efficiency of series hybrid vehicles is usually considerably lower than the parallel or split counterpart. Because of the lower efficiency, series hybrids have not been popular for lightweight vehicles. However, they are suitable for heavy-vehicle applications and, in fact, have been widely adopted for other transportation applications such as locomotives, agricultural machines, and submarines.

For parallel hybrids, a secondary power source (i.e., usually an electric motor) exists in parallel with the ICE. There are many possible configurations for parallel hybrids (Figure 10.2). When the motor is too small for vehicle propulsion (also known as a microhybrid), the existence of the motor (known as the belt alternator starter or the belt starter generator) has only limited effects on the vehicle operation (i.e., engine shutoff and startup and limited regenerative braking). It does not change the way a vehicle is driven. Microhybrids exist only in the form of parallel hybrids and are not of the series or the split type. An electric motor sufficiently large to

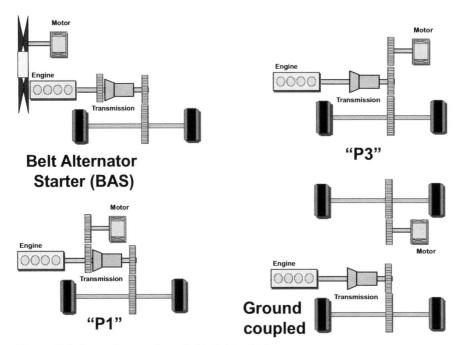

Figure 10.2. Several types of parallel hybrid vehicles.

drive a vehicle, either alone or by assisting the ICE, can be placed in many different locations. Figure 10.2 shows that the motor can be placed either pre-transmission ("P1"), post-transmission ("P3"), or at the originally nondriven axle (i.e., ground coupled). There is a configuration called "P2" that is the same as P1 except that a clutch is placed between the engine and the motor, in which case the torque converter (of the presumably automatic transmission) can be removed for better efficiency.

Because there are many types of parallel hybrids, they share several attributes, but the pros and cons are closely dependent on the configuration. Because the overall architecture is closest to a traditional vehicle, modifications required to obtain a hybrid version are relatively few; thus, development time can be the shortest. The dual power sources also provide better fail-safe characteristics than series hybrids. If they are properly designed, both the engine and the motor can have a much smaller power rating than the original ICE yet achieve the same or better acceleration performance. If a traditional step-gear transmission is used, a parallel hybrid does not have CVT-like function, which results in losses due to the coupling between vehicle and engine speeds. The Honda Integrated-Motor-Assist (IMA) design is an example of a P1 design (Ogawa et al. 2003).

The third type of hybrid vehicle is the power-split type. The best-known examples include the Toyota Hybrid System (Muta et al. 2004) (used in the Prius, Estima minivan, and RX400H) and the Allison Transmission Electric Drives System (Holmes 2001). Both hybrid systems use a planetary gear(s) as the power-summation device as well as to provide torque ratios. Two electric machines (i.e., the motor and the generator, although both can have either role) are used as the secondary power sources to sustain favorable operating conditions for the ICE as well as to augment the engine-driving torque to satisfy a driver's demand. In the case of the original

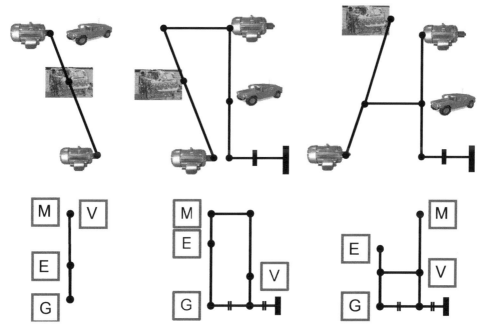

Figure 10.3. Several examples of split hybrid vehicles (Left: original Toyota Hybrid System; Center: original General Motors Design; Right: A Timken design).

Toyota Hybrid Design, one planetary gear is used, whereas in the original General Motors design, two planetary gears are used. Figure 10.3 shows three examples of split hybrid designs, all of which are patented. They are represented in the lever-diagram form in which each "stick" represents a planetary gear with three nodes: sun gear, carrier gear, and ring gear. The small icons connected to the nodes represent the vehicle (V), engine (E), motor (M), and generator (G). From Figure 10.3, it is clear that there are many possible split-vehicle configurations. For split vehicles with two planetary gears, there are in theory 1,152 different ways (Liu and Peng 2008) to connect the four elements, plus possibly a ground clutch, to the 6 nodes, with 288 having the electronic continuously variable transmission (ECVT) function (i.e., the engine and vehicle speeds can be independent). Apparently, in addition to component sizing and control, configuration selection is an important decision in the design of split hybrid vehicles.

In general, split hybrids have high engine efficiency because of the enhanced CVT (ECVT) function. The planetary gear typically is very compact (i.e., only slightly larger than a soft-drink can), has high torque capacity, has simple and robust mechanical design, has very high efficiency, and potentially can have multiple operation modes if clutches are used. However, because it is inherently a mechanical transmission, it is not suitable for certain types of hybrid vehicles. For example, it does not make sense to have a fuel-cell split hybrid. All existing split vehicles use two electric machines and, because of the torque balance on the planetary gears, they are inherently strong hybrids with two sizable electric machines. Therefore, it is more challenging to design a low-cost split hybrid vehicle. Because of the existence of three power sources, the associated sizing and control problem is more complex than for series and parallel hybrids.

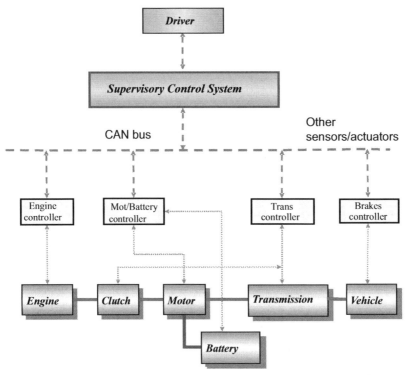

Figure 10.4. An example hierarchical structure of hybrid-vehicle control system: A supervisory control unit on the top interacts with several servo-loop controllers.

10.2 Hybrid Vehicle-Control Hierarchy

Hybrid vehicles require significant modifications to the propulsion and braking systems of ground vehicles. These changes also impact the design and functioning of other subsystems, including electric systems; heating, ventilation, and air conditioning (HVAC); and auxiliary systems. Because hybrid vehicles introduce several additional subsystems (i.e., the traction battery, motor, generator, power converters, and clutches), the control and coordination among all subsystems become more complex. The role of control systems for a hybrid vehicle is similar to an operating system for a computer. Even with the best components, a vehicle with bad control algorithms may perform poorly if the control is poorly designed. It is important to adopt a design process that ensures guaranteed quality of the control system.

Control engineers who design algorithms for a hybrid vehicle usually need to encompass three fundamentally different functions. At the lowest level, there are functions such as initializations, actuator and network monitoring, and sensor diagnosis that must be performed to ensure that all hardware is functioning properly. If the hardware is judged to be functioning properly, then the control features can be initiated. Typically, the hybrid vehicle control system is designed hierarchically (Figure 10.4). At the top is a supervisory control unit, interacting with the human driver as well as other vehicle control systems such as braking, VSC, and HVAC. Under its "supervision" are the ECU, transmission control unit (TCU), battery-management system, and electric motor/drive. The supervisory control unit should be designed assuming that it is fully aware of the capabilities and limitations of the subsystems; thus, all commands sent to the *servo-loop controllers* of the subsystems are achievable. The servo-loop controllers directly regulate the output of the subsystems; for example, a gear-shift command might be sent to the TCU, which then executes the gear shift by manipulating the clutch pressures. Similarly, a command for engine power (or torque) might be communicated to the ECU, which then manipulates the fuel injection, valve, EGR, and so on to achieve reliably and accurately the required engine power (or torque).

The hierarchical structure described herein simplifies the design and improves the modularity and reusability of hybrid vehicle control systems. The supervisory control unit typically is designed based on slow dynamics and focuses on the long-horizon performance of a vehicle (e.g., fuel economy and/or emission during a drive cycle). The servo-loop control units (e.g., ECU and TCU) are designed to achieve accurate regulation or tracking of the commands and must follow quickly the commands from the supervisory unit under a wide range of vehicle load, age, and environmental conditions. Their designs usually are based on classical control theories augmented by extensive calibrations. Once designed, the servo-loop controls can be reused easily on another hybrid vehicle. The supervisory control, conversely, must be redesigned for each hybrid vehicle based on vehicle load and characteristics of the powertrain subsystems. In addition, the supervisory control must strike a balance among multiple objectives, including fuel economy; electric range; emission; and noise, vibration, harshness (NVH). The main challenge in the design of hybrid vehicle control systems is at the supervisory-control level.

The following sections introduce design concepts for the supervisory control functions, focusing on the design for fuel economy while regulating the battery charge. The discussion covers the prevailing design concepts for series, parallel, and split hybrid vehicles. Before delving into the hybrid-vehicle controls and simulations, we first illustrate the basic concepts of drive cycles, driving-power calculation, and fuel/battery state of charge (SOC) calculations using MATLAB examples.

EXAMPLE 10.1: DRIVE CYCLES AND DRIVING POWER. The performance of traditional and hybrid vehicles is evaluated using standard drive cycles that are selected to represent typical driving conditions, such as in an urban or a highway setting. Drive cycles are simply speed profiles as a function of time; when the vehicle weight, rolling resistance, and aerodynamic drag coefficient are specified, the power required to go through these cycles can be calculated. This is illustrated in the following MATLAB example and the plots in Figure 10.5:

```
% Ex10_1.m
% Drive cycle and driving power calculation
clear all, close all
load CYC_HWFET.mat % Load highway cycle
time_highway = cyc_mph(:,1);
speed_highway = cyc_mph(:,2);
figure(1), subplot(221)
plot(time_highway, speed_highway)
xlabel('Time (Sec)')
ylabel('Speed (mph)')
Title('EPA highway cycle')
```

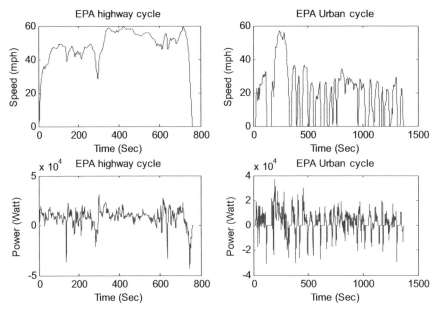

Figure 10.5. Output of Example 10.1. Driving power required for the EPA highway and urban cycles is low, usually less than 20kw for the mid-sized vehicle simulated.

```
clear cyc_mph
load CYC_UDDS.mat  % Load urban cycle
time_urban = cyc_mph(:,1);
speed_urban = cyc_mph(:,2);
subplot(222)
plot(time_urban, speed_urban)
xlabel('Time (Sec)')
ylabel('Speed (mph)')
Title('EPA Urban cycle')


m = 1800;
g = 9.81;
rolling_resistance_coeff = 0.015;
aero_dynamic_drag_coeff = 0.4;
road_grad = 0;   % road grade = 0 rad


N = size(time_highway, 1);
speed_highway = speed_highway*1602/3600;
% Change the unit of speed to m/sec
distance_highway = 0;
for i = 1:N-1,
  accel_highway(i) = (speed_highway(i+1)-speed_highway(i))/1.0;
% delta_T = 1 sec
    F_resistant_highway(i) =
rolling_resistance_coeff*m*g*cos(road_grad) ...
      +0.5*1.202*aero_dynamic_drag_coeff*speed_highway(i)^2 ...
      + m*g*sin(road_grad);
```

```
power_highway(i) = (m*accel_highway(i)+
F_resistant_highway(i))*speed_highway(i);
% Power = F*V
    distance_highway = distance_highway + speed_highway(i);
end
subplot(223)
plot(time_highway(1:N-1), power_highway)
xlabel('Time (Sec)')
ylabel('Power (Watt)')
Title('EPA highway cycle')
N = size(time_urban, 1);
speed_urban = speed_urban*1602/3600;
% Change the unit of speed to m/sec
distance_urban = 0;
for i = 1:N-1,
    accel_urban(i) = (speed_urban(i+1) - speed_urban(i))/1.0;
% delta_T = 1 sec
    F_resistant_urban(i) =
rolling_resistance_coeff*m*g*cos(road_grad) ...
        +0.5*1.202*aero_dynamic_drag_coeff*speed_urban(i)^2 ...
        + m*g*sin(road_grad);
    power_urban(i) = (m*accel_urban(i) +
F_resistant_urban(i))*speed_urban(i);
% Power = F*V (W)
    distance_urban = distance_urban + speed_urban(i);
end
subplot(224)
plot(time_urban(1:N-1), power_urban)
xlabel('Time (Sec)')
ylabel('Power (Watt)')
Title('EPA Urban cycle')
```

EXAMPLE 10.2: EFFICIENT ENGINE OPERATIONS. The ICE has been and likely will remain the primary power source on nearly all hybrid vehicles currently on the market. For these vehicles, efficient engine operation is a necessary (but not quite sufficient) condition for efficient hybrid vehicles. Selection of an efficient operation for an ICE is based on its BSFC, which typically has a "sweet spot" in the area where the engine is most efficient. Some of the basic control concepts start from the simple idea of operating the engine close to the sweet spot, which is highlighted in the plots in Figure 10.6:

```
% Ex10_2 Efficient Engine Operations

load engine_parameters
% Load engine torque (N-m)
% and fuel consumption (g/sec)
% Both of which is a 21 by 28 matrix
% The row (throttle) index is 0:5:100
```

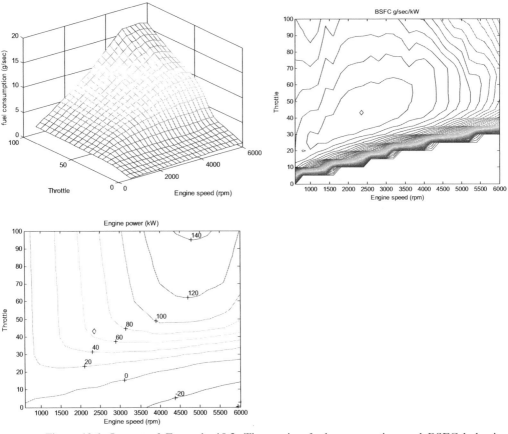

Figure 10.6. Output of Example 10.2. The engine fuel consumption and BSFC behavior become the basis for selecting control rules for efficient engine operations.

```
% The column (engine speed) index is 600:200:6000 (rpm)

engine_speed = [600:200:6000]*2*pi/60;  % rad/sec
Throttle_grid=[0:5:100];

figure(2)
mesh([600:200:6000], Throttle_grid, fuel_map)
xlabel('Engine speed (rpm)')
ylabel('Throttle')
zlabel('fuel consumption (g/sec)')

for k = 1:21,
  for j = 1:28,
      engine_out_power(k,j) =
engine_speed(j)*engine_torque(k,j)/1000;
% engine out power (kW)
      bsfc(k,j) = fuel_map(k,j)/engine_out_power(k,j);
% BSFC = g/sec/kW
      if (bsfc(k,j) < 0)
```

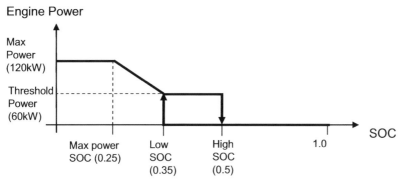

Figure 10.7. "Thermostat"-control concept common for supervisory control of series hybrid vehicles.

```
        bsfc(k,j) = 0.3;
    elseif (bsfc(k,j) > 0.3)
        bsfc(k,j) = 0.3;
      end
    end
  end
  figure(3)
  contour([600:200:6000], Throttle_grid, bsfc, 50)
  xlabel('Engine speed (rpm)')
  ylabel('Throttle')
  title('BSFC g/sec/kW')
  hold on, plot(2350, 43, 'd'), hold off

  figure(4)
  [CS,H]=contour([600:200:6000], Throttle_grid, engine_out_power);
  clabel(CS), xlabel('Engine speed (rpm)')
  ylabel('Throttle')
  title('Engine power (kW)')
  hold on, plot(2350, 43, 'd'), hold off
```

10.3 Control Concepts for Series Hybrids

The supervisory control of series hybrids is relatively straightforward – mainly because the engine can be run independently from vehicle speed and load. This decoupling makes it possible to determine the engine operation while completely disregarding the vehicle states – and, because of this decoupling, it makes no sense to demand engine torque, so a power-based control concept is more appropriate. A commonly used design concept is to determine the engine power based only on the battery SOC. The ICE is idling when there is adequate SOC. When the battery SOC is low, the ICE is turned on and run at a high-efficiency level until the battery SOC reaches a high threshold level. This "thermostat" control concept is illustrated in Figure 10.7. The control algorithm involves only a few parameters: the high/low bounds of the SOC dead band (i.e., 0.5 and 0.35), the threshold power (i.e., 60kW), and the "maximum power SOC" (i.e., 0.25), which are selected based on battery and vehicle characteristics. For modern Li-Ion batteries, the maximum power SOC may need to be higher to improve battery life. If an engine is selected properly based on vehicle load, the engine SOC operates mostly in the hysteresis rectangle. If the vehicle load is very high, it may be necessary for the engine to run at a power level higher than the threshold power. In the special case of when the thermostatic control concept is applied to a fuel-cell–powered series hybrid, the hysteresis rectangle could disappear completely. This is because fuel cells, unlike ICEs, operate efficiently at low power/current level because the efficiency is caused primarily by the Ohmic loss. Therefore, lower stack current is more efficient if the ancillary loads (e.g., compressor and humidifier) are not considered.

The thermostat concept is extremely simple and therefore was a popular concept for early hybrid-vehicle control designs. However, the concept has several drawbacks. First, the engine will be turned on and shut down under conditions that are completely independent of vehicle driving, causing possible confusion and/or discomfort. Second, the engine will experience frequent on/off and deep transients, which may need to be adjusted when considering the effect on emissions. Third, the significant amount of energy stored in and released from the battery incur efficiency loss, sometimes known as the "battery tax." Round-trip efficiency loss is typically in the range of 15 to 30 percent – which is significant and could erase most of the efficiency gains by operating the ICE only around the efficient sweet spot.

EXAMPLE 10.3: THERMOSTATIC CONTROL CONCEPT FOR SERIES HYBRIDS. The engine fuel-consumption and power-generation characteristics are analyzed first. Typical of today's passenger automobiles in the North American market, it can be seen that the engine is somewhat oversized. The sweet spot (denoted by diamond symbols in Figure 10.8) corresponds to a power level of more than 60kW, which is much higher than the vehicle driving load under the EPA highway and urban cycles. Therefore, engine efficiency will be low when it operates far from the sweet spot. In a hybrid configuration, it may be appropriate to downsize the engine. In this example, however, we assume that the original engine is used. Because the engine can be operated completely independently of the vehicle state, we limit the engine to operate in a small region, denoted by thick blue lines in Figure 10.8. From the BSFC plot, it is obvious that operating the engine at a power level much below 40kW would not be efficient, which should be a consideration in designing the control strategy. From the simulation results, it shows that the engine is turned on periodically to charge up the battery, and the SOC vacillates between the selected lower and upper threshold values.

```
% Ex10_3.m
% Compute fuel consumption of a thermostatic controlled
% Series huybrid vehicle for the Highway and Urban Cycles

clear all
ex10_1;  % Load and run ex10_1 to compute driving power
ex10_2;  % Load and run ex10_2 to compute engine map
```

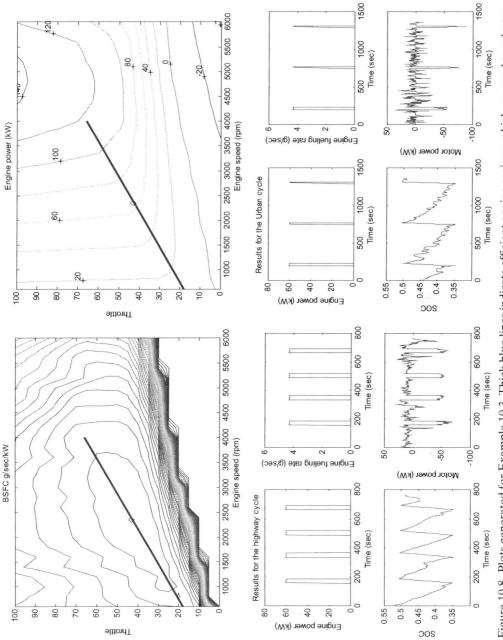

Figure 10.8. Plots generated for Example 10.3. Thick blue lines indicate efficient engine regions, which are used to selectparameters of the thermostat-control algorithm.

159

```
figure(3)
hold on, plot([600;4000], [18; 66], '-', 'LineWidth', 3)
hold off

figure(4)
hold on, plot([600;4000], [18; 66], '-', 'LineWidth', 3)
hold off

% The optimal operation of the engine is characterized
by the end points
% [600rpm, 18% throttle] and [4000rpm, 66% throttle].

opt_eng_speed = [600: 3400/48: 4000];
opt_throttle = [18:1:66];

% Interpolate to find fuel consumption (g/sec) and power (kW)
for each
% point
opt_fuel = interp2([600:200:6000], Throttle_grid, fuel_map,
opt_eng_speed, opt_throttle);
opt_power = interp2([600:200:6000]*2*pi/60, Throttle_grid,
engine_out_power, ...
                    opt_eng_speed*2*pi/60, opt_throttle);
figure(5), subplot(211)
plot(opt_eng_speed, opt_fuel), xlabel('Engine speed (rpm)'),
ylabel('fuel cnsumption (g/sec)')
subplot(212)
plot(opt_eng_speed, opt_power), xlabel('Engine speed (rpm)'),
ylabel('power (kW)')


% Simulate for the two drive cycles, series hybrid, themodtatic
control
SOC_max_power = 0.25;
SOC_low = 0.35;
SOC_high = 0.50;
threshold_power = 60;    % kW
max_power = 120;         % kW
Battery_capacity = 2.5;    % kW-hr

% Simulate for the highway cycle
total_fuel_highway_hev = 0;
SOC(1) = 0.52;
flag = 0;    % flag tracks whether we are on the top or bottom
of the hysteresis rectangle
for i = 1: (size(time_highway, 1)-1),
    if SOC(i) > SOC_high
            P_eng(i) = 0; flag = 0;
```

```
        elseif SOC(i) > SOC_low
            if flag == 0
                P_eng(i) = 0;
            else
                P_eng(i) = threshold_power;
            end
        elseif SOC(i) > SOC_max_power
            flag = 1;
            P_eng(i) = max_power - (SOC(i)-SOC_max_power)/(SOC_low-
SOC_max_power) ...
                    *(max_power-threshold_power);
        else
            flag = 1;    P_eng(i) = max_power;
        end

        P_batt(i) = power_highway(i)/1000 - P_eng(i);
        if P_batt(i) > 0
            SOC(i+1) = SOC(i) -
P_batt(i)*1.0/3600/0.92/Battery_capacity;
        else
            SOC(i+1) = SOC(i) -
P_batt(i)*1.0/3600*0.92/Battery_capacity;
        end

        if P_eng(i) == 0
            fuel_hev(i) = 0;
        else
            fuel_hev(i) = interp1(opt_power, opt_fuel, P_eng(i),
'linear');
        end
        total_fuel_highway_hev = total_fuel_highway_hev + fuel_hev(i);
end

figure(6), subplot(221)
plot([1: size(time_highway,1)-1], P_eng)
xlabel('Time (sec)'), ylabel('Engine power (kW)')
title('Results for the highway cycle')
subplot(222), plot([1: size(time_highway,1)-1], fuel_hev)
xlabel('Time (sec)'), ylabel('Engine fueling rate (g/sec)')
subplot(223), plot([1: size(time_highway,1)], SOC)
xlabel('Time (sec)'), ylabel('SOC')
subplot(224), plot([1: size(time_highway,1)-1], P_batt)
xlabel('Time (sec)'), ylabel('Battery power (kW)')

% Simulate for the urban cycle
total_fuel_urban_hev = 0;
SOC(1) = 0.44;
```

```
flag = 0;      % flag tracks whether we are on the top or bottom
of the hysteresis rectangle
for i = 1: (size(time_urban, 1)-1),
    if SOC(i) > SOC_high
        P_eng(i) = 0; flag = 0;
    elseif SOC(i) > SOC_low
        if flag == 0
            P_eng(i) = 0;
        else
            P_eng(i) = threshold_power;
        end
    elseif SOC(i) > SOC_max_power
        flag = 1;
        P_eng(i) = max_power - (SOC(i)-SOC_max_power)/(SOC_low-
SOC_max_power) ...
                *(max_power-threshold_power);
    else
        flag = 1; P_eng(i) = max_power;
    end

    P_batt(i) = power_urban(i)/1000 - P_eng(i);
    if P_batt(i) > 0
        SOC(i+1) = SOC(i) -
P_batt(i)*1.0/3600/0.92/Battery_capacity;
    else
        SOC(i+1) = SOC(i) -
P_batt(i)*1.0/3600*0.92/Battery_capacity;
    end
    if P_eng(i) == 0
        fuel_hev(i) = 0;
    else
        fuel_hev(i) = interp1(opt_power, opt_fuel, P_eng(i),
'linear');
    end
    total_fuel_urban_hev = total_fuel_urban_hev + fuel_hev(i);
end

figure(7), subplot(221)
plot([1: size(time_urban,1)-1], P_eng)
xlabel('Time (sec)'), ylabel('Engine power (kW)')
title('Results for the Urban cycle')
subplot(222), plot([1: size(time_urban,1)-1], fuel_hev)
xlabel('Time (sec)'), ylabel('Engine fueling rate (g/sec)')
subplot(223), plot([1: size(time_urban,1)], SOC)
xlabel('Time (sec)'), ylabel('SOC')
subplot(224), plot([1: size(time_urban,1)-1], P_batt)
xlabel('Time (sec)'), ylabel('Battery power (kW)')
```

```
% connvert fuel econpmy results from meter/gram to mpg
mpg_highway_HEV =
(distance_highway/1602)/(total_fuel_highway_hev/0.74/3785)
mpg_urban_HEV =
(distance_urban/1602)/(total_fuel_urban_hev/0.74/3785)
```

The thermostat concept can be enhanced by making the ICE power level dependent on the desired driving power and battery SOC (Abthoff et al. 1998; Hayashida and Narusawa 1999). A major reason to have weak coupling between the engine power and the vehicle load is to reduce efficiency losses due to energy conversion. A simple concept is illustrated in Figure 10.9.

In Figure 10.9, the power demand is passed through a low-pass filter to create an engine-power demand. This low-pass filter should be more complicated than a typical low-pass filter because if the power demand is too low, it may not make sense to run the engine at that low level of power, depending on the engine-efficiency and the battery characteristics. If the feed-forward part of the control is achieved properly and if the driving load is adequately high (e.g., on a long uphill), it will reduce energy passing through the battery, thereby reducing battery-efficiency loss. Implementation of the feed-forward algorithm is shown in Example 10.4.

EXAMPLE 10.4: MODIFIED THERMOSTATIC-CONTROL CONCEPT FOR SERIES HY-BRIDS. The feed-forward enhanced thermostatic-control concept is implemented in this example. This concept does not improve fuel economy and is not appropriate when the drive-cycle load is low because the engine would be quite inefficient at low power. Therefore, even considering the battery tax, it may not be justified to run the engine at low power demand. However, if the road load is high (e.g., significant road grade), then the new feed-forward term would help significantly by avoiding the battery loss. In the example, we assume an optimistic battery and power electronic loss of 8 percent (each way), which is close to the best efficiency achieved under ideal conditions. Actual loss can be significantly higher, which means that the feed-forward enhancement may be beneficial in a wider range of conditions.

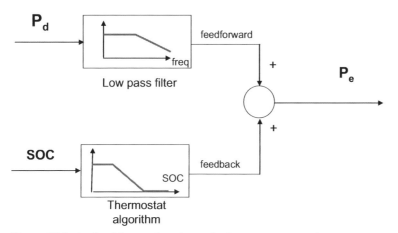

Figure 10.9. A feed-forward enhanced thermostat-control concept to reduce battery efficiency.

```matlab
% Ex10_4.m
% Compute fuel consumption of a thermostatic controlled
% Series hybrid vehicle for the US06 Cycles
% With feedforward control to reduce battery efficiency loss
clear all, close all
load CYC_US06.mat                % Load the highway cycle
time_US06 = cyc_mph(:,1);
speed_US06 = cyc_mph(:,2);
figure(1), subplot(211)
plot(time_US06, speed_US06)
xlabel('Time (Sec)')
ylabel('Speed (mph)')
Title('US06 cycle')

m = 1800;
g = 9.81;
rolling_resistance_coeff = 0.015;
aero_dynamic_drag_coeff = 0.4;
road_grad = 0;                          % road grade = 0 rad

N = size(time_US06, 1);
speed_US06 = speed_US06*1602/3600; % Change the unit of speed to
m/sec
distance_US06 = 0;

for i = 1:N-1,
    accel_US06(i) = (speed_US06(i+1) - speed_US06(i))/1.0;
% delta_T = 1 sec
  F_resistant_US06(i) =
rolling_resistance_coeff*m*g*cos(road_grad) ...
      +0.5*1.202*aero_dynamic_drag_coeff*speed_US06(i)^2 ...
      + m*g*sin(road_grad);
  power_US06(i) = (m*accel_US06(i) +
F_resistant_US06(i))*speed_US06(i);    % Power = F*V
  distance_US06 = distance_US06 + speed_US06(i);
end

subplot(212)
plot(time_US06(1:N-1), power_US06)
xlabel('Time (Sec)')
ylabel('Power (Watt)')
Title('US06 cycle')

ex10_2;    % Load and run ex10_2 to compute engine map

opt_eng_speed = [1500: 2500/35: 4000];
opt_throttle = [31:1:66];
opt_fuel = interp2([600:200:6000], Throttle_grid, fuel_map,
```

```
opt_eng_speed, opt_throttle);
opt_power = interp2([600:200:6000]*2*pi/60, Throttle_grid,
engine_out_power, ...
                    opt_eng_speed*2*pi/60, opt_throttle);
[B_filter,A_filter] = butter(3, 0.02);
P_eng_FF = filtfilt(B_filter,A_filter, power_US06/1000);

% Simulate for the two drive cycles, series hybrid, themodtatic
control
SOC_max_power = 0.25;
SOC_low = 0.35;
SOC_high = 0.50;
threshold_power = 60;      % kW
max_power = 120;           % kW
Battery_capacity = 2.5;    % kW-hr

% Simulate for the US06 cycle
total_fuel_US06_hev = 0;
SOC(1) = 0.52;
flag = 0;     % flag tracks whether we are on the top or bottom
of the hysteresis rectangle
for i = 1: (size(time_US06, 1)-1),
    if SOC(i) > SOC_high
        P_eng_FB(i) = 0; flag = 0;
    elseif SOC(i) > SOC_low
        if flag == 0
            P_eng_FB(i) = 0;
        else
            P_eng_FB(i) = threshold_power;
        end
    elseif SOC(i) > SOC_max_power
        flag = 1;
        P_eng_FB(i) = max_power - (SOC(i)-SOC_max_power)/
(SOC_low-SOC_max_power) ...
                      *(max_power-threshold_power);
    else
        flag = 1;   P_eng_FB(i) = max_power;
    end

    P_eng(i) = P_eng_FF(i) + P_eng_FB(i);

    if P_eng(i) < threshold_power
        P_eng(i) = 0;
    end

    P_batt(i) = power_US06(i)/1000 - P_eng(i); % unit: kW
    if P_batt(i) > 0
        SOC(i+1) = SOC(i) -
```

```
P_batt(i)*1.0/3600/0.92/Battery_capacity;
    else
        SOC(i+1) = SOC(i) -
P_batt(i)*1.0/3600*0.92/Battery_capacity;
    end

    if P_eng(i) == 0
        fuel_hev(i) = 0;
    else
        fuel_hev(i) = interp1(opt_power, opt_fuel, P_eng(i),
'linear');
    end
    total_fuel_US06_hev = total_fuel_US06_hev + fuel_hev(i);
end

figure(5), subplot(221)
plot([1: size(time_US06,1)-1], P_eng)
xlabel('Time (sec)'), ylabel('Engine power (kW)')
title('Results for the US06 cycle')
subplot(222), plot([1: size(time_US06,1)-1], fuel_hev)
xlabel('Time (sec)'), ylabel('Engine fueling rate (g/sec)')
subplot(223), plot([1: size(time_US06,1)-1],
SOC(1:size(time_US06,1)-1))
xlabel('Time (sec)'), ylabel('SOC')
subplot(224), plot([1: size(time_US06,1)-1], P_batt)
xlabel('Time (sec)'), ylabel('Battery power (kW)')

% convert fuel economy results from meter/gram to mpg
mpg_US06_HEV =
(distance_US06/1602)/(total_fuel_US06_hev/0.74/3785)
```

10.4 Control Concepts for Parallel Hybrids

Parallel hybrids have powertrain configurations closest to traditional ICE-only vehicles and therefore are used in many prototype hybrid vehicles. Many different control concepts have been introduced, with different levels of success. Three concepts are introduced in the following sections.

Load-Leveling Concept

This concept is an "engine-centric" idea and focuses only on operating the engine efficiently, within a band of efficient power – similar to the engine-operation analysis described in the series-hybrid discussion. The basic concept can be explained using Figure 10.10, which shows the BSFC of an engine – similar to the plot generated in Example 10.2. However, the y-axis in Figure 10.10 is engine torque. If this engine is used on a parallel hybrid vehicle, then a simple idea to avoid inefficient operations of the engine is to select two power levels (denoted by the green and red lines) and operate the ICE only within these two power levels. When the power requirement is

Table 10.1. *Example load-leveling control concept for parallel hybrid vehicles*

Engine power	SOC $>=$ SOC$_t$	SOC $<$ SOC$_t$
$P_d <= P_{ev}$	$0\ (V < V_{low})$	$P_d + P_{ch}$
$P_{ev} < P_d <= Pe_{max}$	P_d	$P_d + P_{ch}$
$Pd > Pe_{max}$	Pe_{max}	Pe_{max}

lower than the "all-electric power level" (i.e., the green line), the engine will not be used; instead, only the electric motor will be used. When the required power from the driver is higher than the "maximum engine power line" (i.e., the red line), the engine will operate only at that power level, and the deficit will be provided by the battery and electric motor. When the battery SOC is too low, the electric motor may not be able to provide the load-leveling function and the engine power will be modified accordingly. The basic control rules are shown in Table 10.1. To improve engine efficiency, a transmission still should be used to move the operation of the engine along the constant-power lines.

The rules shown in Table 10.1 can be enhanced further by including other factors in the decisions, such as emission, NVH, and vehicle speed.

EXAMPLE 10.5: LOAD-LEVELING CONTROL CONCEPT. The load-leveling concept shown in Table 10.1 is implemented in this example. A noticeable change is that the engine is downsized from the one used in the series hybrid by 30 percent. The engine torque and fuel consumption are assumed to reduce proportionally. The two power-threshold values are selected to be $P_{ev} = 20$kW and $Pe_{max} = 70$kW.

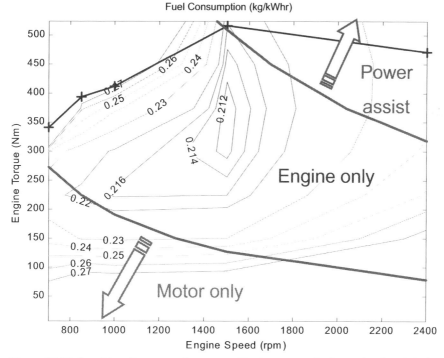

Figure 10.10. Load-leveling-control concept: Engine power is determined mainly based on the desired power level.

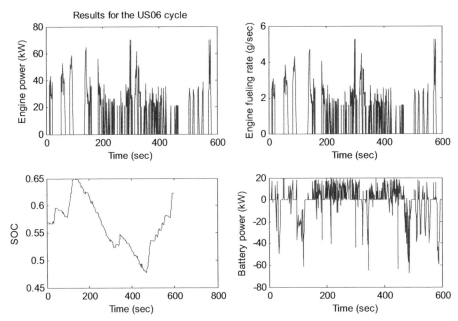

Figure 10.11. Simulation results of a parallel hybrid vehicle based on the load-leveling-control concept on the US06 drive cycle.

Similar to the two previous examples, it is assumed that a CVT is used to maintain the engine operation around its optimal speed point. In this example, the US06 cycle is used, which requires higher power than the urban or the highway cycle. It is observed (see Figure 10.11) that in a parallel-hybrid powertrain, the engine is on most of the time (which is a major difference from a series hybrid).

```
% Ex10_5.m
% Compute fuel consumption of a load levelling
% Parallel hybrid vehicle for the Highway and Urban Cycles

clear all, close all
load CYC_US06.mat                % Load the US06 cycle

time_US06 = cyc_mph(:,1);
speed_US06 = cyc_mph(:,2);
figure(1), subplot(211)
plot(time_US06, speed_US06)
xlabel('Time (Sec)')
ylabel('Speed (mph)')
Title('US06 cycle')

m = 1800;
g = 9.81;
rolling_resistance_coeff = 0.015;
aero_dynamic_drag_coeff = 0.4;
road_grad = 0;                   % road grade = 0 rad

N = size(time_US06, 1);
```

```
speed_US06 = speed_US06*1602/3600;  % Change the unit of speed to
m/sec
distance_US06 = 0;

for i = 1:N-1,
    accel_US06(i) = (speed_US06(i+1) - speed_US06(i))/1.0;
% delta_T = 1 sec
    F_resistant_US06(i) =
rolling_resistance_coeff*m*g*cos(road_grad) ...
        +0.5*1.202*aero_dynamic_drag_coeff*speed_US06(i)^2 ...
        + m*g*sin(road_grad);
    power_US06(i) = (m*accel_US06(i) +
F_resistant_US06(i))*speed_US06(i)/1000;
% Power = F*V, in kW
    distance_US06 = distance_US06 + speed_US06(i);
end

subplot(212)
plot(time_US06(1:N-1), power_US06)
xlabel('Time (Sec)')
ylabel('Power (kW)')
Title('US06 cycle')

load engine_parameters
% Load engine torque (N-m)
% and fuel comsumption (g/sec)
% Both of which is a 21 by 28 matrix
% The row (throttle) index is 0:5:100
% The column (engine speed) index is 600:200:6000 (rpm)

engine_speed = [600:200:6000]*2*pi/60;   % rad/sec
Throttle_grid=[0:5:100];

engine_torque = engine_torque * 0.7;   % Scale down engine by 30%
fuel_map = fuel_map * 0.7;

figure(2)
mesh([600:200:6000], Throttle_grid, fuel_map)
xlabel('Engine speed (rpm)')
ylabel('Throttle')
zlabel('fuel consumption (g/sec)')

for k = 1:21,
    for j = 1:28,
        engine_out_power(k,j) =
engine_speed(j)*engine_torque(k,j)/1000;   % engine out power (kW)
    bsfc(k,j) = fuel_map(k,j)/engine_out_power(k,j);
% BSFC = g/sec/kW
    if (bsfc(k,j) < 0)
```

```
        bsfc(k,j) = 0.3;
      elseif (bsfc(k,j) > 0.3)
          bsfc(k,j) = 0.3;
      end
   end
end

figure(3)
contour([600:200:6000], Throttle_grid, bsfc, 50)
xlabel('Engine speed (rpm)')
ylabel('Throttle')
title('BSFC g/sec/kW')
hold on, plot(2350, 43, 'd'), hold off

figure(4)
[CS,H]=contour([600:200:6000], Throttle_grid, engine_out_power);
clabel(CS), xlabel('Engine speed (rpm)')
ylabel('Throttle')
title('Engine power (kW)')
hold on, plot(2350, 43, 'd'), hold off

figure(3)
hold on, plot([600;4000], [18; 66], '-', 'LineWidth', 3)
hold off

figure(4)
hold on, plot([600;4000], [18; 66], '-', 'LineWidth', 3)
hold off

opt_eng_speed = [600: 3400/48: 4000];
opt_throttle = [18:1:66];
opt_fuel = interp2([600:200:6000], Throttle_grid, fuel_map,
opt_eng_speed, opt_throttle);
opt_power = interp2([600:200:6000]*2*pi/60, Throttle_grid,
    engine_out_power, ...
                        opt_eng_speed*2*pi/60, opt_throttle);
% Simulate for the US06 cycle, parallel hybrid, load leveling
control
P_ev = 20;
Pe_max = 70;
P_ch = 20;
SOC_t = 0.45;
Battery_capacity = 2.5;    % kW-hr

% Simulate for the US06 cycle
total_fuel_US06_hev = 0;
SOC(1) = 0.57;
```

```
for i = 1: (size(time_US06, 1)-1),
    if SOC(i) >= SOC_t
        if power_US06(i) <= P_ev
            P_eng(i) = 0;
        elseif power_US06(i) <= Pe_max
            P_eng(i) = power_US06(i);
        else
            P_eng(i) = Pe_max;
        end
    else
        if power_US06(i) < 0
            P_eng(i) = 0;
        elseif power_US06(i) <= P_ev
            P_eng(i) = power_US06(i) + P_ch;
        elseif power_US06(i) <= Pe_max
            P_eng(i) = power_US06(i) + P_ch;
        else
            P_eng(i) = Pe_max;
        end
    end

    P_batt(i) = power_US06(i) - P_eng(i);   % unit: kW
    if P_batt(i) > 0
        SOC(i+1) = SOC(i) -
P_batt(i)*1.0/3600/0.92/Battery_capacity;
    else
        SOC(i+1) = SOC(i) -
P_batt(i)*1.0/3600*0.92/Battery_capacity;
    end

    if P_eng(i) == 0
        fuel_hev(i) = 0;
    else
        fuel_hev(i) = interp1(opt_power, opt_fuel, P_eng(i),
'linear');
    end
    total_fuel_US06_hev = total_fuel_US06_hev + fuel_hev(i);

end

figure(5), subplot(221)
plot([1: size(time_US06,1)-1], P_eng)
xlabel('Time (sec)'), ylabel('Engine power (kW)')
title('Results for the US06 cycle')
subplot(222), plot([1: size(time_US06,1)-1], fuel_hev)
xlabel('Time (sec)'), ylabel('Engine fueling rate (g/sec)')
subplot(223), plot([1: size(time_US06,1)], SOC)
xlabel('Time (sec)'), ylabel('SOC')
```

```
subplot(224), plot([1: size(time_US06,1)-1], P_batt)
xlabel('Time (sec)'), ylabel('Battery power (kW)')
% convert fuel economy results from meter/gram to mpg
mpg_US06_HEV = (distance_US06/1602)/
    (total_fuel_US06_hev/0.74/3785)
```

Equivalent Consumption Minimization Strategy

The ECMS concept is an instantaneous optimization concept, and the basic idea is that the best engine/battery power split is one that achieves minimum "equivalent fuel consumption." In other words, the battery power is converted to an equivalent fuel consumption, and the sum of the gasoline engine fuel and battery equivalent fuel is minimized at every step. Because the optimal decision is made every sampling time, it is a "greedy" or instantaneous optimization decision process. The ECMS concept relies on the desire that instantaneous optimizations result in overall performance that is close to horizon optimization; that is:

$$\int Min[\dot{m}_f(t)]dt \approx Min \int \dot{m}_f(t)dt \qquad (10.1)$$

In practical applications, the ECMS method usually achieves fuel-consumption performance close to algorithms that are developed based on horizon-optimization methods, although such near-optimality is never guaranteed. In the ECMS concept, the "fuel consumption" to be minimized includes not only the fuel consumed by the engine but also the equivalent fuel consumption by the battery (Paganelli et al. 2000). That is,

$$\dot{m}_f(t) = \dot{m}_{f\_eng} + \dot{m}_{f\_batt} \qquad (10.2)$$

where:

$$\dot{m}_{f\_batt} = \frac{\overline{SC_{eng}} \cdot P_{batt}}{Eff_{electrical}} \qquad (10.3)$$

In other words, the battery power is converted to an equivalent fuel consumption by considering the efficiency of the engine $(\overline{SC_{eng}})$ and the electric-path efficiency $(Eff_{electrical})$. An issue in solving an optimization problem to minimize Eq. (10.2) is that the battery SOC may not be maintained around the desired level and could fluctuate depending on the driving load. To mitigate this problem, Eq. (10.2) is modified to include a multiplier that favors battery SOC around a desired region. The modified fuel-consumption term becomes:

$$\dot{m}_{f\_equi}(t) = \dot{m}_{f\_eng} + f(soc) \cdot \dot{m}_{f\_batt} \qquad (10.4)$$

where the SOC modifier $f(soc)$ may look like the curve shown in Figure 10.12. The characteristics of $f(soc)$ include the following: (1) it is close to 1 around the desired SOC level; and (2) it changes to a lower or higher magnitude to promote charge sustenance. The equation used to generate the curve shown in Figure 10.12 is as follows (Paganelli et al. 2002; Rodatz et al. 2005):

$$f(soc) = 1 - (1 - 0.7x_{soc}) \cdot x_{soc}^3 \quad \text{where} \quad x_{soc} = \frac{SOC - \dfrac{SOC_L + SOC_H}{2}}{SOC_H - SOC_L} \qquad (10.5)$$

Other modifier functions can be used as long as they have the same characteristics stated previously.

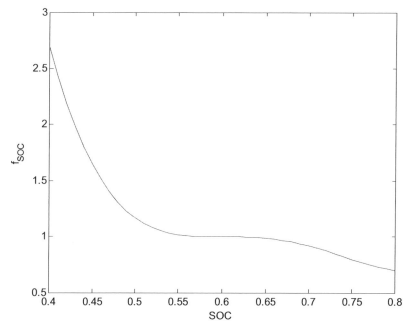

Figure 10.12. An example SOC modifier for the ECMS-control concept.

In general, the total fuel to be minimized depends on the vehicle speed, total desired power, and battery SOC:

$$\dot{m}_{f\_equi}(t) = \dot{m}_{f\_eng}(v, P_d) + f(soc) \cdot \dot{m}_{f\_batt} \qquad (10.6)$$

In other words, the output from such an optimization problem can be represented as a three-dimensional lookup table. When the transmission used in the vehicle has several gears (or when a CVT is used) and the transmission is shifted properly, the vehicle speed is no longer a factor and the engine always operates at the appropriate engine speed for the corresponding engine-power level. In that case, the ECMS algorithm becomes a two-dimensional lookup table, with the desired output power and the battery SOC as the only two independent variables. This special case is illustrated in Example 10.6.

EXAMPLE 10.6: ECMS CONTROL. The ECMS concept is implemented in this example, with the optimal (i.e., minimum fuel consumption) point identified and marked for a given desired output power and battery SOC level. The results are shown in Figure 10.13.

```
%Ex10_6.m
% generate engine power command map using the ECMS concept
% for a parallel HEV
clear all;
close all;

load engine_parameters
Throttle_grid=[0:5:100];
engine_speed = [600:200:6000]*2*pi/60;    % rad/sec
```

```
for k = 1:21,
    for j = 1:28,
      engine_out_power(k,j) =
engine_speed(j)*engine_torque(k,j)/1000; % engine out power (kW)
    end
end


opt_eng_speed = [600: 3400/48: 4000];
opt_throttle = [18:1:66];
opt_fuel = interp2([600:200:6000], Throttle_grid, fuel_map,
opt_eng_speed, opt_throttle);
opt_power = interp2([600:200:6000]*2*pi/60, Throttle_grid,
engine_out_power, ...
                    opt_eng_speed*2*pi/60, opt_throttle);


SC_eng = 0.07;
Eff_elec = 0.95;
soc_L = 0.5;
soc_H = 0.7;
soc = [0.4:0.01:0.8];
x_soc = (soc - (soc_L+soc_H)/2)/(soc_H-soc_L);
f_soc = 1-(1-0.7.*x_soc).*x_soc.^3;                    % 1 by41


figure(1)
plot(soc, f_soc)
xlabel('SOC'), ylabel('f_S_O_C')


Peng = [1:1:120];
fuel_engine = interp1(opt_power, opt_fuel, Peng, 'linear');


for i = 1:120,
  Pd(i) = i*1.0;      % Pd from 1 to 120kw
  P_motor(i,:) = Pd(i)*ones(1,120) - Peng;

  for j = 1:120,
    if P_motor(i,j) > 0
      P_batt(i,j) = P_motor(i,j)/0.92;
    else
      P_batt(i,j) = P_motor(i,j)*0.92;
    end
  end
  for k = 1:41,
    fuel_batt(k,i,:) = f_soc(k)*SC_eng*P_batt(i,:)/Eff_elec;
    fuel_total(k,i,:) = reshape(fuel_engine,1,1,120) +
fuel_batt(k,i,:) ;
  end
end
```

```
% Showing example at a particular case, where Pd = 60kW,
SOC = 0.6

figure(2)
subplot(411), plot(Peng, fuel_engine)
xlabel('Engine Power (kW)'), ylabel('Fuel_e_n_g')
title('Pd = 60kW, SOC = 0.6')
subplot(412), plot(Peng, P_batt(60,:))
xlabel('Engine Power (kW)'), ylabel('Battery power (kW)')
subplot(413), plot(Peng, squeeze(fuel_batt(21,60,:)))
xlabel('Engine Power (kW)'), ylabel('Fuel_b_a_t_t')
subplot(414), plot(Peng, squeeze(fuel_total(21,60,:)))
xlabel('Engine Power (kW)'), ylabel('Total fuel')
hold on
[fuel_min, index] = min(fuel_total(21,60,:));
plot(Peng(index), fuel_min, 'r*'), hold off
```

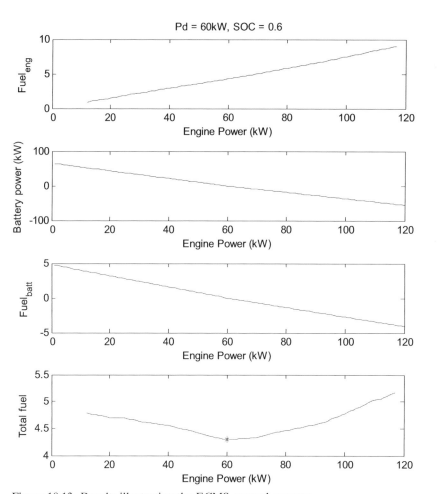

Figure 10.13. Results illustrating the ECMS-control concept.

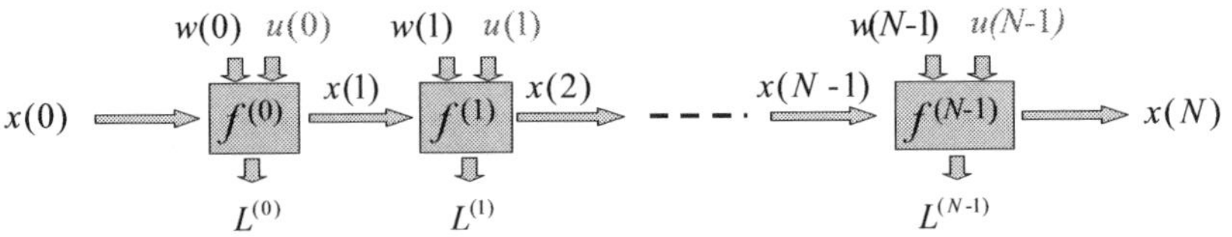

Figure 10.14. A discretized multistage decision process solved by the DP method.

Dynamic Programming

The dynamic programming (DP) method has been used to find an optimal control strategy under either a deterministic drive cycle or a stochastic driving-power distribution. The DP problem is solved numerically, which makes it possible for nonlinear vehicle dynamics and inequality constraints to be considered while also guaranteeing global optimality. The underlying vehicle model incorporates the dynamics, efficiency, and constraints of subsystems to ensure that the overall vehicle performance is optimized as the constraints of the components are satisfied. The continuous vehicle dynamics and the control actions also are discretized, and the DP solution involves searching among discrete control actions that result in the optimal cost function throughout the problem horizon. A multistage decision process solved by the deterministic DP is illustrated in Figure 10.14. The dynamics are discretized so that it has the following form:

$$\mathbf{x}_{k+1} = \mathbf{f}_k(\mathbf{x}_k, \mathbf{u}_k, \mathbf{w}_k) \tag{10.7}$$

where the states, control inputs, and disturbance inputs are represented by $\mathbf{x}, \mathbf{u}$, and $\mathbf{w}$, and respectively. Each box in Figure 10.14 represents mapping during a sampling time, and the total dynamic process has N steps. During each step, different control inputs result in different states and generate different transitional cost L. The DP seeks to find the control actions $u(0), u(1), \ldots, u(N-1)$ that minimize a cost function, which could include both transitional costs at each step and a terminal penalty term. For example:

$$J = \sum_{k=0}^{N-1} (fuel_k + \mu \cdot NOx_k + \upsilon \cdot PM_k) + \alpha(SOC_N - SOC_f)^2 \tag{10.8}$$

which penalizes fuel and emission at each step as well as a terminal penalty on battery SOC.

The solutions from the DP optimization are the global optimal up to the grid accuracy. For deterministic DP problems, the identified DP control solution is not suitable for real-time implementations due to the noncausal nature and heavy computational requirement. However, it is a good design tool for benchmarking and calibrating other real-time control strategies.

Near-optimal power-management algorithms for hybrid vehicles can be extracted from the DP results (Lin et al. 2003a). The results were implemented on prototype vehicles with significant fuel-economy improvement over the more classical rule-based control-design method (Lin et al. 2003b). This fundamental control-design concept later was implemented for several different hybrid vehicles, including hydraulic hybrid vehicles (Wu et al. 2004) and fuel-cell hybrid vehicles (Kim and Peng 2007; Lin et al. 2006).

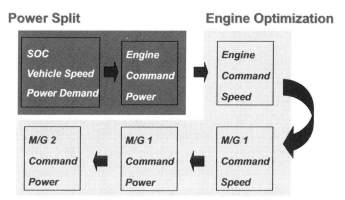

Figure 10.15. Control concept for a single planetary-gear split vehicle, which includes engine optimization, generator-speed control, and motor-torque control.

More recently, a stochastic version of the DP control design procedure was developed (Lin et al. 2006; Tate et al. 2008). The two major benefits of the stochastic dynamic programming (SDP) design technique, in comparison to its deterministic counterpart, are (1) the obtained control results are directly implementable as nonlinear full-state feedback lookup tables; and (2) the control algorithm is not "cycle beating"; instead, it has robust performance because the optimization is over a power-demand transition probability obtained from a rich set of driving conditions. Optimal control results based on deterministic DP and SDP have gained popularity in the hybrid-vehicle field. Many recent hybrid-vehicle designs (Opila et al. 2009) were presented with DP results as a benchmark or are directly motivated by DP results.

10.5 Control Concept for Split Hybrids

As discussed previously, there are many different possible split-hybrid designs and they all behave differently; some also have a mode shift. Therefore, except for systematic and model-based design methods such as DP, other design approaches are likely to require ideas specific to the vehicle configuration. In the following discussion, we explain the control idea for the original Toyota Hybrid System (Harada 2000; Hermance 1999) and the concepts of how to extend that to two-planetary split-vehicle designs.

Figure 10.15 shows a control concept explained in the literature regarding the Toyota Hybrid Systems. First, based on a driver's pedal positions, vehicle speed, and battery SOC, a proper engine-power command is computed that will satisfy a driver's command as well as maintain the proper battery SOC. The engine characteristics then are checked to determine the best engine speed for that engine-power level. Although the ECU is manipulating the engine inputs (e.g., fuel, valve, and EGR) to achieve that engine power, it is also required that the generator (i.e., M/G 1) "hold the line" so that when the engine is trying to generate power to push the vehicle forward, the engine speed does not rev up or down, away from its most efficient operating speed. This desired generator speed is calculated from the vehicle speed and the desired engine speed. A feedback control (e.g., PI control) can be designed to ensure good generator speed, which involves the manipulation of generator torque. Finally, the motor (i.e., M/G 2) is controlled to achieve power balance by ensuring that $P_{M/G2} = P_d - P_e - P_{M/G1}$. In other words, regardless of what action was taken for battery SOC sustenance and generator-speed control, the total power requested by the driver is delivered to the wheel. This ensures that the vehicle is drivable.

The concept described here is "engine centric" and essentially ignores the efficiency losses incurred by the electric machines, power electronics, and battery. When the efficiencies of those components are high, this approach is fine. When DP is used, those subsystem losses are considered automatically, which leads to better efficiency. DP for split vehicles has not been studied extensively and only limited information is available from Liu and Peng (2008).

The control concept illustrated in Figure 10.15 can be used for power-split hybrid vehicles using two planetary gears. However, for those vehicles, two additional factors must be considered: (1) proper timing for mode shifts must be determined, which can involve the consideration for vehicle speed and acceleration; and (2) during the shift, coordination of the oncoming clutch and the outgoing clutch is necessary. Similar to gear shifts in an automatic transmission, it is desirable to have "power-on" mode shifts; that is, propulsion torque still is supplied to the wheels during a mode shift so that drivability is not impacted negatively. Proper calculations of the clutch torque and coordination for the clutch pressure to avoid "tie-up" or "flare," both of which are considered "bad shifts," are required.

10.6 Feedback-Based Supervisory Controller for PHEVs

Recently, many PHEVs have been introduced. These PHEVs have batteries that can be charged from the electrical power grid and can operate for a limited range (i.e., the all-electric range [AER]) as an electric vehicle (Markel and Simpson 2006). At present, supervisory controllers for PHEVs have centralized architectures (Wirasingha and Emadi 2011). A typical PHEV operates in a charge-depleting (CD) or an electric-vehicle (EV) mode before the battery SOC decreases to a certain value; then, it switches to a charge-sustaining (CS) mode and operates like a conventional HEV. The control strategies proposed for HEVs can be applied to the CS mode-controller design for PHEVs. As discussed previously, various control-design methods for HEVs are available. For instance, in Di Cairano et al. (2011), feedback controllers based on model predictive control were evaluated experimentally and showed improved fuel economy compared to two baseline strategies.

In this section, a feedback-based controller for the CS mode is presented. The controller is designed with respect to the EPA US06 cycle; however, simulation results demonstrate that the feedback-based controller also achieves good fuel economy, good driving performance, and charge sustainability over other driving cycles (e.g., the EPA UDDS and HWFET cycles).

Vehicle Model

The control-oriented vehicle model for a series PHEV is shown in Figure 10.16. The model inputs are the wheel-power command, $P_{w,cmd}$, and the reference battery

Table 10.2. *Nominal vehicle configuration*

PHEV	AER (miles)	30
Engine	Engine Power (kW)	50
Generator	Generator Power (kW)	50
Battery	Battery Capacity (kWh)	12
	Battery Maximum Power (kW)	110
Motor	Motor Power (kW)	110
Vehicle	Vehicle Weight (kg)	1,680

SOC, soc_r. The system outputs are the actual wheel power delivered, P_w; engine fuel consumption, *fuel*; and actual battery SOC, *soc*.

The supervisory controller generates the engine/generator-power command and the battery-power command. The wheels are driven by the electric motor. The battery is being charged when the battery power P_b is negative. The component sizes for the vehicle configuration used in this section are listed in Table 10.2. This PHEV is representative of current designs (e.g., the 2011 Chevrolet Volt).

Engine

The engine is modeled using a static fuel-consumption map from ADVISOR (Markel et al. 2002), as shown in Figure 10.16. A combustion engine can achieve a required power (excluding the maximum power) with different combinations of torques and speeds. However, given a required power level, there is usually a unique pair of engine torque and speed that achieves minimum fuel consumption. The Optimal Operating Points Line (OOP-Line) is defined as the curve on which the fuel consumption is minimized for each power level (Konev et al. 2006).

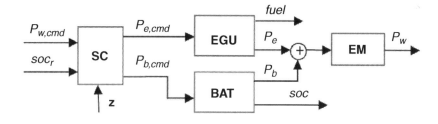

SC – supervisory controller

EGU – internal combustion engine and generator unit

BAT – battery

EM – electric machine

z – feedback state vector

Figure 10.16. Diagram of the PHEV: SC – supervisory controller, EGU – ICE and generator unit, BAT – battery, EM – electric machine, **z** – feedback-state vector.

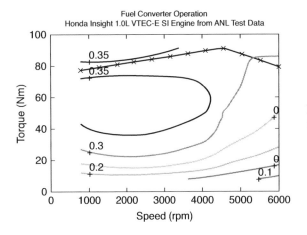

Figure 10.17. Engine-fuel-consumption map (g/W/h).

Using the datapoints of the fuel-consumption map in Figure 10.17, the OOP-Line can be constructed as plotted in Figure 10.18. The blue dots represent the datapoints from the engine fuel-consumption map, and the green line is the OOP-Line consisting of the most efficient points corresponding to each requested power level. For an engine to operate along the OOP-Line, it should avoid large transients; this is based on the perspective that aggressive engine transients and engine operation away from the OOP-Line may degrade fuel economy and emissions. The maximum rate of requested engine-power-output change can be constrained – for example, to 3.5 kW/sec for a system considered in Konev et al. (2006) or to 11 kW/sec for a system considered in Di Cairano et al. (2011). If the rate of engine-power-output change is constrained, the engine can operate smoothly and efficiently along the OOP-Line responding to power commands.

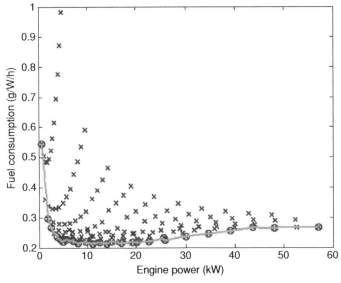

Figure 10.18. Fuel consumption versus engine power along engine OOP-Line.

Battery

Battery efficiency is assumed to be $\eta_b = 0.9$. The battery is modeled as an integrator with parameter $B_s = 2.57\text{e-}5$ (Di Cairano et al. 2011).

$$\Delta \dot{soc} = B_s P_{b,cmd} \tag{10.9}$$

where $\Delta soc = soc_r - soc$. The battery SOC is the only state of the powertrain system.

Electric Machines

The motor and generator are modeled as static with a constant mean efficiency $\eta_m = \eta_g = 0.85$. This assumption can be relaxed easily by incorporating efficiency maps.

Vehicle Dynamics

The vehicle-longitudinal dynamics includes the acceleration force, the rolling resistance force, and the aerodynamic force:

$$F = ma + fF_z + \frac{A\rho C_d v^2}{2} \tag{10.10}$$

where m is the vehicle mass, kg; a is the longitudinal acceleration, m/s^2; f is the rolling-resistance coefficient; F_z is the normal force on the vehicle, N; A is the vehicle cross-sectional area, m^2; ρ is the air density, kg/m^3; C_d is the drag coefficient; and v is the longitudinal velocity, m/s. The vehicle longitudinal dynamics is used to calculate the wheel-power command for the vehicle to follow the driving cycles that are specified with vehicle speed, as in Example 10.1.

Feedback-Based Supervisory Controller

Assume that the PHEV operates in two main modes: (1) the CD mode when the battery SOC is larger than a certain reference value, soc_r; and (2) the CS mode once the battery SOC reaches soc_r. Regenerative braking is activated when the wheel-power command is negative. If the battery SOC exceeds the specified range [soc_{min}, soc_{max}], the control-strategy priority is to drive the battery SOC back to the interval.

In the CD mode, the battery provides the propulsion energy. The engine is used to satisfy the transient-load demand beyond the power capacity of the battery. In the CS mode, the control strategy is to optimize fuel economy and driving performance while sustaining battery charge.

The CD mode and regenerative-braking control are straightforward; therefore, we focus on designing the controller for the CS mode. First, a feedback-based control structure is proposed. Second, the controller gains are obtained through optimization. Third, the obtained controller is tested in realistic simulations during three standard driving cycles – with negative wheel-power command and regenerative braking included – to evaluate fuel economy, driving performance, and battery-charge sustainability.

Controller Structure

Two integrators are introduced to regulate battery SOC and to eliminate wheel-power tracking error in steady state. Recall that the powertrain system is modeled as

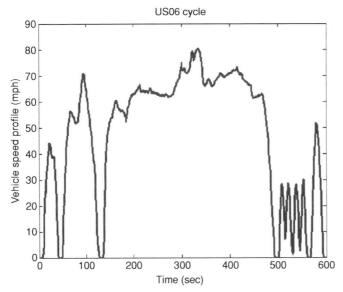

Figure 10.19. EPA US06 driving cycle.

first order with the battery SOC as the single state. The three states of the closed-loop system are as follows:

$$z_1 = \Delta soc = soc_r - soc \tag{10.11}$$

$$z_2 = \int (\Delta soc)dt = \int (soc_r - soc)dt \tag{10.12}$$

$$z_3 = \int (P_{w,cmd} - P_w)dt \tag{10.13}$$

where state z_1 represents the deviation of battery SOC, soc, from the reference value, soc_r; state z_2 represents the integral error of battery SOC; and state z_3 represents the integral error between power command, $P_{w,cmd}$, and actual power delivered, P_w. The actual power equals the engine power P_e plus the battery power P_b.

The control algorithm includes state-feedback control, feed-forward control, and the terms representing information exchange between the engine-power command $P_{e,cmd}$ and the battery-power command $P_{b,cmd}$:

$$P_{e,cmd} = \mathbf{K_1z} + n_1 P_{w,\,cmd} + k_e P_{b,cmd} \tag{10.14}$$

$$P_{b,cmd} = \mathbf{K_2z} + n_2 P_{w,\,cmd} + k_b P_{e,cmd} \tag{10.15}$$

where $\mathbf{z} = [z_1\ z_2\ z_3]^T$ is the state vector; $\mathbf{K_1} = [k_1\ k_2\ k_3]$ and $\mathbf{K_2} = [k_4\ k_5\ k_6]$ are state-feedback-gain vectors; n_1 and n_2 are feed-forward gains; and k_e and k_b are controller gains that represent information exchange.

The regulation of battery SOC should be slow to allow the battery to augment the engine in transients, whereas the wheel-power tracking should be fast and accurate for good driving performance and safety. To achieve different convergence rates for different states, we use eigenstructure assignment (Magni et al. 1997) to decouple the state z_3 from the other two states, z_1 and z_2, which are related to battery SOC.

The aggressive driving cycle of EPA US06 (Figure 10.19) is chosen to generate the controller gains through optimization.

Table 10.3. *Performance results for the vehicle with battery*
($B_s = 2.57e-5$) and corresponding optimal centralized controller

Driving cycle	MPG	$err_{P,max}$	$soc_{dev,p}$
EPA US06	36.05	2.97e-8	−5.33%
EPA HWFET	55.19	4.67e-9	−0.79%
EPA UDDS	57.88	7.56e-9	−1.30%

In the optimization for the controller gains, the cost function J includes three terms: engine-fuel consumption, equivalent fuel consumption from the battery at the end of the driving cycle, and accumulated vehicle-power tracking error:

$$J = \int_0^{T_f} \dot{m}_{ice}(t)dt + K_{eqf}|soc_r - soc_f| + \alpha \int_0^{T_f} |P_{w,cmd}(t) - P_w(t)|dt \quad (10.16)$$

where $\dot{m}_{ice}(t)$ represents the engine-fuel consumption; soc_f is the battery SOC at the end of the driving cycle; K_{eqf} is the equivalent fuel-consumption factor from external charge and α is the penalty weight to drive the power-tracking error to zero. Values of $\alpha = 100$ and $K_{eqf} = 375.44$ are used (Li et al. 2011). The optimization constraints include (1) stability of the closed-loop system with the linear controller for the CS mode, which is enforced by the closed-loop-pole locations; (2) upper and lower power limits of the engine; (3) limit on engine-power rate of change to smooth engine power during continuous engine operation; (4) upper and lower power limits of the battery; (5) upper and lower limits of battery SOC, $soc(t)$; and (6) upper and lower limits of battery SOC at the end of the driving cycle to enforce charge sustainability.

The obtained controller was tested during three standard driving cycles to evaluate fuel economy, driving performance in terms of wheel-power-tracking error, and battery-charge sustainability; detailed results are in Li et al. (2011).

Vehicle performance with optimal controller for the three standard driving cycles is summarized in Table 10.3. Given in the table are the fuel economy in miles per gallon (mpg) considering both fuel consumption from the engine and equivalent fuel consumption from the battery at the end of the driving cycle; the driving performance evaluated by the maximum power-tracking error, $err_{P,max}$, in kW; and the battery- CS performance evaluated by the deviation of the final SOC from the reference SOC, $soc_{dev,p}$.

It is shown in Table 10.3 that the proposed controller provides good fuel economy for each of the different driving cycles. The wheel-power-tracking error, $err_{P,max}$, is less than 2.97e-8 kW for all simulation scenarios, which ensures good driving performance. The maximum deviation of battery SOC at the end of the driving cycles, $soc_{dev,p}$, is −5.33 percent for all simulation scenarios. For the mild EPA UDDS and HWFET cycles, the maximum $soc_{dev,p}$ for all battery applications is −1.3 percent.

For the US06 driving cycle, an example power split between the engine and the battery for the vehicle with battery parameter, $B_s = 2.57e-5$, and the corresponding optimal controller are given in Figure 10.20. When the wheel-power command is positive, the engine provides the slowly changing power and the battery provides

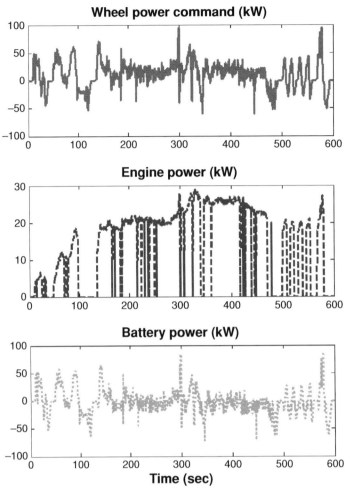

Figure 10.20. Example power split between engine and battery for a vehicle with battery 3 ($B_s = 2.57\text{e-}5$) and corresponding optimal controller over the US06 cycle.

the transient-power command. When the wheel-power command is negative, the engine shuts off without delivering power and the battery is charged by regenerative braking. The SOC trajectory for the same vehicle configuration for the US06 cycle is plotted in Figure 10.21. The initial battery SOC in the simulation is the reference SOC used in the control. Additional simulation results are in Li et al. (2011).

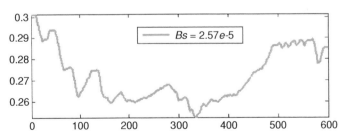

Figure 10.21. Battery SOC profile over the US06 cycle.

PROBLEMS

1. Modify the parameters in Example 10.1 to simulate the power requirements for compact and large passenger automobiles to supplement the results already given for a midsize automobile.

2. What enhancements to the rules in Table 10.1 should be considered for the load-leveling-control strategy for parallel hybrid vehicles?

REFERENCES

Abthoff, J. O., P. Antony, M. Kramer, and J. Seiler, 1998, "The Mercedez-Benz C-Class Series Hybrid," SAE Paper 981123.

Anon, 2005, "Code of Federal Regulations 474.3," http://cfr.vlex.com/source/code-federal-regulations-energy-1059.

Di Cairano, S., W. Liang, I. V. Kolmanovsky, M. L. Kuang, and A. M. Phillips, 2011, "Engine Power Smoothing Energy Management Strategy for a Series Hybrid Electric Vehicle," *Proceedings of the American Control Conference, San Francisco.*

Harada, O., K. Yamaguchi, and Y. Shibata, 2000, "Power Output Apparatus and Method of Controlling the Same," U.S. Patent Number 6,067,801.

Hayashida, M., and K. Narusawa, 1999, "Optimization of Performance and Energy Consumption on Series Hybrid Electric Power System," SAE Paper 1999-01-0922.

Hermance, D., 1999, "Toyota Hybrid System," 1999 *SAE TOPTEC Conference*, Albany, NY, May.

Holmes, A. G., D. Klemen, and M. R. Schmidt, 2001, "Electrically Variable Transmission with Selective Input Split, Compound Split, Neutral and Reverse Modes," U.S. Patent No. 6,527,658.

Kim, M., and H. Peng, 2007, "Power Management and Design Optimization of Fuel Cell/Battery Hybrid Vehicles," *Journal of Power Sources*, Vol. 165, Issue 2, March, pp. 819–32.

Konev, A., L. Lezhnev, and I. Kolmanovsky, 2006, "Control Strategy Optimization for a Series Hybrid Vehicle," SAE Paper 2006–01-0663.

Li, S., I. V. Kolmanovsky, and A. G. Ulsoy, 2011, "Battery Swapping Modularity Design for Plug-in HEVs Using the Augmented Lagrangian Decomposition Method," *Proceedings of the American Control Conference*, San Francisco.

Lin, C., M. Kim, H. Peng, and J. Grizzle, 2006, "System-Level Model and Stochastic Optimal Control for a PEM Fuel Cell Hybrid Vehicle," *ASME Journal of Dynamic Systems, Measurement and Control*, Vol. 128, No. 4, pp. 878–90.

Lin, C., H. Peng, J. W. Grizzle, and J. Kang, 2003a, "Power Management Strategy for a Parallel Hybrid Electric Truck," *IEEE Transactions on Control Systems Technology*, Vol. 11, November, pp. 839–49.

Lin, C., H. Peng, J. W. Grizzle, and M. Busdiecker, 2003b, "Control System Development for an Advanced-Technology Medium-Duty Hybrid Electric Truck," *SAE Transactions Journal of Commercial Vehicles*, Vol. 112–2, pp. 105–13.

Liu, J., and H. Peng, 2008, "Modeling and Control of a Power-Split Hybrid Vehicle," *IEEE Transactions on Control Systems Technology*, Vol. 16, Issue 6, November, pp. 1242–51, Digital Object Identifier 10.1109/TCST.2008.919447.

Magni, J.-F., S. Bennani, J. Terlouw, L. Faleiro, J. de la Cruz, and S. Scala, 1997, "Eigenstructure Assignment," *Robust Flight Control*, Springer, pp. 22–32.

Markel, T., A. Brooker, T. Hendricks, V. Johnson, K. Kelly, B. Kramer, M. O'Keefe, S. Sprik, and K. Wipke, 2002, "ADVISOR: A Systems Analysis Tool for Advanced Vehicle Modeling," *Journal of Power Sources*, Vol. 110, pp. 255–66.

Markel, T., and A. Simpson, 2006, "Plug-In Hybrid Electric Vehicle Energy Storage System Design," NREL/CP-540–39614.

Muta, K., M. Yamazaki, and J. Tokieda, 2004, "Development of New-Generation Hybrid System THS II – Drastic Improvement of Power Performance and Fuel Economy," SAE Paper 2004-01-0064.

Ogawa, H., M. Matsuki, and T. Eguchi, 2003, "Development of a Power Train for the Hybrid Automobile: The Civic Hybrid, *SAE Transactions*, Vol. 112, No. 3, pp. 373–84.

Opila, D. F., X. Wang, R. McGee, J. A. Cook and J. W. Grizzle, 2009, "Performance Comparison of Hybrid Vehicle Energy Management Controllers on Real-World Drive Cycle Data," *Proceedings of the American Control Conference, St. Louis.*

Paganelli, G., T. M. Guerra, S. Delprat, J. J. Santin, M. Delhom and E. Combes, 2000, "Simulation and Assessment of Power Control Strategies for a Parallel Hybrid Car," *IMechE*, Vol. 214, Part D.

Paganelli, G., Y. Guezennec, and G. Rizzoni, 2002, "Optimizing Control Strategy for Hybrid Fuel Cell Vehicle," SAE Paper 2002-01-0102.

Rizoulis, D., J. Burl, and J. Beard, 2001, "Control Strategies for a Series-Parallel Hybrid Electric Vehicle," SAE Paper 2001-01-1354.

Rodatz, P., G. Paganelli, A. Sciarretta, and L. Guzzella, 2005, "Optimal Power Management of an Experimental Fuel Cell/ Supercapacitor-Powered Hybrid Vehicle," *Control Engineering Practice*, Vol. 13, pp. 41–53.

Tate, E. D., J. W. Grizzle, and H. Peng, 2008, "Shortest Path Stochastic Control for Hybrid Electric Vehicles," *International Journal of Robust and Nonlinear Control*, Vol. 18, Issue 14, September, pp. 1409–29.

Wirasingha, S. G., and A. Emadi, 2011, "Classification and Review of Control Strategies for Plug-In Hybrid Electric Vehicles," *IEEE Transactions on Vehicular Technology*, Vol. 60, pp. 111–22.

Wu, B., C. Lin, Z. Filipi, H. Peng, and D. Assanis, 2004, "Optimization of Power Management Strategies for a Hydraulic Hybrid Medium Truck," *Vehicle System Dynamics*, Vol. 42, Nos. 1–2, July–August, pp. 23–40.

11 Modeling and Control of Fuel Cells for Vehicles

11.1 Introduction

In recent decades, various new power-plant alternatives to traditional ICEs have emerged. Because of their packaging and efficiency properties – with typical actual fuel-cell efficiency of approximately 50 to 60 percent compared to typical actual engine efficiency of approximately 20 percent (i.e., gasoline) to 30 percent (i.e., diesel) engines – fuel cells have been studied extensively as an alternative to ICEs (Bernardi and Verbrugge 1992; Boettner et al. 2002; Caux et al. 2005; Department of Energy 2002; Pukrushpan et al. 2004a; Steele and Heinzel 2001; Yang et al. 1998).

Figure 11.1 is a typical configuration of the power flow in a vehicle equipped with a fuel-cell power plant. Hydrogen as fuel is fed to the fuel-cell stack, which generates electricity and water after chemical reactions occur. The electrical energy generated is received by the power-conditioning circuits of the vehicle. Depending on the driver-demand conditions, electronic control units controlling the power electronic circuits, the battery, and the electric motor determine the correct amount of energy that is delivered to the components; the battery is charged to the desired levels and the electrical motor generates the desired torque for vehicle traction. For cases in which the battery is already charged and the energy required for the electrical motor is less than what is generated by the fuel-cell–system electrical load, dumping circuits are used to dissipate the excess energy.

William Grove conducted the first demonstration of a fuel cell in the nineteenth century (Grove 1838). He realized that when water is electrolyzed with a power supply by passing an electric current through it, the output of this reaction is hydrogen and oxygen. When the power supply is replaced with an ammeter, a small current is obtained; that is, the electrolysis is reversed and the hydrogen and oxygen recombine, and an electric current is produced as a result.

The reaction given in Eq. (11.1) uses hydrogen as fuel and is the fundamental reaction for a fuel-cell system:

$$2H_2 + O_2 \rightarrow 2H_2O \tag{11.1}$$

At the end of the reaction described in Eq. (11.1), electrical energy is produced. The experiment mentioned previously reasonably describes the basic principle of the fuel cell. To amplify the generation of electricity, certain conditions should be improved.

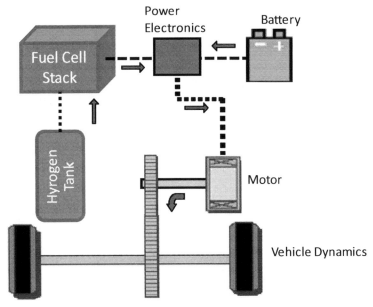

Figure 11.1. Typical fuel-cell hybrid-electric vehicle layout.

Primary reasons for the insufficient or small current include the low contact area and the distance between the electrodes (i.e., resistance). As shown in Figure 11.2, the electrodes usually are made from flat and porous material, covered with a thin layer of electrolyte. Porosity of the material helps the process for the electrolyte and the gas penetration to maximize the contact among the electrode, electrolyte, and gas.

Various details are different for different types of fuel cells and for the reaction between hydrogen and oxygen, which produces an electric current. For example, to

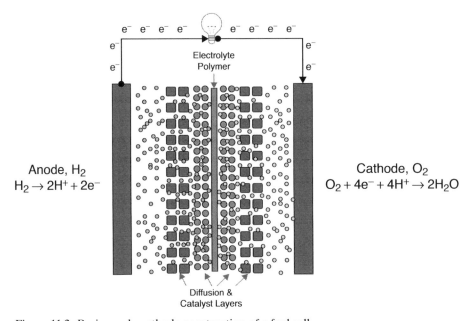

Figure 11.2. Basic anode-cathode construction of a fuel cell.

generate the electrons, reactions at each electrode shown in Figure 11.2 should be considered in the following two parts:

1. The hydrogen gas ionizes, releases electrons, and creates H^+ ions at the anode, as shown in Eq. (11.2):

$$H_2 \rightarrow 2H^+ + 2e^- \tag{11.2}$$

2. Oxygen reacts with electrons taken from the electrode and H^+ ions from the electrolyte, forming water at the cathode, as shown in Eq. (11.3).

$$O_2 + 4e^- + 4H^+ \rightarrow 2H_2O \tag{11.3}$$

The ionization reaction described in Eq. (11.2) releases energy. For continuous reaction, electrons produced at the anode (i.e., Part 1) should go through an electrical circuit to the cathode and H^+ ions concurrently should pass through the electrolyte (i.e., Part 2). Special materials such as proton exchange membranes (PEMs) can be made to possess mobile H^+ ions (i.e., protons) and are used extensively in fuel-cell systems.

With increasing focus and attention on these activities, the fuel-cell research community is active and has been working on many interesting topics in recent years. One topic is fuel-cell–performance prediction so that better fuel-cell components can be designed and optimal fuel-cell operating points can be better understood. An example in this category is PEM fuel cells without any external humidification (Bernardi and Verbrugge 1992) and special membranes with "self-humidification." There also are studies regarding the steady-state–system models that were developed for component sizing and cumulative fuel consumption or hybridization studies. There are numerous publications on fuel-cell modeling, and a few are suitable for control studies. The control-oriented model presented here is based on Pukrushpan et al. (2004b), which is one of the few models suitable for fuel-cell–control design.

The remainder of this chapter is structured as follows. In the next section, a method for obtaining the dynamic model of the fuel-cell systems is presented. Then, suitable control methods for fuel-cell systems are studied. The chapter concludes with a discussion of supervisory control algorithms and parametric design considerations for fuel-cell electric vehicles.

11.2 Modeling of Fuel-Cell Systems

In this section, modeling of fuel-cell systems for developing control algorithms is discussed. Principal ideas, terminology, and structure of the control-oriented model presented in Pukrushpan et al. (2004b) is followed. A computer-simulation model of a dynamic system based on mathematical equations and empirical data is an invaluable tool for control-algorithm development purposes. In these types of models, the correct amount of physical and empirical modeling with the correct structure is crucial for model accuracy and simulation effort.

The diagram in Figure 11.3 is an example of a fuel-cell system for automotive applications, as presented in Pukrushpan et al. (2004a). A fuel-cell stack must be

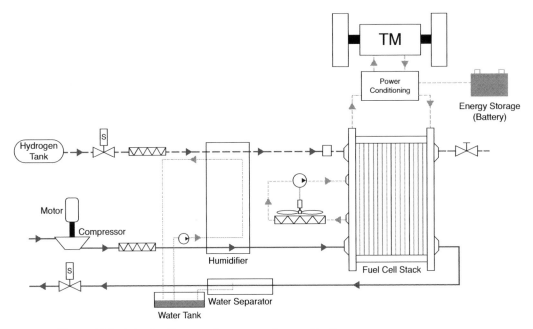

Figure 11.3. A fuel-cell system for automotive applications.

integrated with several auxiliary components to form a complete fuel-cell system. Typically, the fuel-cell stack is augmented by the following four auxiliary systems:

(1) *Hydrogen supply system.* This is a common approach to supply fuel to the fuel-cell stack using a hydrogen tank. The hydrogen flow must be controlled using a supply valve so that the desired flow or pressure is obtained for favorable energy-conversion conditions.

(2) *Air-supply system.* The oxygen (i.e., air) is supplied by a compressor, which is used to increase the power density of the overall system. The air-supply system consists of an air compressor, an electric motor, and pipes or manifolds between the components. The compressor not only achieves desired airflow, it also increases air pressure that significantly improves the reaction rate at the membranes – thus, the overall efficiency and power density of the stack.

(3) *Cooling systems.* Cooling systems generally are used to maintain the temperature of the fuel-cell stack at a desired temperature, given the outside conditions. Because the pressurized airflow leaving the compressor is at a higher temperature, an air cooler may be needed to reduce the temperature of the air entering the stack.

(4) *Humidification system.* To obtain ideal reaction conditions, external humidification systems for both anode and cathode gases also are used. Although alternatives exist, external humidification generally is preferred because it provides more control on the fuel-cell system. The water level in the tank is maintained by collecting water generated in the stack, which is carried out with the airflow. The excessive heat released in the fuel-cell reaction is removed by the cooling system, which circulates de-ionized water through the fuel-cell stack and removes the excess heat via a heat exchanger.

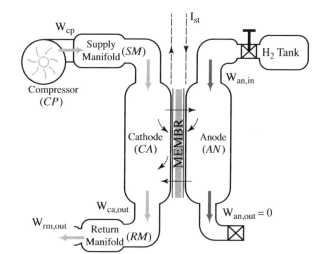

Figure 11.4. Fuel-cell stack layout.

The power of the fuel-cell stack is a function of the current drawn from the stack and the resulting stack voltage. The cell voltage is a function of the stack current, reactant partial pressure inside each cell, cell temperature, and membrane humidity. Most of the fuel-cell mathematical models used in the literature (e.g., Boettner et al. 2002; Caux et al. 2005; Golbert and Lewin 2004; and Pukrushpan et al. 2004b) assume that the stack is well designed and that the cells perform similarly and can be combined as a stack. Therefore, at a single stack temperature, the same set of polarization curves for all cells can be used.

The fuel-cell stack studied in this chapter and based on Dicks and Larminie (2000) and Pukrushpan et al. (2004a) is shown in Figure 11.4. The inputs to this system can be considered to be the pressurized air (i.e., oxygen) and hydrogen; the outputs can be considered to be the electric current and the water generated. A power-conditioning circuit usually is needed because the voltage of the fuel-cell stack varies significantly. The nomenclature used throughout this section and in the equations is given in Table 11.1.

For the model presented here, it is assumed that the cathode and anode volumes of the multiple fuel cells are combined as a single-stack cathode and anode volumes. The anode supply and the return manifold volumes are small, which allows the combination of these volumes into one "anode" volume. The model is based primarily on physics; however, several phenomena are described using empirical equations. In the following subsections, models for the compressor, manifolds, fuel-cell stack, air cooler, and humidifier are presented.

Compressor and Manifolds

The rotational dynamics and a flow map are used to model the compressor. The law of conservation of mass is used to track the gas species in each volume. The principle of mass conservation is applied to calculate the properties of the combined gas in the supply and return manifolds. A combined rotational model used to represent

Table 11.1. *Nomenclature*

A_{fc}	Fuel-cell active area (cm^2)	x	Mass fraction or system state vector
A_T	Valve-opening area (m^2)	y	Mole fraction or system measurements
C_D	Throttle-discharge coefficient	γ	Ratio of the specific heats of air
C_p	Specific heat (J/kg/K)	η	Efficiency
D_w	Membrane-diffusion coefficient	λ	Excess ratio or water content
	(cm^2/sec)	τ	Torque (Nm)
E	Fuel-cell open-circuit voltage		
F	Faraday's number		**Subscripts**
I	Stack current (A)	*act*	Activation
J	Rotational inertia (kgm^2)	*air*	Air
M	Molecular mass (kg/mol)	*an*	Anode
P	Power (Watt)	*ca*	Cathode
R	Gas constant or electrical	*conc*	Concentration
	resistance (Ω)		
T	Temperature (K)	*cp*	Compressor
V	Volume (m^3)	*fc*	Fuel cell
W	Mass flow rate (kg/sec)	*gen*	Generated
a	Water activity	*in*	Inlet
c	Water concentration (mol/cm^3)	*m*	Membrane
d_{cp}	Compressor diameter (m)	*membr*	Across membrane
i	Current density (A/cm^2)	*ohm*	Ohmic loss
m	Mass (kg)	*out*	Outlet
n	Number of cells	*rm*	Return manifold
n_d	Electro-osmotic drag coefficient	*sm*	Supply manifold
p	Pressure (Pa)	*st*	Stack
t	Time (sec)	*v*	Vapor
t_m	Membrane thickness (cm)	*w*	Water
u	System input	*des*	Desired
ω	Rotational speed (rad/sec)	*dmd*	Demand
v	Voltage (V)	*regen*	Regeneration

the dynamic behavior of the compressor is given in Eq. (11.4):

$$J_{cp}\frac{d\omega_{cp}}{dt} = \frac{1}{\omega_{cp}}(P_{cm} - P_{cp}) \tag{11.4}$$

where $P_{cm}(v_{cm}, \omega_{cp})$ is the compressor motor (cm) power and P_{cp} is the load power. In addition, the compressor-motor torque can be calculated using a static-motor equation.

The law of conservation of energy is applied to the air in the supply manifold to account for the effect of temperature variations. The cathode supply manifold (s_m) includes pipe and stack manifold volumes between the compressor and the fuel cells. The supply manifold pressure, p_{sm}, is governed by the mass continuity and energy conservation equations in Eq. (11.5):

$$\begin{aligned}
\frac{ds_{sm}}{dt} &= W_{cp} - W_{ca,in} \\
\frac{dp_{sm}}{dt} &= \frac{\gamma R_a}{V_{sm}}(W_{cp}T_{cp,out} - W_{ca,in}T_{sm})
\end{aligned} \tag{11.5}$$

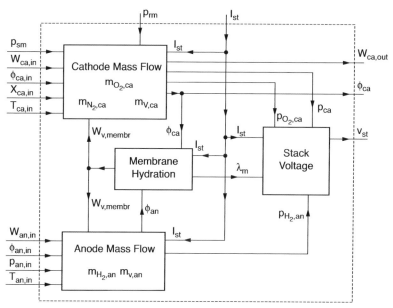

Figure 11.5. Dynamic model of a fuel-cell stack.

where R_a is the air-gas constant, V_{sm} is the supply-manifold volume, and T_{sm} is the temperature of the flow inside the manifold calculated from the Ideal Gas Law. The supply-manifold exit flow, $W_{sm,out}$, can be calculated as a function of p_{sm} and p_{ca} using the linearized nozzle-flow equation.

Unlike the supply manifold, in which temperature changes must be considered, the temperature in the return manifold, T, is assumed to be constant and equal to the temperature of the flow leaving the cathode. The return manifold pressure, p_{rm}, is governed by the principle of mass conservation and the Ideal Gas Law through isothermal assumptions:

$$\frac{ds_{rm}}{dt} = \frac{R_a T_{rm}}{V_{rm}} (W_{ca,out} - W_{rm,out}) \tag{11.6}$$

Fuel-Cell Stack Model

The electrochemical reaction at the membranes is assumed to occur instantaneously. The fuel-cell stack (st) model contains the following four interacting submodels, as shown in Figure 11.5:

(1) stack-voltage model
(2) anode-flow model
(3) cathode-flow model
(4) membrane-hydration model

We assume that the stack temperature is constant at T. The voltage model contains an equation to calculate stack voltage based on fuel-cell pressure, temperature, reactant gas partial pressures, and membrane humidity.

Cathode Mass Flow Model

This model captures the cathode airflow behavior and is developed using the mass conservation principle and the thermodynamic and psychometric properties of air. Several assumptions are made, as follows:

(1) All gases obey the Ideal Gas Law.
(2) The temperature of the air inside the cathode is equal to the stack temperature.
(3) The properties of the flow exiting the cathode, such as temperature, pressure, and humidity, are assumed to be the same as those inside the cathode.
(4) When the relative humidity of the gas exceeds 100 percent, vapor condenses into liquid form. The liquid water does not leave the stack and either evaporates when the humidity drops below 100 percent or accumulates in the cathode.
(5) The flow channel and cathode-backing layer are combined into one volume; that is, the spatial variations are ignored.

The mass continuity is used to balance the mass of the three elements — oxygen, nitrogen, and water – inside the cathode volume, as shown in Eq. (11.7):

$$\frac{dm_{O_2}}{dt} = W_{O_2,in} - W_{O_2,out} - W_{O_2,react}$$

$$\frac{dm_{N_2}}{dt} = W_{N_2,in} - W_{N_2,out} \tag{11.7}$$

$$\frac{dm_{w,ca}}{dt} = W_{v,ca,in} - W_{v,ca,out} + W_{v,gen} + W_{v,membr}$$

Anode-Mass-Flow Model

This model is similar to the cathode-flow model. Hydrogen partial pressure and anode-flow humidity are determined by balancing the mass of hydrogen, m_{H2}, and water in the anode and can be calculated as shown in Eq. (11.8):

$$\frac{dm_{H_2}}{dt} = W_{H_2,in} - W_{H_2,out} - W_{H_2,react}$$

$$\frac{dm_{w,an}}{dt} = W_{v,an,in} - W_{v,an,out} - W_{v,membr} \tag{11.8}$$

Assumptions for this mathematical model are as follows:

(1) Pure hydrogen gas is assumed to be supplied to the anode from a hydrogen tank.
(2) It is assumed that the hydrogen flow rate can be adjusted instantaneously using a valve while maintaining a minimum pressure difference across the membrane.
(3) The inlet hydrogen flow is assumed to have 100 percent relative humidity.
(4) The anode outlet flow represents possible hydrogen purge and is currently assumed to be zero (W).
(5) The anode hydrogen temperature is assumed to be equal to the stack temperature.

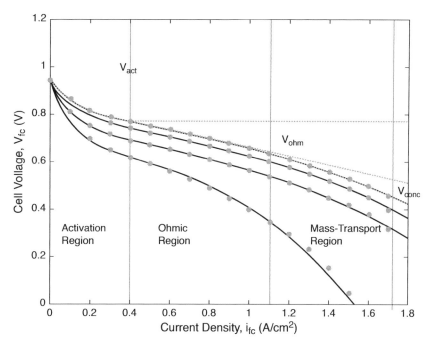

Figure 11.6. Fuel-cell polarization curve.

Membrane-Hydration Model

The membrane-hydration model captures the effect of water transport across the membrane. Both water content and mass flow are assumed to be uniform over the surface area of the membrane, and they are functions of stack current and relative humidity of the gas in the anode and cathode.

Combining the two water-transport mechanisms – the electro-osmotic drag phenomenon and the "back-diffusion" of water – the water flow across the membrane from anode to cathode can be calculated as shown in Eq. (11.9):

$$W_{v,membr} = M_v A_{fc} n \left(n_d \frac{i}{F} - D_w \frac{(c_{v,an} - c_{v,ca})}{t_m} \right) \qquad (11.9)$$

The dynamically varying pressure and relative humidity of the reactant gas flow inside the stack flow channels are calculated in the cathode- and the anode-flow models. The process of water transfer across the membrane is governed by the membrane-hydration model. These subsystem models are discussed in the following subsections.

Stack-Voltage Model

Stack voltage is calculated as a function of stack current, cathode pressure, reactant partial pressures, fuel-cell temperature, and membrane humidity. The current–voltage relationship is commonly given in the form of the polarization curve, which is plotted as cell voltage, v_{fc}, versus cell-current density, i_{fc}, as shown in Figure 11.6 for a specific cell temperature and various pressures.

Because the fuel-cell stack consists of multiple fuel cells connected in series, the stack voltage, V_{st}, is obtained as the sum of the individual cell voltages, and the stack current, I_{st}, is equal to the cell current. The current density then is defined as stack current per unit of cell active area, i_{fc}. Stack voltage can be calculated by multiplying the cell voltage, V_{fc}, by the number of cells – assuming that all cells are identical. The fuel-cell voltage is calculated using a combination of physical and empirical relationships, as shown in Eq. (11.10):

$$V = E - V_{act} - V_{ohm} - V_{conc}$$
$$= E - \left[V_0 + V_a \left(1 - e^{-c_1 i} \right) \right] - \left[i R_{ohm} \right] - \left[i \left(c_2 \frac{i}{i_{max}} \right)^{c_3} \right] \qquad (11.10)$$

The open-circuit voltage, E, is calculated from the energy balance between the reactants and products and the Faraday Constant. The activation overvoltage, V_{act}, arises from the need to move electrons and to break and form chemical bonds at the anode and cathode, which also depends on temperature and oxygen partial pressure.

Based on the nonlinear mathematical model developed in Eqs. (11.1) through (11.10) and outlined in Figures 11.4 and 11.5, a dynamic model of the fuel-cell stack can be developed with computer-aided engineering (CAE) tools such as MATLAB/Simulink to analyze the system response. A downloadable version of the model outlined here is available at www.springer.com/engineering/control/book/ 978-1-85233-816-9 (Pukrushpan et al. 2004a). The top-level structure of the model is shown in Figure 11.7.

For the model given in Pukrushpan et al. (2004b) and shown in Figure 11.7, simulation results when a series of step changes in stack-current and compressor-motor input voltage are applied to the stack are shown in Figures 11.8a and b, respectively, at a nominal stack operating temperature of 80 degrees Celsius.

During the first four steps, the compressor voltage is controlled so that the optimal oxygen excess ratio (approximately 2.0) is maintained. This can be achieved with a simple static feed-forward controller. The remaining steps then are applied independently, resulting in different levels of oxygen-excess ratios (Figure 11.8e). During a positive-current step, the oxygen-excess ratio drops due to the depletion of oxygen (Figure 11.8e), which causes a significant drop in the stack voltage (Figure 11.8c). When the compressor voltage is controlled by the feed-forward algorithm, there is still a noticeable transient effect on the stack voltage (Figure 11.8c) and oxygen partial pressure at cathode exit (Figure 11.8f). The step at $t = 18$ seconds shows the response of giving a step increase in the compressor input while maintaining the constant stack current. An opposite case is shown in seconds; the response between 18 and 22 seconds shows the effect of running the system at an excess ratio higher than the optimum value. Although the stack voltage (i.e., power) increases, the net power actually decreases due to the increased parasitic loss.

11.3 Control of Fuel-Cell Systems

The previous sections discussed supporting systems (i.e., hydrogen supply, air supply, cooling, and humidification) that help the fuel-cell stack system work effectively, as shown in Figure 11.3. All of these systems have desired operating points based on

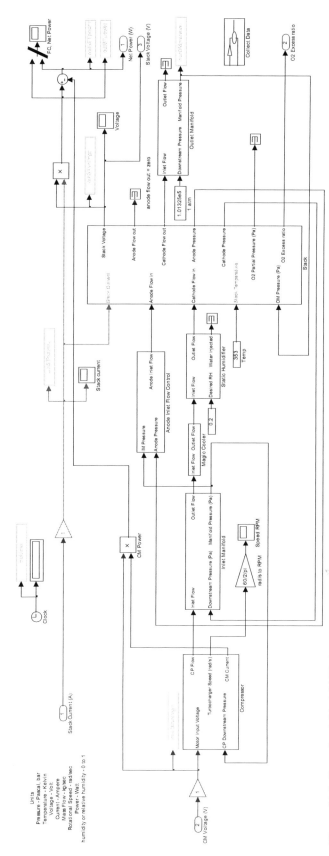

Figure 11.7. Downloadable MATLAB/Simulink fuel-cell model.

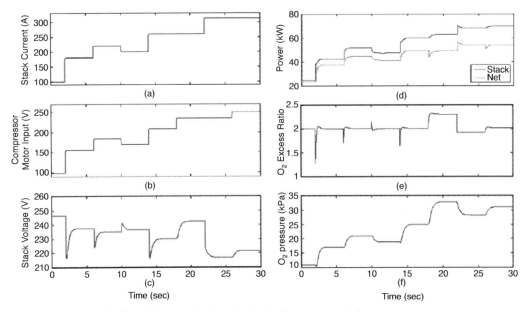

Figure 11.8. Simulation results for the fuel-cell system model.

vehicle conditions and, when regulated, they effectively help the fuel-cell stack generate energy at optimum levels. Typical operations for these systems are summarized as follows:

(1) *Hydrogen-supply system.* The hydrogen flow must be controlled using a supply valve so that the desired flow or pressure is obtained for favorable energy-conversion conditions. The hydrogen supply valve is the actuator for this system, verifying whether the proper amount of hydrogen is delivered and monitored via the fuel-stack measurements.

(2) *Air-supply system.* The oxygen (i.e., air) is supplied by a compressor, which is used to increase the power density of the overall system. The air-supply actuation is controlled by the air-compressor motor.

(3) *Cooling systems.* Cooling systems generally are used to keep the temperature of the fuel-cell stack at a desired temperature, given the outside conditions. The cooling actuation is controlled by changing the speed of the fan motor, and the temperature is monitored using the stack-temperature outputs.

(4) *Humidification system.* To obtain ideal reaction conditions, external humidification systems for both anode and cathode gases are used. The actuation of the humidifier is controlled by changing the air-blower speed, and the humidity is monitored by using fuel-cell stack temperature and pressure outputs.

The block diagram in Figure 11.9 shows a control system that is composed of multiple SISO controllers. This closed-loop system uses desired stack voltage ($v_{st,des}$) as the input and actual stack voltage (v_{st}) as the output. Desired stack-voltage input is used in the electrical bus controller to determine the proper stack current, i_{st}. It also is mapped to a desired compressor-speed signal, $\omega_{cp,des}$, using a feed-forward controller that is representative of a proper operating point for the fuel-cell system.

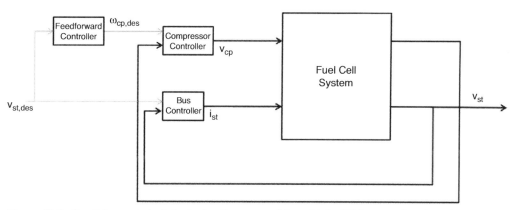

Figure 11.9. Possible control of the fuel-cell system with multiple SISO controllers based on desired stack voltage.

The compressor-speed response then can be controlled by its own SISO controller, as shown in the figure.

The controller structure proposed in Figure 11.9 is composed of a series of simple SISO feedback and feed-forward controllers. These types of controllers usually are preferred in control practice because of their simplicity and vast experience with the design. However, for cases in which the plant dynamics (i.e., the fuel-cell system) are closely coupled, implementation of the individual (or loosely coupled) SISO controllers may not provide the most effective performance, and more complex controller structures are recommended. Based on the nonlinear mathematical model developed in Eqs. (11.1) through (11.10) and outlined in Figures 11.1 and 11.4, the linear time-invariant (LTI) system analysis in the MATLAB/Simulink control system toolbox can be used to design MIMO controllers after linearization is performed for specific operating conditions.

A linear model based on this analysis is given in state-space form in Eq. (11.11). For the linear system presented, the nominal operating point is chosen as $P_{net} = 40$ kW, $\lambda_{O2} = 2$, which corresponds to nominal inputs of $I_{st} = 191$A and $v_{cm} = 164$ Volt.

$$\dot{\mathbf{x}} = \mathbf{A}\mathbf{x} + \mathbf{B}\mathbf{u}$$
$$\mathbf{y} = \mathbf{C}\mathbf{x} + \mathbf{D}\mathbf{u}$$

$$(11.11)$$

where:

$$\mathbf{A} = \begin{bmatrix} -6.3091 & 0 & -10.9544 & 0 & 83.7446 & 0 & 0 & 24.0587 \\ 0 & -161.0830 & 0 & 0 & 51.5292 & 0 & -18.0261 & 0 \\ -18.7858 & 0 & -46.3136 & 0 & 275.6592 & 0 & 0 & 158.3471 \\ 0 & 0 & 0 & -17.3506 & 193.9373 & 0 & 0 & 0 \\ 1.2996 & 0 & 2.9693 & 0.3977 & -38.7024 & 0.1057 & 0 & 0 \\ 16.6424 & 0 & 38.0252 & 5.0666 & -479.3840 & 0 & 0 & 0 \\ 0 & -450.3869 & 0 & 0 & 142.2084 & 0 & -80.9472 & 0 \\ 2.0226 & 0 & 4.6212 & 0 & 0 & 0 & 0 & -51.2108 \end{bmatrix}$$

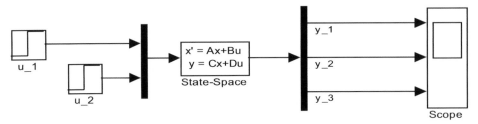

Figure 11.10. Simulink model to simulate the linearized MIMO model.

$$
\mathbf{B} = \begin{bmatrix}
0 & -0.0316 \\
0 & -0.0040 \\
0 & 0 \\
3.9467 & 0 \\
0 & 0 \\
0 & 0 \\
0 & -0.0524 \\
0 & 0
\end{bmatrix}
$$

$$
\mathbf{C} = \begin{bmatrix}
0 & 0 & 0 & 5.0666 & -116.4460 & 0 & 0 & 0 \\
0 & 0 & 0 & 0 & 1 & 0 & 0 & 0 \\
12.9699 & 10.3253 & -0.5693 & 0 & 0 & 0 & 0 & 0
\end{bmatrix}
$$

$$
\mathbf{D} = \begin{bmatrix}
0 & 0 \\
0 & 0 \\
0 & -.2966
\end{bmatrix}
$$

and $\mathbf{x} = \begin{bmatrix} m_{O_2} & m_{H_2} & m_{N_2} & \omega_{cp} & p_{sm} & m_{sm} & m_{w,an} & p_{rm} \end{bmatrix}^T$,

$\mathbf{u} = \begin{bmatrix} v_{cm} & I_{st} \end{bmatrix}, \quad \mathbf{y} = \begin{bmatrix} W_{cp} & p_{sm} & v_{st} \end{bmatrix}$

In this linear model, masses, pressure, rotational speed, flow rate, power, voltage, and current are given in grams, bar, kRPM, g/sec, kW, V, and A, respectively. Figure 11.10 shows a MATLAB/Simulink block diagram, with which the linear system given in Eq. (11.11) can be used to perform control design studies.

Figures 11.11 and 11.12 present the dynamic response of the linear system in Eq. (11.11) when a step input is applied to the compressor input voltage, v_{cm}, and stack current, I_{st}, respectively.

These types of studies can be used to determine the coupling between inputs and outputs for plants to identify control-design opportunities. For example, Figure 11.12 shows that the stack-current input is weakly coupled with the supply-manifold pressure p_{sm}, whereas the other two plant outputs are meaningfully responsive. Using results of the linearized-model analysis, simple MIMO controllers can be designed using conventional techniques (see Problem 10.4). The design

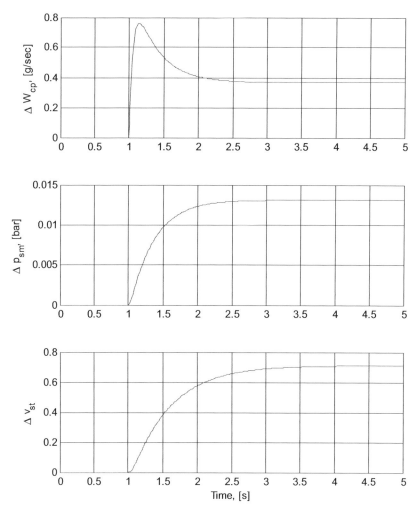

Figure 11.11. Response of the linearized model to a step compressor voltage input.

process for such multivariable controllers is a well-studied field in control theory, and many results exist for the design of MIMO systems. For example, in Boettner et al. (2002), Caux et al. (2005), Guzzella and Sciarretta (2007), and Pukrushpan et al. (2004a), related topics such as dynamic simulation, the observability of the model developed, and possible linearization techniques are discussed.

Figure 11.13 shows a vehicle control system controller that uses the driver-demand pedal inputs to control simultaneously the fuel-cell system and the traction at the electric motor. Because the controller is based on a linearized model, more than one MIMO controller should be designed and implemented. During operation of the vehicle based on current conditions, the "state" of the vehicle can be detected and the corresponding MIMO controller can be designed based on the linearized-system model at those operating conditions.

11.4 Control of Fuel-Cell Vehicles

The powertrain configuration for a typical fuel-cell–powered HEV is shown in Figure 11.1. This configuration includes an electric motor; a battery; a fuel-cell

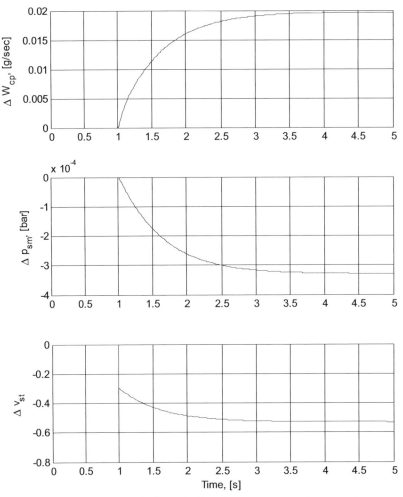

Figure 11.12. Response of the linearized model to a step stack current input.

system; and power electronics to facilitate energy flow among the battery, fuel-cell system, and electric motor. Although it is not shown in the figure, the electric motor, the battery, and the fuel-cell system have controller modules. The coordination of these controllers is performed by the vehicle-supervisory control algorithm. During typical operation conditions for a fuel-cell vehicle, the driver torque demand is obtained via the accelerator- or the brake-pedal sensor signals as well as monitored vehicle conditions.

To design a control system for a vehicle equipped with a fuel-cell power plant, various control-design approaches can be followed. As a first approach, we can consider each system as an individual control entity and design a single control algorithm for each system with set points calculated from the driver's instantaneous commands based on the torque at the wheels' response of the vehicle. This is similar to the controller structure presented in Figure 11.9, with all of the vehicle functions included. Although this approach results in simpler individual control problems to solve, the strong interaction among these systems results in poor overall system performance (i.e., vehicle longitudinal response). As discussed in the previous section on modeling of fuel-cell systems, the interaction between the supporting systems is

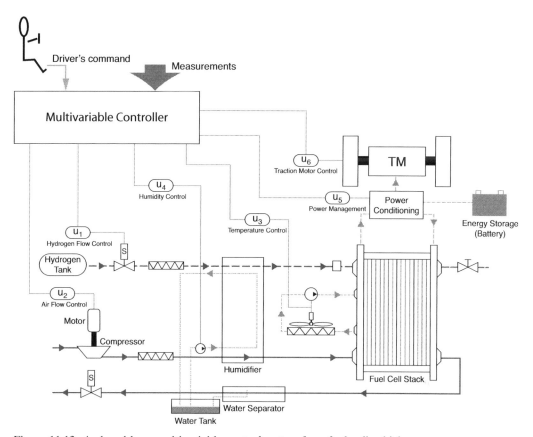

Figure 11.13. A closed-loop multivariable control system for a fuel-cell vehicle.

strong. The design of separate SISO controllers is not preferred and not used except in the test benches and academic studies for replicating certain conditions.

In Figure 11.13, a closed-loop multivariable control system is integrated with the fuel-cell stack system shown. Although the LTI model analysis and suitable control design methods are useful for the initial design phase, more detailed control algorithms typically are necessary to run the complex fuel-cell systems discussed herein. Figure 11.14 shows in detail typical inputs and outputs for such a multivariable controller, shown in Figure 11.13. This represents a case – which we can classify as the second control-design approach – in which the fuel-cell stack supporting systems and the vehicle longitudinal dynamics are considered a single system with interacting subsystems. It is important to note that some of the inputs in Figure 11.14 in fact can be used to decide which alternative MIMO control to use rather than being used directly in the controller. Also, some outputs from the controller (e.g., blower command) can be an on/off signal rather than a continuous command.

Another approach to develop a controller for the system, using the supervisory control development techniques through optimization, has been studied by many researchers (Ehsani et al. 2004; Gao and Ehsani 2001; Schell et al. 2005; Yang et al. 1998). The vehicle supervisory controller controls the energy flow among the fuel-cell system, the battery, and the electric motor by issuing the appropriate set-point commands to component controllers. During a typical drive cycle, both the fuel-cell

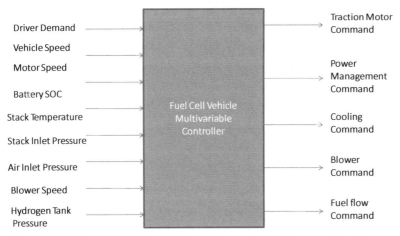

Figure 11.14. More realistic fuel-cell vehicle multivariable controller I/O.

system and the battery can be used to provide the energy that is needed to meet the power demand by the driver. The battery also can be used to store the energy when the demand is less than the produced energy or when the electric motor is used in generator mode for braking. The objective of controls is to utilize the energy flow such that the fuel economy is maximized while the drivability requirements are met and the vehicle components are operated at their recommended operating conditions.

To meet the desired fuel-cell vehicle-control strategy objectives, the supervisory control strategy should direct the components in the system such that the electric motor always meets the power demanded by the driver (i.e., the drivability condition). This condition also should be accompanied by the goals of the fuel-cell system operating at its optimum operating region and maintaining the battery power level within an optimum range to ensure maximized charging and discharging efficiency (i.e., the energy-efficiency condition).

When a driver gives an accelerator-pedal or brake-pedal command, it is used to calculate a driver-power-demand command, P_{dmd}. For traction cases and for the fuel-cell electric vehicle whose powertrain configuration is described in Figure 11.1, this calculated value corresponds to power that the electric motor is commanded to produce. Using the transmission efficiency, η_m, the electric power input to the motor, $P_{m,in}$, may be modified as $P_{m,in} = P_{dmd}/\eta_m$ at the electric motor controller for a certain motor speed. During regenerative braking through the motor, it functions as a generator, and the electric-power output from the motor, $P_{m,out}$, may be calculated as $P_{m,out} = P_{regen,dmd}*\eta_m$. It is important to note that $P_{regen,dmd}$ is the braking-power command to the motor, which may be different from the power demanded, P_{dmd}, because part of it can come from the mechanical brake at the wheels.

Based on the motor-power command and other vehicle information – such as energy level in the battery and minimum operating power of the fuel-cell system, below which the efficiency of the fuel cell decreases significantly – the fuel-cell system and the battery can be requested to provide the power to the system. This is usually accomplished by defining a power-split policy based on system conditions and various operating modes of the drivetrain, which can be described for a fuel-cell electric vehicle as follows:

Neutral: The fuel-cell system or the battery does not supply power to the drivetrain. The fuel-cell system may operate at idle for quick start and warm-up conditions.

Braking: The fuel-cell system operates at idle and the battery absorbs the regenerative braking energy supplied according to the brake-system operating characteristics. The amount of regenerative braking typically is calculated based on the amount of braking reported by the brake-controller module.

Hybrid Traction: If the commanded motor-input power is greater than the rated power of the fuel-cell system, the hybrid-traction mode is used, in which the fuel-cell system operates with its rated power and the remaining power demanded is supplied by the battery. The rated power of the fuel-cell system can be set as the maximum level of the optimal operating region of the fuel cell.

Battery Charge: If the commanded motor-input power is smaller than the preset minimum power of the fuel-cell system and the battery needs charging (i.e., the energy level is less than the minimum value), the fuel-cell system operates with its rated power, part of which goes to the drivetrain and the other part goes to the battery.

Battery Only: If the battery does not need charging (i.e., the energy level is near its maximum value), the fuel-cell system operates at idle and the battery alone drives the vehicle. In the latter case, the peak power that the battery can produce is greater than the commanded motor-input power.

Fuel Cell Only: If the load power is greater than the preset minimum power and less than the rated power of the fuel cell, and the battery does not need charging, the fuel-cell system alone drives the vehicle. Otherwise, if the battery does need charging, the fuel-cell system operates with its rated power, part of which goes to the drivetrain to drive the vehicle and the other part is used to charge the battery.

Figure 11.15 shows a baseline supervisory control algorithm chart for a typical fuel-cell vehicle and for the operating conditions described herein. This is a basic strategy that can be used as a starting point for simulations in which P_{dmd} is power demanded by the driver (i.e., braking or acceleration based on the sign); P_{FC} is the power output of the fuel-cell system; $P_{FC,op}$ is the current operating power of the fuel-cell system; P_{batt} is the battery power (i.e., charge or discharge depending on the sign); and SOC is the battery state of the charge with maximum and minimum limits, SOC_{min} and SOC_{max}, respectively.

11.5 Parametric Design Considerations

The parametric design of the fuel-cell–powered hybrid drivetrain includes the design of the electric-motor power, the fuel-cell–system power, and the battery power and energy capacity. The motor power is to meet the acceleration performance of a vehicle, as discussed in previous chapters. Required motor power can be calculated

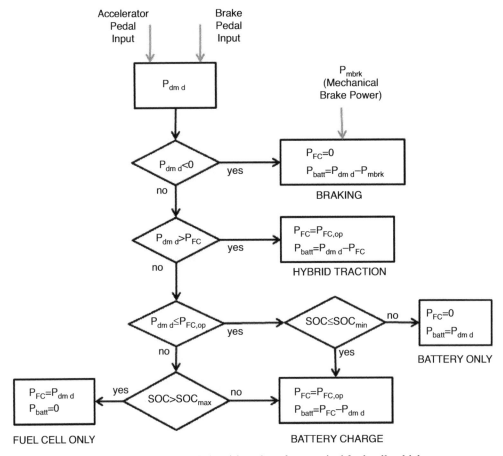

Figure 11.15. Supervisory-control algorithm chart for a typical fuel-cell vehicle.

by looking at desired maximum acceleration and constant speed on various road grades.

The fuel-cell system selected for a vehicle must supply sufficient power to support it at highway speeds (e.g., high constant speeds for a long time) and to support it when medium speeds are desired for medium time conditions (i.e., city driving) without the help of the battery. Because it has a limited amount of energy, the battery should be used only to supply peak power in short periods. Once the maximum power of the motor is determined by the specified acceleration performance and the rated power of the fuel-cell system is determined by the constant-speed driving, the power requirements of the battery can be determined as shown in Eq. (11.12):

$$P_{batt} = P_{motor} * \eta_m - P_{FC} \tag{11.12}$$

where P_{batt} is the rated power of the battery, P_{motor} is the maximum electric-motor power, η_m is the efficiency of the motor drive, and P_{FC} is the operating power of the fuel-cell system. The energy change in the battery during a driving cycle can be expressed as given in Eq. (11.13):

$$E = \int_{t=0}^{t=t_f} P_{batt} dt \tag{11.13}$$

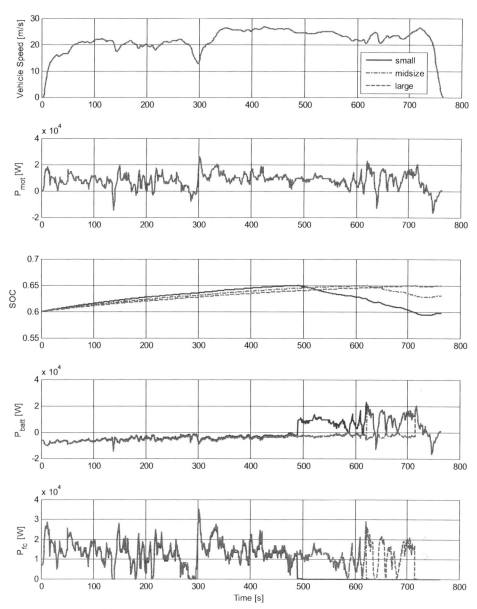

Figure 11.16. Candidate fuel-cell vehicle configuration simulated for the Federal Test Procedure drive cycle.

where P_{batt} is the discharge and charge power of the battery (based on the sign) and t_f is the duration of the drive cycle. The energy changes, E, in the battery depend on the size of the fuel-cell system selected, the vehicle control strategy described in Figure 11.15, and the drive cycle to which the vehicle is subjected.

A parametric design for these systems can be analyzed via simulation. First, using a CS battery strategy, different drive cycles that are representative of the target market and vehicle dynamics are run to obtain the results shown in Figures 11.16 and 11.17. By analyzing the maximum fuel-cell power and its utilization ratio (i.e., on/off

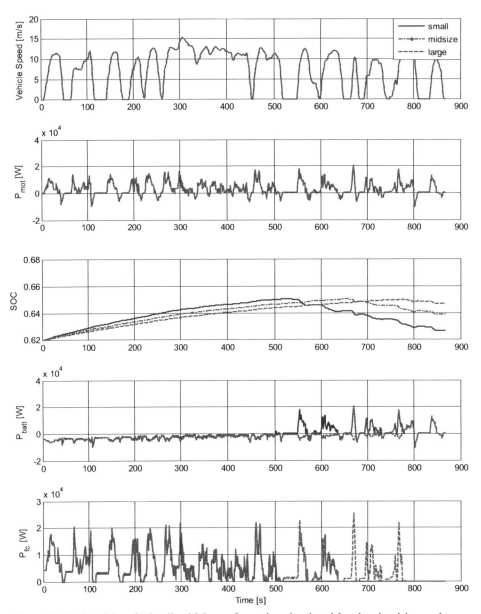

Figure 11.17. Candidate fuel-cell vehicle configuration simulated for the city drive cycle.

ratio) and the corresponding maximum battery power required to run the system and its maximum energy-difference sweep, the size of the battery and the fuel-cell system can be established. In addition to the power considerations, it is important to consider that when a larger battery and the fuel-cell system capacities are selected, the vehicle total weight and cost are usually heavier and more expensive. For the conceptual design stage, various computer simulation tools can be used for sizing and fuel-efficiency analysis (Guzzella and Amstutz 1999; Rousseau and Pasquier 2001).

Table 11.2. *Vehicle properties*

Vehicle Mass	1,600 kg
Electric-Motor Power	65 kW
Fuel-Cell Power	80 kW
Battery SOC Limits	0.3–0.7
Battery Maximum Power/Energy	40 kW/1.5 hWh

PROBLEMS

1. In Eqs. (11.1) through (11.3), the electrode reactions and charge flow for an acid-electrolyte fuel cell are presented. There also are other ways to obtain the same reaction, such as using an alkaline-electrolyte fuel cell. For this type of fuel cell, explain the reactions and discuss the possibility of using such systems in fuel-cell HEVs.

2. In Figure 11.1, a typical fuel-cell electric-vehicle powertrain layout is shown. Discuss the advantages and disadvantages of this configuration. Also compare the operation principle of this vehicle to typical HEV applications such as the Ford Escape Hybrid and Toyota Prius.

3. Consider the linear state-space model given in Eq. (11.11) and use the stack-current and compressor-voltage input data given in Pukrushpan et al. (2004a) to obtain the plant output traces. Compare the two sets of results and explain the differences.

4. Using the state-space linear model given in Eq. (11.11), propose an optimal MIMO controller that will regulate the plant outputs to desired values, as shown in the following block diagram. Investigate the possibility of using a state-feedback controller; if not, provide a controller structure accordingly. Use suitable results from Pukrushpan et al. (2004a) as necessary.

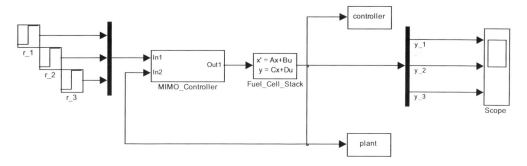

5. Using the control-algorithm chart in Figure 11.15, develop a supervisory control algorithm for the fuel-cell–system model given in Eqs. (11.1) through (11.10) and in vehicle longitudinal dynamics equations from Chapter 4. Use the vehicle parameters listed in Table 11.2.

6. By comparing the plots in Figures 11.16 and 11.17, determine the effects of using different battery sizes and possible supervisory control, for both highway and city driving, to better utilize fuel-cell power.

REFERENCES

Bernardi, D. M., and M. W. Verbrugge, 1992, "A Mathematical Model of the Solid Polymer Electrolyte Fuel Cell," *Journal of the Electrochemical Society*, Vol. 139, 1992, p. 2477.

Boettner, D. D., G. Paganelli, Y. G. Guezennec, G. Rizzoni, and M. J. Moran, 2002, "Proton Exchange Membrane Fuel Cell System Model for Automotive Vehicle Simulation and Control," *Journal of Energy Resources Technology*, Vol. 124, March, pp. 20–7.

Caux, S., J. Lachaize, M. Fadel, P. Shott, and L. Nicod, 2005, "Modelling and Control of a Fuel Cell System and Storage Elements in Transport Applications," *Journal of Process Control*, Vol. 15, 2005, pp. 481–91.

Department of Energy (U.S.), 2002, *Fuel Cell Handbook*, Morgantown, PA.

Dicks, A., and J. Larminie, 2000, *Fuel Cell Systems Explained*, Wiley & Sons, Inc.

Ehsani, M., Y. Gao, S. E. Gay, and A. Emadi, 2004, *Modern Electric, Hybrid Electric, and Fuel Cell Vehicles: Fundamentals, Theory, and Design*, CRC Press.

Gao, Y., and M. Ehsani, 2001, "Systematic Design of Fuel Cell Powered Hybrid Vehicle Drive Train," *IEEE Electric Machines and Drives Conference*, 604–11, DOI: 10.1109/IEMDC.2001.939375.

Golbert, J., and D. R. Lewin, 2004, "Model-Based Control of Fuel Cells: (1) Regulatory Control," *Journal of Power Sources*, Vol. 135, September 2004, pp. 135–51.

Grove, W. R., 1838, "On a new voltaic combination." *Philosophical Magazine and Journal of Science*, Vol. 13, p. 430.

Guzzella, L., and A. Amstutz, 1999, CAE Tools for Quasi-Static Modeling and Optimization of Hybrid Powertrains, *IEEE Transactions on Vehicular Technology*, Vol. 48, No. 6, pp. 1762–9.

Guzzella, Lino, and A. Sciarretta, 2007, *Vehicle Propulsion Systems: Introduction to Modeling and Optimization*, Springer Publishing Co.

Pukrushpan, J. T., A. G. Stefanopoulou, and H. Peng, 2004a, *Control of Fuel Cell Power Systems: Principles, Modeling, Analysis, and Feedback Design*, Springer-Verlag.

Pukrushpan, J. T., H. Peng, and A. G. Stefanopoulou, 2004b, "Control-Oriented Modeling and Analysis for Automotive Fuel Cell Systems," *ASME Journal of Dynamic Systems, Measurement, and Control*, Vol. 126, p. 14.

Rousseau, A., and M. Pasquier, 2001, "Validation of a Hybrid Modeling Software (PSAT) Using its Extension for Prototyping (PSAT-PRO)," *Proceedings of the 2001 Global Powertrain Congress*, Detroit, MI, pp. 1–9.

Schell, Andreas, H. Peng, D. Tran, E. Stamos, C.-C. Lin, and M. J. Kim. 2005. "Modelling and Control Strategy Development for Fuel Cell Electric Vehicles," *Annual Reviews in Control* 29(1): 159–68. DOI: 10.1016/j.arcontrol.2005.02.001.

Springer-Verlag, Web site for Control of Fuel-Cell Power Systems Book, available at www.springer.com/engineering/control/book/978–1-85233–816-9. (January 11, 2012).

Steele, B. C. H., and A. Heinzel, 2001, "Materials for Fuel-Cell Technologies," *Nature*, Vol. 414, 2001, pp. 345–52.

Yang, W., B. Bates, N. Fletcher, and R. Pow, 1998, "Control Challenges and Methodologies in Fuel Cell Vehicle Development," *Proceedings of the SAE Conference*, Issue 328, pp. 363–70.

PART III

VEHICLE CONTROL SYSTEMS

12 Cruise and Headway Control

One of the most widely adopted and visible control systems available on contemporary vehicles sold in the United States is the cruise control, which automatically regulates the vehicle longitudinal velocity by throttle adjustments. Typically, a vehicle cruise-control system is activated by a driver who wants to maintain a constant speed during long highway driving. This relieves the driver from having to continually adjust the throttle. The driver activates the cruise controller while driving at a particular speed, which then is recorded as the desired (or set-point) speed to be maintained by the controller.

Intelligent cruise control systems – also known as autonomous or adaptive cruise control (ACC) systems – are the next-generation product for cruise control. When no lead vehicle is within sight, an ACC vehicle behaves like a conventional cruise-control vehicle by maintaining a constant (i.e., target) speed. However, an ACC vehicle also has a headway-control mode. When the vehicle, using a range sensor, detects that it is close to the in front vehicle, the controller switches to headway-control mode and adjusts the speed to maintain a desired (i.e., safe) headway. Many field tests have been conducted to assess the real-life performance of ACC vehicles and consumers generally are receptive to the convenience provided by them. A rapidly growing number of luxury vehicles now offer ACC as an option.

12.1 Cruise-Controller Design

The main goal in designing the cruise-control algorithm is to maintain vehicle speed smoothly but accurately, even under large variations in plant parameters and road grade. In the case of cruise control for heavy trucks, the vehicle mass can change by up to several hundred percentage points (e.g., from a tractor with no trailer to towing a fully loaded trailer). Furthermore, a truck transmission may have numerous gears and frequent gear shifting may occur. Therefore, powertrain behavior may vary significantly, which causes additional complexity.

In the case of passenger automobiles or SUVs/light trucks, vehicle mass still may change noticeably but it is within a much smaller range. Also, the cruise-control function usually is activated only above a threshold speed (e.g., 30 mph); therefore, gear shifting is less of a problem. In this chapter, we assume that the engine and transmission characteristics can be inverted easily through engine-map inversion.

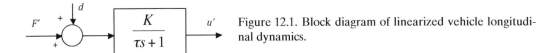

Figure 12.1. Block diagram of linearized vehicle longitudinal dynamics.

In other words, the desired traction force can be obtained easily through table lookup and commanding proper throttle angle or fuel-injection rate. In addition, we assume that tire friction is not a limiting factor. Therefore, we assume that we can directly manipulate traction force. Under these assumptions, a cruise-control design can be based on the equation of motion for longitudinal vehicle dynamics presented in Chapter 3 and the free-body diagram in Figure 4.5. Thus, from Eq. (4.4), the vehicle longitudinal motion is described by:

$$m\frac{du}{dt} = F_x - mg\sin\Theta - fmg\cos\Theta - 0.5\rho AC_d(u + u_w)^2 \tag{12.1}$$

Equation (12.1) is nonlinear in the forward velocity, $u(t)$, but otherwise is a simple dynamic system: it only has one state variable. So, what are the main challenges in cruise-control design problems? The difficulties arise mainly from two factors: (1) plant uncertainty due to change of vehicle weight, and (2) external disturbances due to road grade. Thus, a good cruise-control algorithm must work well under these uncertainties. We can linearize Eq. (12.1) about a nominal operating condition. At equilibrium (i.e., when $du/dt = 0$), Eq. (12.1) can be solved for:

$$F_{x0} = mg\sin\Theta_0 + fmg\cos\Theta_0 + 0.5\rho AC_d(u_0 + u_w)^2 \tag{12.2}$$

For example, using the numerical values:

$$g = 9.81\,\text{m/s},\ u_0 = 20\,\text{m/s},\ \Theta_0 = 0,\ m = 1000\,\text{kg},$$
$$\rho = 1.202\,\text{kg/m}^3,\ A = 1\,\text{m}^2,\ C_d = 0.5,\ f = 0.015,\ u_w = 2\,\text{m/s};$$

this yields $F_{x0} = 292.6\,N$. Linearizing Eq. (12.1) about the specified operating (i.e., equilibrium) state by using a Taylor series expansion yields the linearized equation:

$$\tau\dot{u}' + u' = K(F' + d) \tag{12.3}$$

where the incremental, or perturbed, variables are defined as:

$$u = u_0 + u';\quad F_x = F_{x0} + F';\quad \Theta = \Theta_0 + \Theta'\quad d = mg(f\sin\Theta_0 - \cos\Theta_0)\Theta';$$
$$\tau = (m/(\rho C_d A(u_0 + u_w)));\quad K = (1/(\rho C_d A(u_0 + u_w)));$$

Using the parameter values given here, we obtain $\tau = 75.632$ sec and $K = 0.0756$ (m/s)/N. Based on the linearized equations, the process model for control-system design can be represented as shown in the block diagram in Figure 12.1.

Using this model, a PI controller can be designed, which has the transfer function:

$$G_c(s) = \frac{F'(s)}{e(s)} = K_P + K_I/s \tag{12.4}$$

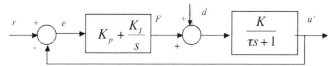

Figure 12.2. Closed-loop system with PI controller.

As shown in the block diagram in Figure 12.2, the closed-loop-system equations become:

$$u'(s) = \frac{K(K_I + K_P s)}{\tau s^2 + (KK_P + 1)s + KK_I} r + \frac{Ks}{\tau s^2 + (KK_P + 1)s + KK_I} d \quad (12.5)$$

Thus, the controller gains (i.e., parameters K_P and K_I) must be selected to achieve good performance for the closed-loop system. This requires consideration of a number of factors, such as stability, steady-state errors, and transient response, which are discussed in Example 12.1.

EXAMPLE 12.1: PI CRUISE CONTROLLER DESIGN (FIRST-ORDER PLANT). A PI cruise-controller design requires consideration of several performance requirements (e.g., stability, steady-state error, and transient response) in selecting the control gains, K_P and K_I. Various control-design techniques, including pole placement, root-locus, and Bode plots, can be used. These techniques are made considerably easier to apply by the availability of computer-aided control-system design programs such as those available in MATLAB. The controller design is based on the plant-model parameters $K = 0.0756$ (m/s)/N and $\tau = 75.632$ seconds given previously. The closed-loop system has a second-order denominator, and its performance can be designed by manipulating the poles of the closed-loop characteristic equation. For example, the controller gains can be selected to achieve no overshoot ($M_p = 0$ or $\zeta = 1$) and a settling time of $t_s = 46$ seconds. The program also generates for the fixed value of $T_I = (K_p/K_I) = 18.686$ a root-locus plot with K_P as a parameter (see Figure 12.3). Finally, for the selected

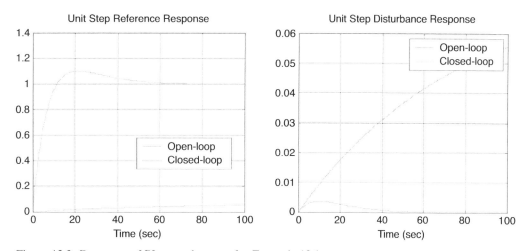

Figure 12.3. Response of PI control system for Example 12.1.

$K_P = 186.86$ and $K_I = 10$ values, the closed-loop system performance is simulated for step changes in reference speed and disturbance inputs, as shown
in Figure 12.3. For these selected gains, the closed-loop system exhibits zero
steady-state error and a settling time of approximately 45 seconds, as expected.
However, despite the value of $\zeta = 1$ the response exhibits some overshoot. This
is because the settling time and overshoot criteria used here are for second-order
systems with no zeros; the closed-loop system is second-order but has a zero at
-0.0535. If the response of the closed-loop system was determined to be not fast
enough, a smaller T_I could be selected. For example, the step response of the
system with $T_I = 9.34$ and $K_P = 401$ shows a 50 percent reduction in rising and
settling times.

```
% Ex12_1.m
% Define the linearized model parameters
u0=20; g=9.81; m=1000; f=0.015; Theta=0;
rho=1.202; A=1; Cd=0.5; uw=2;
% Calculate the equilibrium force, Fx0:
Fx0 = m*g*sin(Theta) + f*m*g + ...
              0.5*rho*A*Cd*(u0+uw)^2;
% The time constant and dc gain are:
Tau=(m/(rho*A*Cd*(u0+uw))); K=Tau/m;
% Fix the Ti gain (Ki=Kp/Ti) for the PI control:
Ti=186.86/10.0;
% The loop transfer function for PI control is
% Kp*(num/den) where num/dec = (1+1/Ti*s)*K/(1+Tau*s)
num_o=K*[Ti 1]; den_o=[Ti*Tau Ti 0];
% Calculate open loop zeros and poles:
z=roots(num_o);
p=roots(den_o);
% Obtain the root locus for Kp:
K_root=[0:10:300];
R=rlocus(num_o,den_o,K_root);
plot(R,'*'); title('RootLocus Plot');
axis([-0.2 0 -0.05 0.05]), pause;
[K_root, POLES]=rlocfind(num_o,den_o)

% Simulate the reference unit step response
% with and w/o control:
t=[0:0.5:100];
sys = tf([K],[Tau 1]);
y=step(sys,t);
Kp=186.86; KI = Kp/Ti;
numc=K*[Kp KI]; denc=[Tau K*Kp+1 K*KI];
% Calculate the closed-loop zeros and poles:
zc=roots(numc)
pc=roots(denc)
sys_c = tf(numc, denc);
yc=step(sys_c, t);
```

```
plot(t,y,'-r',t,yc,'--b'); grid;
xlabel('Time (sec)')
title('Unit Step Reference Response');
legend('Open-loop', 'Closed-loop'); pause;
% Simulate the disturbance unit step response
% with and w/o control:
yd=step(sys,t);
sys_disturbance = tf([K 0], denc);
ycd=step(sys_disturbance,t);
plot(t,yd,'-r',t,ycd,'--b'); xlabel('Time (sec)'),grid;
title('Unit Step Disturbance Response');
legend('Open-loop', 'Closed-loop');
```

The simple model used for the cruise-control design in Example 12.1 has several shortcomings. First, it was assumed that the tractive force can be changed instantaneously by the controller – that is, the throttle/engine system dynamics has been neglected. Second, the controller design is based on linearized equations of longitudinal motion, and the time constant and gain are assumed to be constant. It reality, the vehicle-forward dynamics are nonlinear and the corresponding linear dynamics vary depending on operating conditions. Both issues have been addressed by various studies in the literature (e.g., Liubakka et al. 1991; Oda et al. 1991; and Tsujii et al. 1990). Example 12.2 illustrates the design of a PID cruise controller based on a more complex but still linear model, which accounts for the throttle/engine dynamics. It also has been proposed to develop adaptive versions of the cruise controller to estimate the actual parameters K and τ under various operating conditions. An adaptive version of the PI cruise-control system is discussed in Example 12.3.

The closed-loop transfer function of the vehicle shown in Figure 12.4 is:

$$u'(s) = \frac{KK_a(K_Ds^2 + K_Ps + K_I)}{s^2(\tau s + 1)(\tau_a s + 1) + KK_a(K_Ds^2 + K_Ps + K_I)}r$$
$$+ \frac{Ks^2(\tau_a s + 1)}{s^2(\tau s + 1)(\tau_a s + 1) + KK_a(K_Ds^2 + K_Ps + K_I)}d$$

EXAMPLE 12.2: PID CRUISE-CONTROLLER DESIGN (SECOND-ORDER ACTUATOR). The process model in Eq. (12.3) neglects any dynamics associated with the actuator, which adjusts the throttle angle. In Tsujii et al. (1990), a *dc* motor throttle actuator model is given as:

$$G_a(s) = \frac{K_a}{s(\tau_a s + 1)}$$

where the input is the motor-drive duty cycle (percent) and the output is the tractive force. Parameter values used here are the same as those in Example 12.1, plus

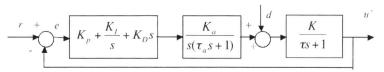

Figure 12.4. Closed-loop system with PID controller and throttle dynamics.

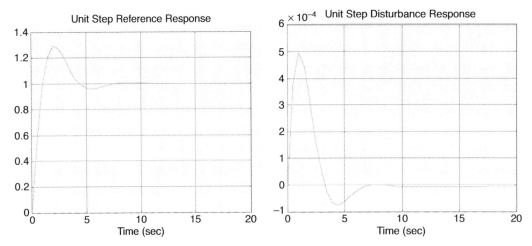

Figure 12.5. Response of PID control system for Example 12.2.

$K_a = 10$ (Newton-sec/%) and $\tau_a = 0.05$ seconds. Clearly, the actuator dynamics usually is much faster than the vehicle dynamics (i.e., $\tau_a \ll \tau$), and it is reasonable to neglect actuator dynamics as in Example 12.1. However, in this example, the throttle dynamics is considered and the controller is a PID controller rather than the PI controller used in Example 12.1. The closed-loop block diagram is shown in Figure 12.4. With the controller gains, $K_P = 120$, $K_I = 12$, and $K_D = 120$, the closed-loop-system block diagram and responses to step reference and disturbance inputs are illustrated in Figure 12.5. The response is faster than the PI controller shown in Example 12.1 but has a larger overshoot. The P gain is selected for a damping ratio of approximately 0.707.

```
% Ex12_2.m
% Define the linearized model parameters
clear
u0=20; g=9.81; m=1000; f=0.015; Theta=0;
rho=1.202; A=1; Cd=0.5; uw=2;
% Calculate the equilibrium force, Fx0:
Fx0 = m*g*sin(Theta) + f*m*g + ...
           0.5*rho*A*Cd*(u0+uw)^2;
% Time constant & dc gain are then:
Tau=(m/(rho*A*Cd*(u0+uw))); K=Tau/m;
% The throttle actuator is first-order with integrator:
Ka=10; TauA=0.05;
% Fix the Ti gain (Ki=Kp/Ti) for the PID control:
Ti = 100.0;
% Fix the Td gain (Td=Kd/Kp) for PID control:
Td = 1.0;
% The open loop transfer function for PID control is
% Kp*(num/den) where:
num_o=K*Ka*[Td 1 1/Ti];
den_o=[Tau*TauA Tau+TauA 1 0 0];
```

```
% Calculate and display open loop zeros and poles:
z=roots(num_o)
p=roots(den_o)
% Obtain the root locus for Kp:
K_locus=0:5:160;
R=rlocus(num_o,den_o,K_locus);
plot(R,'*'); title('Root locus'), pause
% Simulate the unit step response with PID control:
t=[0:0.5:20];
Kp=120.0;
numc=Kp*num_o;
denc=den_o+[0 0 numc];
% Calculate the closed-loop zeros and poles:
zc=roots(numc)
pc=roots(denc)
sys_c = tf(numc, denc);
yc=step(sys_c, t);
plot(t,yc,'r'); xlabel('Time (sec)'), grid;
title('Unit Step Reference Response'); pause;
% Simulate disturbance unit step
% response with and w/o control:
sys_disturbance = tf([K*TauA K 0 0], denc);
ycd=step(sys_disturbance,t);
plot(t,ycd,'--b'); xlabel('Time (sec)'),grid;
title('Unit Step Disturbance Response');
```

EXAMPLE 12.3: ADAPTIVE PI CRUISE-CONTROLLER DESIGN. As mentioned previously, the gain, K, and time constant, τ, vary with operating conditions (e.g., mass, speed, and wind velocity). In this example, the model in Eq. (12.3) is used (i.e., actuator dynamics is neglected) and an estimation algorithm is developed to identify the parameters. In other words, the parameters K and τ are estimated on-line from the measured signals. The controller gains then are updated by using the estimated parameters. This "adaptive" cruise-control (ACC) system structure is illustrated in Figure 12.6, which consists of two major

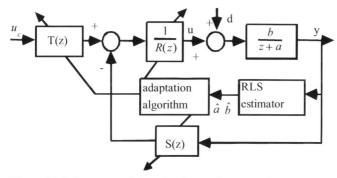

Figure 12.6. Structure of the adaptive cruise control system.

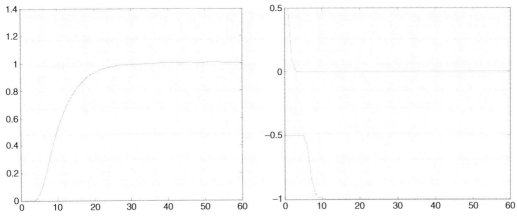

Figure 12.7. Adaptive PI control for Example 12.3.

parts: (1) a recursive least squares (RLS) parameter-estimation algorithm; and (2) an indirect (or self-tuning) PI controller strategy. The MATLAB program shown here is used to generate the results shown in Figure 12.7. The implementation is digital, so a discrete-time process model is used, obtained for a sampling rate of 1 second. The parameter estimates converge quickly (i.e., in about 5 to 10 seconds) and the response is comparable to that shown in Figure 12.2 (somewhat slower and with more overshoot), despite the fact that the unknown parameters are initialized to zero. In the MATLAB program, the RLS estimation algorithm is used to estimate the plant parameters. The basic equations of the RLS algorithm are summarized as follows:

$$y(k) = \phi^T(k)\theta$$
$$\hat{y}(k) = \phi^T(k)\hat{\theta}$$
$$\hat{\theta}(k+1) = \hat{\theta}(k) + K(k)[y(k+1) - \phi^T(k+1)\hat{\theta}]$$
$$P(k+1) = [I - K(k)\phi^T(k+1)]P(k)/\lambda$$
$$K(k) = P(k)\phi(k+1)[\lambda + \phi^T(k+1)P(k)\phi(k+1)]^{-1}$$

In this example,

$$y(k) = -ay(k-1) + bu(k-1)$$

Therefore,

$$\phi^T(k) = [-y(k-1)\,u(k-1)] \qquad \hat{\theta} = [\hat{a}\ \ \hat{b}]^T$$

```
% Ex12_3.m
  K = 0.0756; Tau = 75.632;
  A=-(1/Tau); B=(K/Tau); T=1.0;
  [F,G]=c2d(A,B,T);
  C=1; D=0; [num,den]=ss2tf(F,G,C,D,1);
  b=num(2)/den(1); a=den(2)/den(1);
% initialization:
```

```
   kmax=60; n=2; lambda=0.995;
   theta = [-0.5;0.5]; P = 1000*eye(n);
   u_old = 0.0; y_old = 0.0; uc_old = 1.0;
   phi = [-y_old; u_old];
   uc=1.0+0.01*randn(1);
   Y = zeros(kmax+1,n+3);
   Y(1,:)=[u_old uc y_old theta'];
% System with a ''PI'' controller
% designed for given z, wn
% and parameter estimation using RLS:
   z=0.95; wn=0.1;
% Controller Ru = Tuc - Sy
for k=1:kmax,
   uc = 1.0 + 0.02*randn(1);
   y = - a*y_old + b*u_old;
   [theta,P] = rls(y,theta,phi,P,lambda);
   a1 = theta(1); b1=theta(2);
   % we use estimated values
   % (a1, b1) of the
   % process for the controller
   % gains and control:
   s0 = (exp(-2*z*wn*T)+a1)/b1;
   s1 = (1-a1-2*cos(wn*T* ...
       (1-z^2)^0.5)*exp(-z*wn*T))/b1;
   t0 = s0 + s1;
   u = u_old + t0*uc_old - ...
       s0*y_old -s1*y;
   % update the measurement vector phi
   % for next iteration:
   phi = [-y;u];
   u_old = u; y_old = y; uc_old = uc;
   Y(k+1,:)=[u uc y a1 b1];
end
% Display the results:
k=[0:kmax]'; u_uc_y_ai_bi = Y;
plot(k,Y(:,3),'r'); title('Response');
xlabel('Sampling Interval, k'); ylabel('y'); pause;
plot(k,Y(:,4:n+3)); title('Parameter Estimates');
xlabel('Sampling Interval, k'); ylabel('Estimated a & b');
% function saved in
% rls.m
function [theta, P] = rls(y, theta,
                phi, P, lambda)
K=P*phi*inv(lambda+phi'*P*phi);
theta=theta+K*(y-phi'*theta);
P=(eye(size(P))-K*phi')*P/lambda;
```

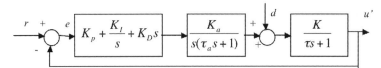

Figure 12.8. Block diagram of the adaptive cruise control system.

EXAMPLE 12.4: ADAPTIVE PID CRUISE-CONTROLLER DESIGN. In this example, we revisit Example 12.2, assuming that some kind of parameter-identification method has been implemented and that we have access to accurate estimation of vehicle parameters (i.e., τ and K). How then can we take full advantage of the three design variables (i.e., the PID gains: K_p, K_I, and K_D) for superior performance? In previous examples, we used the root-locus technique, which helps to select one control parameter; the other two parameters were specified in advance. Thus, a form of trial and error or engineering judgment is needed to find a set of good control parameters. In this design, we show an alternative design method based on pole placement. First, we derive the closed-loop transfer function:

$$\frac{u'}{r} = \frac{K_D K_a K s^2 + K_P K_a K s + K_I K_a K}{\tau \tau_a s^4 + (\tau + \tau_a) s^3 + (1 + K_D K_a K) s^2 + K_P K_a K s + K_I K_a K}$$

The closed-loop poles are the roots of the denominator polynomial (i.e., the characteristic equation), which can be rewritten in the following monic form:

$$s^4 + \frac{\tau + \tau_a}{\tau \tau_a} s^3 + \frac{1 + K_D K_a K}{\tau \tau_a} s^2 + \frac{K_P K_a K}{\tau \tau_a} s + \frac{K_I K_a K}{\tau \tau_a} = 0$$

Because we only have three design variables, we can place only three of the four closed-loop poles. The remaining pole then is determined by the fixed coefficient $(\frac{\tau + \tau_a}{\tau \tau_a})$ of the closed-loop characteristic equation. Notice that the actuator-time constant is much faster than the vehicle-time constant (i.e., 0.2 versus 55~140). If we select three slow (i.e., dominant) closed-loop poles, the last pole, which may move a little when τ changes, will be far from the imaginary axis and thus does not noticeably affect the system performance. Based on the damping ratio and bandwidth of the complex conjugate pairs, we determine that the three poles should locate at $-0.3 \pm 0.1 j$ and -0.4. The closed-loop characteristic equation then becomes:

$$(s^2 + 0.6s + 0.1)(s + 0.4)\left(s + \frac{\tau + \tau_a}{\tau \tau_a} - 1\right)$$

$$= s^4 + \frac{\tau + \tau_a}{\tau \tau_a} s^3 + \left(\frac{\tau + \tau_a}{\tau \tau_a} - 0.66\right) s^2 + \left(0.34 \frac{\tau + \tau_a}{\tau \tau_a} - 0.3\right) s$$

$$+ 0.04\left(\frac{\tau + \tau_a}{\tau \tau_a} - 1\right)$$

The PID gains to achieve these pole assignments are then:

$$K_D = \frac{\tau + \tau_a - 0.66 \tau \tau_a - 1}{K_a K} \qquad K_P = \frac{0.34(\tau + \tau_a) - 0.3 \tau \tau_a}{K_a K}$$

$$K_I = \frac{0.04(\tau + \tau_a) - 0.04 \tau \tau_a}{K_a K}$$

The unplaced pole will be located at $-\frac{1}{\tau_a} - \frac{1}{\tau} + 1$, which is about 10 times faster than the three slower poles. In other words, the performance of the adaptive controller is not affected much even when the vehicle-operation conditions vary. In the following example program, we assume that the vehicle is accelerating from 15 to 30 m/sec in 30 seconds and a 2 percent road gradient also is assumed to be present. It can be seen that the adaptive PID control algorithm is tracking the desired speed command with small error under these two uncertainties.

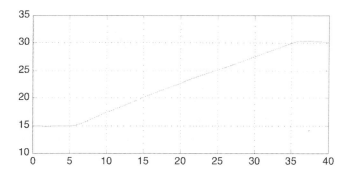

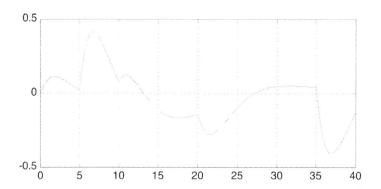

In this MATLAB program, we used the Euler approximation for integration. This may introduce some error but it is more flexible because all of the intermediate variables (e.g., engine force F and command to actuator c) can be examined. Sometimes it is desirable to keep track of these intermediate variables. For example, the force (F) in the following figure shows that the controller design is reasonable because the force command is not excessively large:

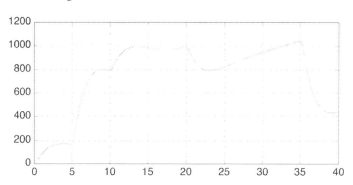

```
% Ex12_4.m
g=9.81; m=1000; f=0.015; Theta=0;
rho=1.202; A=1; Cd=0.5; uw=0;
Ka=10; TauA=0.2; T=0.01; u(1)=15.0;
error=0.0; error_old=0.0; int_error=0.0;
F_dot=0.0; F=0.0; t=0:T:40;
r=[15*ones(1,5/T+1),15+(1:30/T)*0.5*T,30*ones(1,5/T)];
d=[zeros(1,10/T+1), -0.02*m*g*ones(1,10/T), zeros(1,20/T)];
for i=1:40/T,
    Tau=(m/(rho*A*Cd*(u(i)+uw))); K=Tau/m;
    Kd=(TauA+Tau-0.66*TauA*Tau-1)/(Ka*K);
    Kp=(0.34*(TauA+Tau)-0.3*TauA*Tau)/(Ka*K);
    Ki=(0.04*(TauA+Tau)-0.04*TauA*Tau)/(Ka*K);
    error_old=error;
    error=r(i)-u(i);
    int_error=int_error+error*T;
    c=Kd*(error-error_old)/T+Kp*error+Ki*int_error;
    F_dotdot=(Ka*c-F_dot)/TauA;
    F_dot=F_dot+F_dotdot*T;
    F=F+F_dot*T;
    u_dot=(K*(F+d(i))-u(i))/Tau;
    u(i+1)=u(i)+u_dot*T;
end
plot(t(1:40/T),u(1:40/T),'r'), grid
xlabel('time (sec)'); title('True speed (m/sec)')
pause; plot(t(1:40/T),r(1:40/T)-
u(1:40/T),'r'); grid
xlabel('time (sec)'); title('Speed error (m/sec)')
```

12.2 Autonomous Cruise Control: Speed and Headway Control

A vehicle cruise-control system regulates vehicle longitudinal velocity. As an inter-mediate step between platooning (see Chapter 19) and cruise control, the control of the vehicle longitudinal motion (i.e., velocity and position) is considered. The idea is to allow the vehicle to regulate speed when there are no automobiles in front of it. However, when a vehicle is detected in front, then the controller regulates relative distance between the two of them (i.e., "headway"). Such a system (i.e., ACC or intelligent cruise control) can be effective in stop-and-go traffic and on the highway and can be used in conjunction with the auto-cruise system described in Chapter 19. The vehicle cruise-control system uses the desired speed set by the driver (e.g., the speed limit) as a reference input when there is no vehicle in sight. When a vehicle is detected in front, the system switches to the headway-control mode until the lead vehicle exceeds the reference speed or vanishes from sight, when it then switches back to speed control. In this scenario, the driver does not need to be concerned about longitudinal-motion control under conditions ranging from stop-and-go traffic to highway driving. A formulation of the control-system design problem for ACC is presented in the following example.

Figure 12.9. Headway control problem.

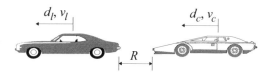

EXAMPLE 12.5: AUTONOMOUS CRUISE CONTROL SYSTEM DESIGN. To differentiate from the adaptive cruise control problem presented in the previous section, we name the range-regulation enhanced cruise-control system autonomous cruise control (ACC). In the literature, this system is commonly referred as intelligent cruise control (ICC), adaptive cruise control (ACC), or autonomous intelligent cruise control (AICC). When no lead vehicle is in sight, the performance of an ACC vehicle is similar to a traditional cruise-control vehicle. When a lead vehicle is present, we must model the interaction with the lead vehicle and keep track of the range variable, which is one of the most important performance variables.

It is assumed that the measurement of the vehicle longitudinal velocity (v_c) as well as of the distance (R) between the first vehicle (i.e., controlled vehicle) and the vehicle in front (i.e., lead vehicle) are available. This is illustrated in Figure 12.9. Using a linearized description of the longitudinal velocity of each vehicle, as discussed in a previous section, we can obtain the following state equations:

$$\dot{x}_1 = x_3 - x_2$$
$$\dot{x}_2 = -(1/\tau_c)x_2 + (K_c/\tau_c)u + (K_c/\tau_c)w_c$$
$$\dot{x}_3 = -(1/\tau_l)x_3 + (K_l/\tau_l)w_l$$

where $x_1 = R$, $x_2 = v_c$, $x_3 = v_l$ and u is the control input (i.e., traction force) applied to the controlled vehicle (i.e., Vehicle c). The inputs w_c and w_l represent disturbances (e.g., grade disturbances for both vehicles and driver inputs for the lead vehicle). The open-loop-system block diagram is shown in Figure 12.10a, and an equivalent block diagram is shown in Figure 12.10b. The two block diagrams result in a different number of state variables because of the difference in the selection of disturbance inputs and the fact that the vehicle displacements are ignored in the second representation. When an ACC algorithm is implemented using differential GPS, for example, then it may be interesting to keep vehicle displacements as state variables. That extension is readily achieved by simply integrating the vehicle speeds.

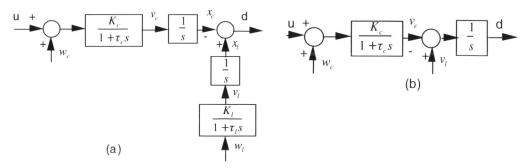

Figure 12.10. Open-loop block diagram for ACC.

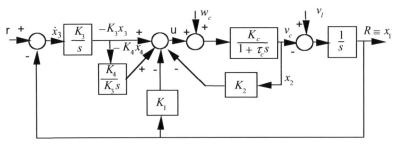

Figure 12.11. Closed-loop block diagram of the ACC system.

Consider the design of a controller to maintain a desired headway, r, based on measurements of R and v_c only. This requires only a range sensor mounted on the controlled vehicle, in addition to measurement of the controlled variable (i.e., vehicle longitudinal speed) and no information from other vehicles or from the roadway infrastructure. The control problem is illustrated in the block diagram shown in Figure 12.11. The reference input, r, should be computed as a function of the vehicle forward velocity to maintain safe headway (e.g., $r = t_h v_c$, where t_h is the desired headway time). Also, the velocity of the lead vehicle (i.e., Vehicle l) is not known and acts as a disturbance input. The control, u, on Vehicle c must be calculated to maintain the desired headway. A feedback controller can be designed for this system based on the following state equations:

$$\dot{x}_1 = -x_2 + v_l$$
$$\dot{x}_2 = -(1/\tau_c)x_2 + (K_c/\tau_c)(u + w_c)$$
$$\dot{x}_3 = (R - r) = (x_1 - r)$$
$$\dot{x}_4 = x_3$$
$$u = -k_1 x_1 - k_2 x_2 - k_3 x_3 - k_4 x_4$$

This controller feeds back the measured values $x_1 = R$ and $x_2 = v_c$; x_3, the integral of the error, $e = d - r$; and x_4, the double integral of the error. The two integrals are used because it is assumed that the lead vehicle velocity (which acts as a disturbance to be rejected by the control law) can be modeled as a ramp-type disturbance input. With the double integrator built into the control algorithm, the controlled vehicle can follow the lead vehicle even when it accelerates. The controller gains, k_i, can be determined by a variety of methods, and a pole-placement design technique is used in the MATLAB program. The closed-loop poles are placed at $s_{1,2} = -\zeta\omega_n \pm \omega_n\sqrt{1 - \zeta^2}j$, and $s_{3,4} = -\alpha\zeta\omega_n$. The values $\zeta = 0.9$, $\omega_n = 0.4$, and $\alpha = 3.0$ are used in the MATLAB simulations for Example 12.5.

In this setup, the lead-vehicle speed (v_l) is treated as a disturbance to be rejected. It may be desired to treat w_l (instead of v_l) as the unknown disturbance and include v_l as a state variable (x_5), which can be used for the control/warning purposes. In this case, the dynamic equations are as follows:

$$\dot{x}_1 = x_5 - x_2$$
$$\dot{x}_2 = -(1/\tau_c)x_2 + (K_c/\tau_c)u + (K_c/\tau_c)w_c$$
$$\dot{x}_3 = (d - r) = (x_1 - r)$$

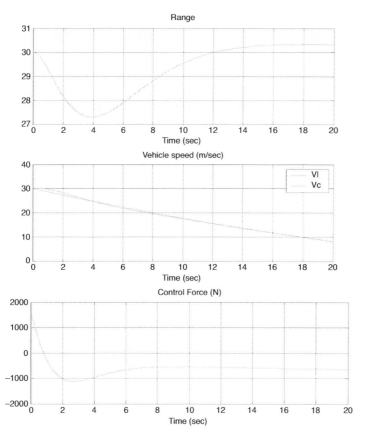

Figure 12.12. Simulation results of the ACC system.

$$\dot{x}_4 = x_3$$
$$\dot{x}_5 = -(1/\tau_c)x_5 + (K_c/\tau_c)w_l$$
$$u = -k_1x_1 - k_2x_2 - k_3x_3 - k_4x_4$$

which, for the closed-loop system, can be written in state-space form as:

$$\mathbf{x} = \begin{bmatrix} 0 & -1 & 0 & 0 & 1 \\ -k_1K_c/\tau_c & -(1+k_2K_c)/\tau_c & -k_3K_c/\tau_c & -k_4K_c/\tau_c & 0 \\ 1 & 0 & 0 & 0 & 0 \\ 0 & 0 & 1 & 0 & 0 \\ 0 & 0 & 0 & 0 & -1/\tau_l \end{bmatrix} \mathbf{x}$$

$$+ \begin{Bmatrix} 0 \\ 0 \\ -1 \\ 0 \\ 0 \end{Bmatrix} r + \begin{Bmatrix} 0 \\ K_c/\tau_c \\ 0 \\ 0 \\ 0 \end{Bmatrix} w_c + \begin{Bmatrix} 0 \\ 0 \\ 0 \\ 0 \\ K_l/\tau_l \end{Bmatrix} w_l$$

The simulation results are shown in Figure 12.12 for a moderate deceleration case. The controller maintains the desired range (i.e., 30 meters), despite changes in the lead-vehicle velocity.

```
% Ex12_5.m
g=9.81; uw=0.0; u0=30.0; rho =1.202;
Theta=0.0; ThetaPrime=0.0;
% Controlled vehicle parameters:
mc=1000.0; Cdc=0.5; Arc=1.5; fc=0.015;
Kc=(1/(rho*Cdc*Arc*(u0+uw)));Tc=mc*Kc;
wc0=mc*g*(fc*sin(Theta)-cos(Theta))*ThetaPrime; Fc=(u0/Kc);
% Lead vehicle parameters (typical):
ml=1500.0;Cdl=0.6;Arl=1.95;fl=0.015;
Kl=(1/(rho*Cdl*Arl*(u0+uw)));Tl=ml*Kl;
Fl=(u0/Kl);
wl0=(ml*g*(fl*sin(Theta)-cos(Theta))*ThetaPrime);
t=[0:0.1:20]';
U0=Fc*ones(size(t));          % Nominal control force
wl= -800*(1+0.01*t);          % Ramp function
wc=wc0*ones(size(t));
disturbance=[U0 wc wl];
% 4-state system for controller-design:
Aa=[0 -1 0 0;
    0 -1/Tc 0 0;
    1 0 0 0;
    0 0 1 0];
Ba=[0;Kc/Tc;0;0];
% Controller design:
pc=[roots([1 2*0.9*0.4 0.4^2]); -1.08; -1.18];
K=place(Aa,Ba,pc);
% Closed-loop simulation (5 states, keep track of vl (x5)):
Ac=[0 -1 0 0 1;
    -K(1)*Kc/Tc -(1+K(2)*Kc)/Tc -K(3)*Kc/Tc -K(4)*Kc/Tc 0;
    1 0 0 0 0;
    0 0 1 0 0;
    0 0 0 0 -1/Tl];
Bc=[0 0 0;
    0 Kc/Tc 0;
    -1 0 0;
    0 0 0;
    0 0 Kl/Tl];
% outputs: x1 (range), vc and vl
Cc=[1 0 0 0 0;0 1 0 0 0; 0 0 0 0 1]; Dc=zeros(3,3);
r=30.0*ones(size(t));
disturbance=[r U0+wc wl]; xc0=[30 u0 0 -
(u0/Kc+K(1)*30+K(2)*u0)/K(4) u0];
[yc,xc]=lsim(Ac,Bc,Cc,Dc,disturbance,t,xc0);
subplot(211), plot(t,yc(:,1)); title('Range');
xlabel('Time (sec)'); grid;
subplot(212), plot(t,yc(:,3), 'r',t,yc(:,2),'-.b');
title('Vehicle speed (m/sec)');
```

```
xlabel('Time (sec)'); grid;
legend('Vl', 'Vc'); pause;
clf, subplot(211)
u= U0-K(1)*xc(:,1)-K(2)*xc(:,2)-K(3)*xc(:,3)-K(4)*xc(:,4);
plot(t, u'); title('Control Force (N)');
xlabel('Time (sec)'); grid
```

PROBLEMS

1. Grades on interstate highways usually are limited to about 4 percent or less (i.e., rise over run, or the $\tan\theta$ is about 0.04). On major roads, grades may occasionally reach 10 to 20 percent. In the simulation in Example 12.1, a unit-step disturbance input was used and caused a transient change in the speed with the PI control. Rerun the simulation in Example 12.1 for the PI cruise controller with grade-disturbance inputs of (a) 15 percent, and (b) −15 percent. Use the following general expression for the grade-disturbance input:

$$d = mg(f \sin\theta_0 - \cos\theta_0)\theta'$$

Use the same parameter values as in Example 12.1. Are the observed deviations from the desired speed "acceptable" for this range of extreme disturbance inputs?

2. In this problem, carry out several calculations to assess whether an adaptive version of the fixed-gain PI controller in Example 12.1 is desirable. Based on the results of your calculations in Parts (a) and/or (b), do you recommend the design of an adaptive PI cruise controller?

    (a) Consider the fixed-gain PI cruise controller in Example 12.1. This was designed for $m = 1{,}000$ kg, $u_w = 2$ m/s, and $u_o = 20$ m/s. Apply this controller to the system with $m = 1{,}250$ kg, $u_w = 5$ m/s, and $u_o = 30$ m/s. Compared to the performance specifications used in Example 12.1, how does this controller perform?

    (b) For cruise control (see Example 12.1), linearization was used to obtain the following time constant and gain:

$$\tau = (m/(\rho C_d A(u_o + u_w)));$$
$$K = (1/(\rho C_d A(u_o + u_w)));$$

For the nominal values given in this chapter and following Eq. (12.2), plot the values of τ and K for each of the following cases:

(1) $m = [1{,}000{:}1{,}250]$ (i.e., 25 percent variation in m from curb weight to loaded weight)

(2) $u_w = [-10{:}10]$ (i.e., ±10m/s [about ±22 mph] head/tail wind)

(3) $u_o = [20{:}30]$ (i.e., 10m/s change in forward velocity)

3. Example 12.5 considers an ACC to maintain a desired headway based on measurements of d and v_c only (i.e., no inputs from other vehicles or from the roadway infrastructure). The control problem is illustrated in block-diagram form in Figure 12.11. Note that the reference input, r, would be computed as a function of the vehicle forward velocity to maintain safe headway. Also note that the velocity, v_l, of the lead vehicle is not known and acts as a disturbance input. The control u on Vehicle c must be calculated to maintain the desired headway. A feedback controller is designed for this system based on the following state equations:

$$\dot{x}_1 = -x_2 + v_l$$
$$\dot{x}_2 = -(1/\tau_c)x_2 + (K_c/\tau_c)(u + w_c)$$
$$\dot{x}_3 = (d - r) = (x_1 - r)$$
$$\dot{x}_4 = x_3$$
$$u = -k_1 x_1 - k_2 x_2 - k_3 x_3 - k_4 x_4$$

This controller feeds back the measured values $x_1 = d$ and $x_2 = v_c$, as well as x_3, the integral of the error, $e = d - r$, and x_4, the double integral of the error. The two integrals of error are used because it is assumed that the lead-vehicle velocity, v_l, (which acts as a disturbance to be rejected by the control law) is well represented as a ramp-type function in time. The controller gains, k_i, can be determined by a variety of methods, and a pole-placement design technique is used in the MATLAB program in Example 12.5. The closed-loop poles are placed at $s_{1,2} = -\zeta\omega_n \pm \omega_n\sqrt{1 - \zeta^2}j$ and $s_{3,4} = -\alpha\zeta\omega_n$. The values $\zeta = 0.9$, $\omega_n = 0.4$, and $\alpha = 3.0$ are used in the MATLAB simulations in Example 12.5.

Repeat Example 12.5 but with a simpler control law that does not feed back the double integral of the error (i.e., x_4) and has the form $u = -k_1 x_1 - k_2 x_2 - k_3 x_3$. Compare the results for a ramp-disturbance input, v_l, to those obtained using the previous control law, $u = -k_1 x_1 - k_2 x_2 - k_3 x_3 - k_4 x_4$.

REFERENCES

Druzhinina, M., A. G. Stefanopoulou, and L. Moklegaard, 2002a, "Speed Gradient Approach to Longitudinal Control of Heavy-Duty Vehicles Equipped with Variable Compression Brake," *IEEE Transactions on Control System Technology*, Vol. 10, No. 2, March, pp. 209–21.

Druzhinina, M., A. G. Stefanopoulou, and L. Moklegaard, 2002b, "Adaptive Continuously Variable Compression Braking Control for Heavy-Duty Vehicles," *ASME Journal of Dynamic Systems, Measurement, and Control*, Vol. 124, No. 3, September.

Fancher, P., H. Peng, and Z. Bareket, 1996, "Comparison of Three Control Algorithms on Headway Control for Heavy Trucks," *Vehicle System Dynamics*, Vol. 25 Suppl., pp. 139–51.

Fancher, P., H. Peng, Z. Bareket, C. Assaf, and R. Ervin, 2001, "Evaluating the Influences of Adaptive Cruise Control Systems on the Longitudinal Dynamics of Strings of Highway Vehicles," *Proceedings of the 2001 IAVSD Conference*, Copenhagen, Denmark, August.

Himmelspach, T., and W. Ribbens, 1989, "Radar Sensor Issues for Automotive Headway Control Applications," *Fall 1989 Summary Report for the Special Topics Course in IVHS, Appendix G*, University of Michigan.

Ioannou, P. A., and C. C. Chien, 1993, "Autonomous Intelligent Cruise Control," *IEEE Transactions on Vehicular Technology*, Vol. 42, No. 4, pp. 657–72.

Klein, R. H., and J. R. Hogue, 1980, "Effects of Crosswinds on Vehicle Response – Full-Scale Tests and Analytical Predictions," *SAE Paper No. 800848*. Detroit, MI.

Liang, C., and H. Peng, 1999, "Optimal Adaptive Cruise Control with Guaranteed String Stability," *Vehicle System Dynamics*, Vol. 32, No. 4–5, November, pp. 313–30.

Liubakka, M. K., D. S. Rhode, J. R. Winkelman, and P. V. Kokotovic, 1993, "Adaptive Automotive Speed Control," *IEEE Transactions on Automatic Control*, Vol. 38, No. 7, July, pp. 1011–20.

Liubakka, M. K., J. R. Winkelman, and P. V. Kokotovic, 1991, "Adaptive Automotive Speed Control," *Proceedings of the American Control Conference*, Boston, MA, June, pp. 439–40.

Oda, K., H. Takeuchi, M. Tsujii, and M. Ohba, 1991, "Practical Estimator for Self-Tuning Automotive Cruise Control," *Proceedings of the American Control Conference*, Boston, MA, June, pp. 2066–71.

Okuno, A., A. Kutami, and K. Fujita, 1990, "Towards Autonomous Cruising on Highways," SAE Paper No. 901484.

Rajamani, R., and Chunyu Zhu, "Semi-Autonomous Adaptive Cruise Control Systems," 1999, *Proceedings of the American Control Conference*, Vol. 2, June, pp. 1491–5.

Swaroop, D., and K. R. Rajagopal, 1999, "Intelligent Cruise Control Systems and Traffic Flow Stability," *Transportation Research*, Part C, Vol. 7, pp. 329–52.

Tsujii, M., H. Takeuchi, K. Oda, and M. Ohba, 1990, "Application of Self-Tuning to Automotive Cruise Control," *Proceedings of the American Control Conference*, San Diego, CA, May, pp. 1843–8.

Xu, Z., and P. A. Ioannou, 1994, "Adaptive Throttle Control for Speed Tracking," *Vehicle System Dynamics*, Vol. 23, No. 4, May, pp. 293–306.

Yoshimoto, K., H. Tanabe, and M. Tanaka, 1994, "Speed Control Algorithm for an Automated Driving Vehicle," *Proceedings of the AVEC '94*, pp. 408–13.

13 Antilock Brake and Traction-Control Systems

Antilock brake systems (ABS) were first introduced on railcars at the beginning of the 20th century. The original motivation was to avoid flat spots on the steel wheels; however, it soon was noted that stopping distance also was reduced by the ABS. Robert Bosch received a patent for ABS in 1936. In 1948, a Boeing B-47 was equipped with ABS to test its effectiveness in avoiding tire blowout on dry concrete and spinouts on icy runways. It used a "bang-bang" (i.e., dump brake pressure to zero, then rebuild) control strategy. Fully modulating ABS control strategies were introduced in the 1950s (e.g., Ford Lincoln, Goodyear, and HydroAire). A rear-wheels-only ABS was first available in luxury automobiles in the late 1960s. The systems used in the 1960s and 1970s were developed by Bendix, Kelsey-Hayes, and AC Electronics, among others. Legal concerns then delayed further development in the United States, and European companies took the lead in the next two decades. Demand skyrocketed in the early 1990s when the benefits of ABS for vehicle-steering control and shorter stopping distances were recognized and accepted widely. Most new passenger vehicles sold in the United States today are equipped with ABS. It is important to note that ABS will not work properly if the user input or road condition varies quickly. For example, according to a recent test report by the NHTSA (Forkenbrock et al. 1999), all of the test vehicles equipped with ABS stop within a longer distance than those without ABS on loose-gravel roads. Therefore, improvements still are needed in this relatively mature technology.

In 1987, Bosch combined a traction-control function with its ABS system. A traction-control system (TCS), generally speaking, is the counterpart of ABS and works in the case of acceleration. Both systems aim not only to avoid wheel lockup or spin but also to maintain wheel slip near an optimal value, which achieves maximum possible tire longitudinal forces and, in the meantime, avoids significant reduction in tire lateral forces. TCS frequently is advertised as a device to help drivers pull off from icy patches or slopes; however, a complete TCS actually could work under all vehicle speeds. Unlike ABS, the major barrier to its acceptance is not legal issues but rather competition with existing systems that achieve similar functions (e.g., limited-slip differentials and 4WD). Because the TCS function can be achieved with minimal extra hardware requirements, and most of the work involves software modification of ABS and engine-control programs, it maintains a good competitive position with respect to those alternative systems.

ABS/TCS for 2-axle commercial vehicle with TCS brake controller

1 Wheel-speed sensor, 2 Pulse ring, 3 ABS/TCS electronic control unit, 4 Pressure-control valve,
5 Shuttle valve, 6 Service-brake valve, 7 Combination brake cylinder, 8 Air reservoir, 9 Drain valve,
10 Diaphragm actuator, 11 TCS solenoid valve, 12 Relay valve.

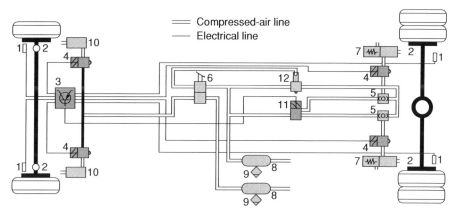

Figure 13.1. ABS/TCS system (Bosch Driving Safety Systems).

ABS, shown in Figure 13.1, are commercially available on many mass-market production vehicles, but traction control (i.e., anti-spin) systems are less widespread. The ABS modulate the brake force (i.e., pressure), much like a good driver does when braking on ice, to maintain braking forces near the peak of the $\mu - \lambda$ curve (Figure 13.2). In ABS, only the braking force is modulated by the controller, whereas in TCS (i.e., anti-spin), both the braking force and the traction force (through engine control) are manipulated. Wheel speeds and brake pressures are measured, vehicle speed and road friction are estimated, and brake pressure and engine torque are modulated to maintain a desired slip ratio.

As shown in Figure 13.2a, the lateral force on a wheel is a maximum near-zero slip, and there is "sufficient" lateral force when the traction force is at its maximum level, at about 10 to 30 percent slip. The lateral force goes to zero as the slip approaches 100 percent (i.e., the wheel is fully locked). This explains the loss of steering control or directional stability in the case of locked front or rear wheels, respectively. These $\mu - \lambda$ curves depend strongly on road-surface conditions, as

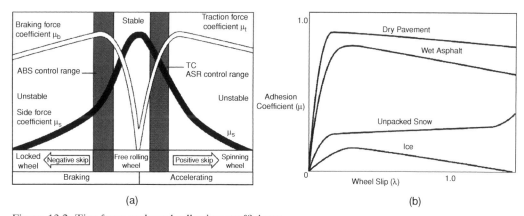

Figure 13.2. Tire-force and road-adhesion coefficients.

illustrated in Figure 13.2b. ABS attempt to maintain the braking force near the peak value. When a road surface changes, however, identifying even a proper target slip ratio can be a challenging task. The regulation of tire-slip ratio is accomplished by modulating the brake pressure in a fast apply–hold–release type of cyclic action. Because the road friction is unknown, the ABS cannot be guaranteed to achieve the exact peak braking-force value. Rather, it tries to maintain a slip ratio that averages around the target slip, which is expected to be close to where the peak force can be obtained.

When the target slip ratio is near the optimal slip (i.e., maximum force), it is clear from Figure 13.2 that ABS can maximize braking forces in the longitudinal direction (i.e., minimize stopping times and distances) while also maintaining large lateral forces for directional stability when steering in braking maneuvers. There are other potential benefits of ABS; for example, Chapter 4 discusses the fore–aft shift of normal forces on the tires due to acceleration or deceleration (see Example 4.2). Without an ABS, efficient braking distribution requires proportioning the brake pressure between the front and rear wheels. However, ABS can adjust continuously individual wheel-brake pressures to maintain each wheel near its optimal adhesion coefficient. This same principle applies not only for fore–aft load variations but also for load and adhesion (i.e., μ) variations between the right and left wheels.

13.1 Modeling

A mathematical model for traction-controller design is based on a one-wheel description of that rotational dynamics and a model (see Chapter 3) for the vehicle longitudinal dynamics. The application of Newton's second law to the rotational dynamics of a wheel gives:

$$\dot{\omega}_w = [(T_e - T_b - r_w F_x - T_w(\omega_w)]/J_w \tag{13.1}$$

where ω_w is the wheel rotational velocity, T_e is the engine torque, T_b is the braking torque, r_w is the wheel radius, F_x is the traction (or braking when negative) force, and $T_w(\omega_w)$ is a wheel-friction torque that is a function of the wheel rotational speed. In the following discussion, we assume dry-friction and viscous-friction models at the wheel; that is, $T_w(\omega_w) = f_w F_z + b_w \omega_w$. The effective wheel inertia, J_w, is the wheel inertia, I_w, in braking. During acceleration, $J_w = I_w + r_g^2 I_e$, where I_e is the engine inertia and r_g is the gear ratio.

The wheel slip, which is due to deformation of the tread on the pneumatic tire (Figure 13.3) during acceleration, is defined as:

$$\lambda = \frac{\omega_w - \omega_v}{\omega_w} \quad \text{when accelerating } (\omega_w > \omega_v) \tag{13.2}$$

and during deceleration as:

$$\lambda = \frac{\omega_w - \omega_v}{\omega_v} \quad \text{when decelerating } (\omega_w > \omega_v) \tag{13.3}$$

where $\omega_v = u/r_w$ and u is the vehicle longitudinal velocity. The vehicle longitudinal dynamics is governed by Eq. (4.4), which is repeated here:

$$\dot{u} = [-0.5\rho C_d A(u + u_w)^2 - fmg\cos\Theta - mg\sin\Theta + N_w F_x]/m \tag{13.4}$$

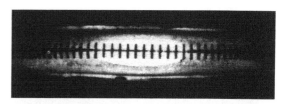

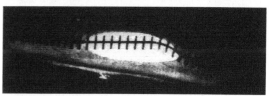

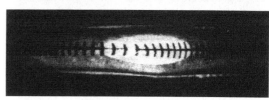

Figure 13.3. Slip in a pneumatic tire. Top to bottom: free rolling, turning, braking, and accelerating.

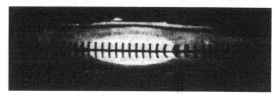

The tire traction force is defined by:

$$F_x = \mu F_z \qquad (13.5)$$

where the surface-adhesion coefficient, μ, is a function of λ, as illustrated in Figure 13.2, and F_z is the normal force at each tire. In this book, we use positive F_z, which is less confusing. Therefore, $\mu > 0$ for acceleration and $\mu < 0$ for deceleration (i.e., braking) for this sign convention. The number of drive wheels (for acceleration) is N_w; or, in the case of braking, N_w is the total number of wheels.

It is now convenient to rewrite these equations in standard state-variable form. Define $x_1 = \omega_v$ and $x_2 = \omega_w$ to rewrite the equations of motion as:

$$\dot{x}_1 = [-(0.5\rho C_d A)(r_w x_1 + u_w)^2 - fmg\cos\Theta - mg\sin\Theta + N_w F_z \mu(\lambda)]/(mr_w)$$
$$\dot{x}_2 = [-f_w F_z - b_w x_2 - F_z r_w \mu(\lambda) + T]/J_w \qquad (13.6)$$

where $T = (T_e - T_b)$ and $\lambda = (x_2 - x_1)/x_1$ for braking $= (x_2 - x_1)/x_2$ for accelerating. Equation (13.6) is highly nonlinear, primarily due to the $\mu - \lambda$ relationship but also to the friction terms in Eqs. (13.1) and (13.4). In the following discussion, it is assumed that the wind velocity, $u_w = 0$, and the road-grade angle, $\Theta = 0$.

For simplicity, we assume that the actuator dynamics can be neglected. In actual systems, of course, this may not be a good assumption. The brake actuator usually needs to incorporate a time delay (on the order of 100 ms) plus a first-order time-lag

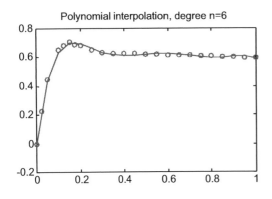

Figure 13.4. Polynomial curve fitting for tire $\lambda - \mu$ curve.

system. The throttle actuator typically is faster. However, it is connected in series with the engine, which requires a nonlinear manifold dynamics in the form of:

$$\frac{dm_{ai}(\omega_e, P_m)}{dt} = \beta \cdot f\left(\frac{P_m}{P_a}\right) \cdot TC(\alpha)$$

where:

m_{ai} = air mass flow rate	ω_e = engine speed
P_m = intake manifold pressure	β = maximum airflow rate
f = pressure-ratio influence function	P_a = atmospheric pressure
α = throttle angle	TC = throttle-valve characteristic function

and the engine-torque generation introduces a non-negligible lag. In this book, however, we ignore these actuator dynamics.

13.2 Antilock Braking Systems

For the ABS, $\omega_w < \omega_v$, so $\lambda = \frac{\omega_w - \omega_v}{\omega_v} < 0$. Example 13.1 illustrates the polynomial fit for the $\lambda - \mu$ curves, Example 13.2 introduces the so-called Magic Formula for nonlinear tire modeling, Example 13.3 gives the nonlinear simulation for braking, and Example 13.4 provides a linear ABS controller.

EXAMPLE 13.1: CURVE FITTING OF A LAMBDA-MU CURVE. In this MATLAB program, a polynomial fit is obtained for a set of $\lambda - \mu$ data given at the beginning of the example program. The polynomial curve is represented in the following form:

$$\mu = \sum_{i=1}^{n} c_i \lambda^{(n+1-i)}$$

where the coefficient values, c_i, are to be identified. The results of this polynomial fit are shown in Figure 13.4, which seems to work reasonably well. Other functional forms (i.e., nonpolynomial) could be used to represent the $\lambda - \mu$ curve. For example, the Magic Formula proposed by Pacejka and co-workers (1987, 1989) could be used to obtain a smooth fit. The Magic Formula is used widely in the automotive industry (see Example 13.2).

```
% Ex13_1.m The mu-lambda curve
% Define the mu-lambda data:
Data = ...
  [ 0.0000    0.0000
    0.0250    0.2250
    0.0500    0.4500
    0.1000    0.6500
    0.1250    0.6850
    0.1500    0.7050
    0.1750    0.6900
    0.2000    0.6800
    0.2500    0.6500
    0.3000    0.6350
    0.3500    0.6300
    0.4000    0.6275
    0.4500    0.6250
    0.5000    0.6225
    0.5500    0.6200
    0.6000    0.6175
    0.6500    0.6150
    0.7000    0.6125
    0.7500    0.6100
    0.8000    0.6075
    0.8500    0.6050
    0.9000    0.6000
    0.9500    0.5975
    1.0000    0.5950];
lambda = Data(:,1); mu = Data(:,2);
% We would like to fit the function
% mu=c(1)*lambda^n+c(2)*lambda^(n-1)
%        + c(3)*lambda^(n-2)
%        +...+c(n)*lambda+c(n+1);
% to the data. In the following, n=6
[c,S]=polyfit(lambda, mu, 6);
c
pause; t = (0: 0.05: 1)';
y = polyval(c,t);
plot(lambda,mu,'or',t,y,'-b')
title('Polynomial interpolation, degree n=6')
```

EXAMPLE 13.2: MATLAB IMPLEMENTATION OF THE MAGIC FORMULA (PURE LONGI-TUDINAL SLIP). The most famous special-function–based tire model is the Magic Formula proposed by Bakker, Nyborg, and Pacejka (1987). The special function looks like the following:

$$F_x = D_x \sin(C_x \tan^{-1}(B_x \phi_x)) + S_{vx} \tag{13.7}$$

$$F_y = D_y \sin(C_y \tan^{-1}(B_y \phi_y)) + S_{vy} \tag{13.8}$$

where:

$$\phi_x = (1 - E_x)(S_x + S_{hx}) + \frac{E_x}{B_x} \tan^{-1}(B_x(S_x + S_{hx})) \qquad (13.9)$$

$$\phi_y = (1 - E_y)(\alpha + S_{hy}) + \frac{E_y}{B_y} \tan^{-1}(B_y(\alpha + S_{hy})) \qquad (13.10)$$

The meanings of the coefficients are as follows:

B = stiffness factor
C = shape factor
D = peak factor (i.e., determines peak magnitude)
E = curvature factor
Sh = horizontal offset
Sv = vertical offset

and α = slip angle (in degrees) and S_x = longitudinal slip. The special-function form also applies to the self-aligning moment, the formulas for which are exactly the same as those shown in Eqs. (13.7) through (13.10). In general, the coefficients are functions of the tire normal force F_z.

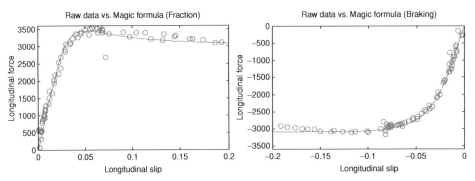

```
% Ex13_2.m
% Magic Formula empirical tire model
% Tire3 contains [slip Fx] data for Fz=3225 Newton
load tire3
s = tire3(:,1);  % Longitudinal slip
fx = tire3(:,2); % Longitudinal force in Newton


% First figure: Traction case
N=4; % resample because too many data points
plot(s(1:N:size(s,1)), fx(1:N:size(fx,1)), 'ob')
xlabel('Longitudinal slip')
ylabel('Longitudinal force Fx (N)')
title('Raw data')
hold on
pause


% The Magic Formula for traction case
Bx = 25;
```

```
Cx = 1.35;
Dx = 3444;
Ex = -2.9;
Shx = 0;
Svx = 0;

for i=1:101,
  s2(i) = (i-1)*0.002;        % slip between 0 and 0.2
  phi_x = (1-Ex)*(s2(i)+Shx) + Ex/Bx*atan(Bx*(s2(i)+Shx));
  fx2(i) = Dx*sin(Cx*atan(Bx*phi_x))+Svx;
end

plot(s2,fx2,'-r'), axis([0 0.2 0 3600])
title('Raw data vs. Magic formula (Traction)')
hold off
pause

% Second figure: Braking case
N=4;
plot(s(1:N:size(s,1)), fx(1:N:size(fx,1)), 'ob')
xlabel('Longitudinal slip')
ylabel('Longitudinal force Fx (N)')
hold on

% The Magic Formula for braking case
Bx = 25;
Cx = 1.15;
Dx = 3100;
Ex = -0.4;
Shx = 0;
Svx = 0;

for i=1:101,
  s2(i) = (i-1)*0.002;        % slip between 0 and 0.2
  phi_x = (1-Ex)*(s2(i)+Shx) + Ex/Bx*atan(Bx*(s2(i)+Shx));
  fx2(i) = Dx*sin(Cx*atan(Bx*phi_x))+Svx;
end

plot(-s2,-fx2,'-r'), axis([-0.2 0 -3600 0])
title('Raw data vs. Magic formula (Braking)')
hold off
```

In the following discussion, the vehicle parameter values are assumed to be:

$m = 1,400$ kg	$f = 0.01$	$I_w = 0.65$ kg-m^2	$r_w = 0.31$ m
$F_z = 3,560$ N	$C_d = 0.5$	$\rho = 1.202$ kg/m^3	$A = 1.95$ m^2
$b_w = 0$ kg-m/s	$f_w = 0$ $N_w = 4$	$T_e = 0.0$ N-m	$T_b = 1,000$ N-m

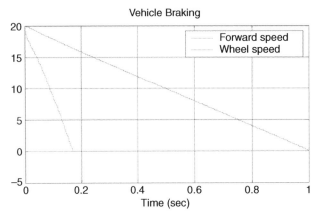

Figure 13.5. Simulation results of nonlinear braking.

EXAMPLE 13.3: NONLINEAR VEHICLE-BRAKING BEHAVIOR. The coefficients calculated from the $\mu - \lambda$ data in Example 13.1 are used in this example to simulate the nonlinear system described by Eq. (13.6) for the parameter values listed previously. The results are shown in Figure 13.5. Notice from the figure that wheel lock occurs rapidly (i.e., in about 0.2 seconds) and that the vehicle takes about 1 second to come to a stop. Because the wheel is locked, the lateral stability of the vehicle can be expected to be poor.

```
% Ex13_3.m
ti=0.0; tf=1.0; xi=[20.0,20.0];
tol=1.0e-4; trace=1;
[t,x]=ode45('Ex13_3a',ti,tf,xi,tol,trace);
plot(t,x); title('Vehicle Braking');
xlabel('Time (sec)'); ylabel('x'); grid;
legend('Forward speed','Wheel speed')

% A separate file Ex13_3a.m
function xdot=Ex13_3a(t,x);
% Define the simulation parameters:
m=1400; rho=1.202; Cd=0.5; A=1.95; g=9.81;
Theta=0.0; bw=0.0; f=0.01; uw=0.0; fw=0.0;
Iw=2.65; rw=0.31; Nw=4; Fz=3560.0;
% Define the mu-lambda polynomial coefficients:
c=[-68.593, 238.216,-324.819,219.283,
 -75.58, 12.088, -0.0068];
if x(1) >= x(2),
 lambda=(x(2)-x(1))/x(1);
else
 lambda=(x(2)-x(1))/x(2);
end;
al = abs(lambda);
if al > 1.0, al =1.0; end;
```

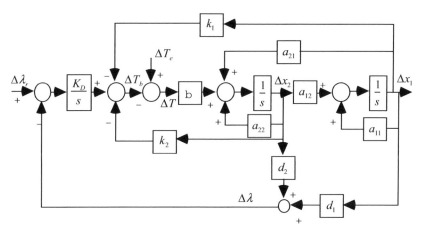

Figure 13.6. Integral plus state feedback form of linear ABS control.

```
mu=sign(lambda)*c*[al^6;al^5; ...
    al^4;al^3;al^2;al;1];
% Define the torque input T = Te - Tb;
Te =0.0; Tb=1000.0;
T = Te - Tb;
% Define the state equations:
if x(1) < 0.0, x(1) = 0.0; end;
if x(2) < 0.0, x(2) = 0.0; end;
xdot=[(-(0.5*rho*Cd*A)*(uw+rw*x(1))^2+
    Nw*Fz*mu-f*m*g*cos(Theta)-
    m*g*sin(Theta))/(m*rw);
      (fw*Fz-bw*x(2)-Fz*rw*mu+T)/Iw];
if x(1) <= 0.0, xdot(1)=0.0; end;
if x(2) <= 0.0, xdot(2)=0.0; end;
```

EXAMPLE 13.4: A LINEAR CONTROLLER DESIGN FOR ABS. A controller is designed for an ABS based on the linearization of Eq. (13.6). The parameter values given in the previous examples are used. The linearized controller block diagram is shown in Figure 13.6. The controller is the integral-plus-state-feedback type, and this controller, when applied to the linearized system, is expected to give good performance. The linearized system performance as shown in Figure 13.7a, however, is not very good. This is due to the presence of the zero near the origin of the s-plane, which leads to a major overshoot in the response. We expect the performance of this controller, when applied to the actual nonlinear system, to be even worse due to the strong nonlinearities in Eq. (13.6).

$$a_{11} = \frac{-\rho C_d A r_w x_{10}}{m} + \frac{N_w F_z}{m r_w}\left(\frac{\partial \mu}{\partial \lambda}\right)_0 \left(\frac{\partial \lambda}{\partial x_1}\right)_0 \quad a_{12} = \frac{N_w F_z}{m r_w}\left(\frac{\partial \mu}{\partial \lambda}\right)_0 \left(\frac{\partial \lambda}{\partial x_2}\right)_0$$

$$a_{21} = \frac{-F_z r_w}{I_w}\left(\frac{\partial \mu}{\partial \lambda}\right)_0 \left(\frac{\partial \lambda}{\partial x_1}\right)_0$$

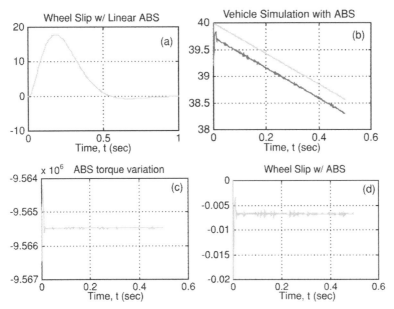

Figure 13.7. Simulation results of linear ABS.

$$a_{22} = \frac{-b_w}{I_w} - \frac{F_z r_w}{I_w} \left(\frac{\partial \mu}{\partial \lambda}\right)_0 \left(\frac{\partial \lambda}{\partial x_2}\right)_0, \quad b = \frac{1}{I_w} \quad d_1 = \left(\frac{\partial \lambda}{\partial x_1}\right)_0 = -\frac{x_{20}}{x_{10}^2}$$

$$d_2 = \left(\frac{\partial \lambda}{\partial x_2}\right)_0 = \frac{1}{x_{10}}$$

This is illustrated in Figures 13.7b–d. Thus, the linear controller shown here is not effective in practice. Furthermore, in an actual application, the $\mu - \lambda$ curve characteristics are not known, as was assumed here (other parameters also are uncertain). The results for both the linear and nonlinear models are shown in Figure 13.7. A piecewise linear control design strategy, with gain scheduling, can be expected to improve these results.

```
% Ex13_4.m
% Define the parameters:
m=1400; rho=1.202; Cd=0.5; A=1.95; g=9.81;
Theta=0.0; bw=0.0; f=0.01; uw=0.0;
Iw=0.65; Jw=Iw; rw=0.31; Nw=4; Fz=3560.0;
x10=40.0; x20=32.0; lref=(x20-x10)/x10;
al=abs(lref);
c=[-68.593,238.216,-324.819,219.2837,
     -75.58, 12.0878, -0.0068];
dmdl=c*[6*al^5;5*al^4;4*al^3;3*al^2;
         2*al;1;0];
dldx1=-(x20/x10^2); dldx2=1/x10;
% Define the A,B,C, and D matrices:
```

```
A=[(-rho*Cd*A*(rw^2)*x10+ ...
    Nw*Fz*dmdl*dldx1)/(m*rw), ...
    (Nw*Fz*dmdl*dldx2)/(m*rw),0;
      (-rw*Fz*dmdl*dldx1)/Jw, ...
      (-bw -rw*Fz*dmdl*dldx2)/Jw,0;
 dldx1,dldx2,0;];
Bu=[0;1/Jw;0]; Br=[0;0;-1];
C=[dldx1 dldx2 0]; D=0.0;
z=0.707; wn=10.0;
p=[-z*wn+i*wn*sqrt(1-z^2);
   -z*wn-i*wn*sqrt(1-z^2); -2*z*wn];
Co = det(ctrb(A,Bu));
Ob = det(obsv(A,C));
K = place(A,Bu,p);
[num,den]=ss2tf((A-Bu*K),Br,C,D,1);
zeros=roots(num);
poles=roots(den);
t=[0:0.01:1.0]; r0 = 0.05;
[y,x]=step((A-Bu*K),r0*Br,C,r0*D,1,t);
plot(t,y+lref);
title('Wheel Slip w/ Linear ABS'); grid;
xlabel('Time (sec)'); ylabel('lambda');

% Ex13_4a.m
ti=0.0; tf=0.5; xi=[40.0;40.0;0.0];
tol=1.0e-4; trace=1;
[t,x]=ode45('Ex13_4b',ti,tf,xi,tol,trace);
plot(t,[x(:,1) x(:,2)]);
title('Vehicle Simulation with ABS');
xlabel('Time, t (sec)'); ylabel('x'); grid; pause;
K=[2.3913e+05 3.0993e+01 3.5418e+06];
u=-K*x';
plot(t,u); title('vehicle ABS torque variation');
xlabel('Time, t (sec)'); grid; pause;
lambda=(x(:,2)-x(:,1))./x(:,1);
plot(t,lambda);
title('Wheel Slip w/ ABS');
xlabel('Time, t (sec)'); grid;

% A separate file Ex13_4b.m
function xdot=Ex13_4b(t,x);
% Define the simulation parameters:
m=1400; rho=1.202; Cd=0.5; A=1.95; g=9.81;
Theta=0.0; bw=0.0; f=0.01; uw=0.0;
Iw=0.65; rw=0.31; Nw=4; Fz=3560.0;
% Initialize:
Te = 0.0;
```

```
% Define the mu-lambda polynomial
c=[-68.593,238.216,-324.8197,219.2837,
    -75.5800, 12.0878, -0.0068];
% Define wheel slip lambda:
if x(1) >= x(2),
    lambda=(x(2)-x(1))/x(1);
else
    lambda=(x(2)-x(1))/x(2);
end;
al = abs(lambda);
if al > 1.0, al =1.0; end;
mu=sign(lambda)*c*[al^6;al^5;al^4;al^3;al^2;al;1];
% Define torque input T=Te - Tb;
K=[2.3913e+05 3.0993e+01 3.5418e+06];
x10=40; x20=32; lref=(x20-x10)/x10;

Tb=7.594e+2+K*[x(1)-x10;x(2)-x20;x(3)];
if Tb<0.0, Tb=0.0; end;
T=Te-Tb;
xdot=[(-(0.5*rho*Cd*A)*(uw+rw*x(1))^2+
      Nw*Fz*mu-f*m*g*cos(Theta)-
      m*g*sin(Theta))/(m*rw);
      (-f*Fz-bw*x(2)-Fz*rw*mu+T)/Iw;
      lambda-lref];
if x(1) <= 0.0,
    x(1)=0.0; xdot(1)=0.0; end;
if x(2) <= 0.0,
    x(2)=0.0; xdot(2)=0.0; end;
```

The previous example shows that the linear controller design is not effective for this application because the nominal operating point is not an equilibrium (i.e., $\dot{x} \neq 0$). A nonlinear control strategy is typically used in practice (e.g., Tan and Chin 1991; Tan and Tomizuka 1990a, 1990b). One approach is to construct a rule-based or table-lookup type of control algorithm. For example, a rule to be used in a TCS might be as follows:

$$\Delta T = 0.0$$
$$\text{IF } \omega_w > \omega_v \text{ THEN } \lambda = \frac{\omega_w - \omega_v}{\omega_w} \text{ ELSE } \lambda = \frac{\omega_w - \omega_v}{\omega_v} \text{ END}$$
$$\text{IF } |\lambda(t)| < \lambda \min \text{ THEN } DT = \text{sgn}(\lambda)\, \alpha_1\, |T(t-1)| \text{ END}$$
$$\text{IF } |\lambda(t)| > \lambda\max \text{ THEN } DT = -\text{sgn}(\lambda)\, \alpha_2\, |T(t-1)| \text{ END}$$
$$T(t) = T(t-1) + \Delta T$$

Thus, the rule-based controller calculates a change (ΔT) in the torque (T) for both braking and accelerating cases. A control system of this type is intuitively appealing but the development of the rules, tuning, and calibration depend on experience and trial and error. For example, the values of λ_{min}, λ_{max}, and α_i in the previous algorithm

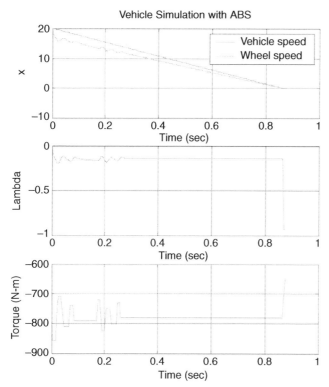

Figure 13.8. Simulation results of rule-based ABS control system.

must be chosen (e.g., $\lambda_{min} = 0.13$, $\lambda_{max} = 0.17$, $\alpha_1 = 0.07$, and $\alpha_2 = 0.09$). Furthermore, there are no guarantees regarding the behavior of such a system in terms of stability or performance.

EXAMPLE 13.5: A RULE-BASED CONTROLLER FOR ABS. The previous rule-based controller is implemented in this ABS simulation. The vehicle and wheel speeds are shown in Figure 13.8a, and in Figure 13.8b for the slip versus time. The slip during braking is maintained at a value near 0.15. Although the stopping time is approximately the same as in Example 13.3 (i.e., about 0.9 seconds), it is expected that vehicle handling should be improved over that case (see Figure 13.2a). In this simulation, the $\mu - \lambda$ curve is relatively flat (see Figure 13.4); thus, we do not observe a significant difference in the stopping distance.

```
% Ex13_5.m
% Initialize Torque (at Te=0, Tb=800 for braking and
% Te=1800, Tb=0 for accelerating)
global Torque;
Te = 0.0; Tb = 800.0; Torque = Te - Tb;
% Define the mu-lambda polynomial coefficients:
c = [-68.5937, 238.2160, -324.8197, 219.2837, ...
      -75.5800, 12.0878,  -0.0068];
```

```
% Perform the closed-loop simulation:
ti=0.0; tf=1.0; t1 = ti;
xi=[20.0;20.0];

% Set the rule-based control parameters:
alph1 = 0.07; alph2=0.09; lmin=0.13; lmax=0.17;
x_log(1,:)=xi';
t_log(1)=ti;
Torque_log(1) = Torque;
OPTIONS=odeset('MaxStep',0.001,'RelTol',1.0e-3);

for i=1:175,
  i
  t2=ti+i*(tf-ti)/200;
  tspan=[t1:0.001:t2];
  [t,x]=ode23('Ex13_5a',tspan,xi,OPTIONS);
  x_log(i+1,:)=x(size(x,1),:);   % data log
  xi=x(size(x,1),:)';
  t1 = t(size(t,1));
  t_log(i+1)=t(size(t,1));
  % Define wheel slip lambda:
  if xi(1) >= xi(2),
    lambda=(xi(2)-xi(1))/xi(1);
  else
    lambda=(xi(2)-xi(1))/xi(2);
  end;

  % Calculate the coefficient mu:
  al = abs(lambda);
% if al > 1.0 al=1.0; end;
% mu = -sign(lambda)*c*[al^6;al^5;al^4;al^3;al^2;al;1];
  % Define the change in torque delT:
  if al<lmin,
    delT = sign(lambda)*alph1*abs(Torque);
  elseif al>lmax,
    delT = -sign(lambda)*alph2*abs(Torque);
  else
    delT = 0.0;
  end;
  Torque = Torque + delT;
  Torque_log(i+1) = Torque;
end

% Plot the results:
subplot(311), plot(t_log,[x_log(:,1) x_log(:,2)]);
title('Vehicle Simulation with ABS');
xlabel('Time (sec)'); ylabel('x');
```

```
legend('Vehicle speed', 'wheel speed'); grid;
%lambda=(x_log(:,2)-x_log(:,1))./x_log(:,2);% TCS
lambda=(x_log(:,2)-x_log(:,1))./x_log(:,1); % ABS
subplot(312), plot(t_log,lambda,'r');
xlabel('Time (sec)'); ylabel('Lambda'); grid;
subplot(313), plot(t_log,Torque_log,'r');
xlabel('Time (sec)'); ylabel('Torque (N-m)'); grid

%Ex13_5a.m
function xdot = Ex13_5a(t,x);
global Torque;
% Define the simulation parameters:
m=1400; rho=1.202; Cd=0.5; A=1.95; g=9.81;
Theta=0.0; bw=0.0; f=0.01; uw=0.0; fw=0.0;
Iw=0.65; rw=0.31; Fz=3560.0;
% For braking: Ie=0, Nw=4; and for accel: Ie=0.429, Nw=2.
rg=9.5285; Ie=0.0; Nw=4;
Jw = Iw + (rg^2)*Ie;
% Define the mu-lambda polynomial coefficients:
c = [-68.5937, 238.2160, -324.8197, 219.2837, ...
     -75.5800, 12.0878, -0.0068];

% Define wheel slip lambda:
if x(1) >= x(2),
  lambda=(x(2)-x(1))/x(1);
else
  lambda=(x(2)-x(1))/x(2);
end;

% Calculate the coefficient mu:
al = abs(lambda);
if al > 1.0 al=1.0; end;
mu = sign(lambda)*c*[al^6;al^5;al^4;al^3;al^2;al;1];

% Define the state equations:
xdot=[(-(0.5*rho*Cd*A)*(uw+rw*x(1))^2+Nw*Fz*mu-
      f*m*g*cos(Theta)-m*g*sin(Theta))/(m*rw);
      (fw*Fz-bw*x(2)-Fz*rw*mu+Torque)/Jw];
if x(1) <= 0.0, x(1)=0.0; xdot(1)=0.0; end;
if x(2) <= 0.0, x(2)=0.0; xdot(2)=0.0; end;
```

13.3 Traction Control

The preceding section discusses manipulation of the brake torque to control the wheel slip during deceleration. Here, the combined manipulation of the engine and brake torques to achieve anti-slip acceleration and anti-skid braking is discussed. For acceleration, we use the same relationships given in Eq. (13.6) except that

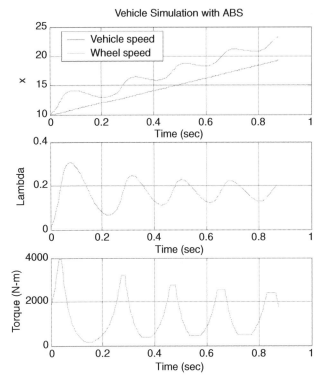

Figure 13.9. Simulation results of a rule-based TCS.

N_w is the number of drive wheels, the effective wheel inertia is $J_w = I_w + r_g^2 I_e$, and the wheel slip is $\lambda = (x_2 - x_1)/x_2$. Notice that during acceleration, $x_2 > x_1$; therefore, $\lambda > 0$. Thus, a controller for the acceleration case can be developed in a manner similar to the deceleration case. The two can be combined based on the measured values of the wheel and vehicle speeds and the resulting sign of λ.

EXAMPLE 13.6: A RULE-BASED TRACTION-CONTROL SYSTEM Simulations for the acceleration case can be performed in a manner similar to the ABS simulations shown in Examples 13.1 through 13.5. Additional parameter values for the acceleration case are as follows:

$$I_e = 0.429 \text{ kg-m}^2 \quad r_g = 9.5285 \quad N_w = 2 \quad T_e = 1,800.0 \text{ N-m}$$

A rule-based TCS simulation can be performed using the same MATLAB program as in Example 13.5 with changes in the parameter values as indicated previously. The controller parameters were set at the values $\lambda_{min} = 0.13$, $\lambda_{max} = 0.17$, $\alpha_1 = 0.14$, and $\alpha_2 = 0.15$. Also, there must be an additional limit so that the applied torque does not exceed that of the engine torque. The simulation results are shown in Figure 13.9. The slip is maintained at a value of approximately 0.15, which is the center of the two limits set in the controller.

The vehicle-speed measurement problem, which is necessary to determine λ, is difficult (Anonymous 1988c). Typically, lambda is estimated based on wheel-speed measurements and by considering the change in wheel speed (or wheel acceleration/deceleration). Notice also that actuation dynamics are neglected; that is, we assumed that commanded changes in brake or engine torque occur instantaneously. A complete controller design also must consider these effects.

Another more systematic nonlinear-control approach is based on the variable-structure control or the sliding-mode control approach. The basis for the sliding-mode-control or variable-structure-control approach is the determination of a switching logic, which switches between two nonlinear control functions. The switching logic depends on the sign of a switching function, S, and the system state moves (or slides) along this function – thus, the name *sliding mode* control. For example, given $\ddot{x} \pm x = 0$, we can have either system (S1) or (S2):

$$\dot{x}_1 = x_2$$
$$\dot{x}_2 = -x_1 \quad \text{(S1)}$$

or

$$\dot{x}_1 = x_2$$
$$\dot{x}_2 = +x_1 \quad \text{(S2)}$$

Define a switching surface (sliding mode) by $S = \dot{x} + c(x - x_d)$; then, select the control input $u(t)$ to achieve $S = 0$. For example, if $S > 0$, select u to make the system (S1); if $S < 0$, select u to make the system (S2). The resulting trajectories will look like:

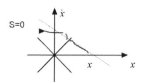

This method, which is equally applicable to the ABS or the TCS problem, is briefly described here (Tan and Chin 1991). The basis for the sliding-mode control approach is the determination of a switching logic, which switches between two nonlinear control functions. The switching logic depends on the sign of a switching function, S, and the system-state moves (or slides) along this function – thus, the name *sliding-mode control*. The switching functions for accelerating are as follows:

$$S_1(\lambda, \dot{\lambda}) = \dot{\lambda} + c_1(\lambda - \lambda_1)$$

and for braking:

$$S_2(\lambda, \dot{\lambda}) = \dot{\lambda} + c_2(\lambda - \lambda_2)$$

where $\lambda_1 = 0.15$ and $\lambda_2 = -0.15$. The control action is calculated for accelerating from:

If $S_1 < 0$, then $\dot{T} = 30{,}000.0$
If $S_1 > 0$, then $\dot{T} = -H(T)c_1 T - [P_1 + E_c]$

and for braking from:

If $S_2 > 0$,   then $\dot{T} = -30,000.0$
If $S_2 < 0$,   then $\dot{T} = -H(-T)c_2 T - P_2$

where the reference slip $\lambda_r = 0.15$ and the other functions and parameters are defined in Tan and Chin (1991).

Example 13.7 illustrates results of traction control from laboratory and road tests. It was mentioned previously that maintaining a value of $\lambda = 0.15$–0.20 is desirable not only for good traction in deceleration and acceleration but also for maintaining high lateral forces during such maneuvers. Example 13.7 presents the performance of a sliding-mode traction controller and gives results from Tan and Chin (1991) and from Tan and Tomizuka (1990a, 1990b).This example includes an adaptive sliding-mode controller based on estimation of the unknown $\mu - \lambda$ characteristics. Estimation of the tire–road friction coefficient is an important research area not only for traction control but also for vehicle-stability control and other active safety systems (Rajamani et al. 2010). Researchers also studied the effects of ABS on combined longitudinal and lateral vehicle motions (Kimbrough 1991; Taheri and Law 1991).

EXAMPLE 13.7: SLIDING-MODE TRACTION CONTROLLER. A nonlinear traction-controller design is described in a series of papers by Tan and co-authors (Tan and Tomizuka 1990a, 1990b; Tan and Chin 1991). These papers describe a variable-structure control (or sliding-mode control) approach to TCS problems (Tan and Chin 1991). An adaptive version of the approach is described in Tan and Tomizuka (1990b) and includes estimation of the $\mu - \lambda$ characteristics in addition to a sliding-mode controller. Finally, in Tan and Tomizuka (1990a), the digital (or discrete-time) implementation of this controller is described. A detailed discussion of these control strategies is beyond the scope of this chapter and readers are referred to the references for further details. The adaptive version, which estimates the $\mu - \lambda$ relationship from measurements, uses online identification much like the ACC example in Chapter 12. However, the function to be identified here is nonlinear, as are the controlled system and the controller. Finally, the digital implementation must consider the effects of sampling and the actuator dynamics (i.e., application of brake torque in ABS and combined braking and engine torque for traction control).

Trajectories from simulation results for anti-spin acceleration are illustrated in Figure 13.10, which shows the convergence of the trajectory in the $\lambda - \dot{\lambda}$ plane to the sliding-mode surface, S. The control torque is shown for this same case in Figure 13.10b. Figure 13.10c shows the resulting simulated wheel and vehicle speeds during this acceleration maneuver. Figure 13.11 illustrates a dynamometer test-cell setup and evaluation test results for this controller for ABS applications. Results in Figures 13.11b–d show the effects of sampling time and drum surface (i.e., wet or dry). Shown in Figure 13.12 are the results of dynamometer tests using the adaptive version of the ABS sliding-mode controller. Figures 13.12a–c show results from tests for various slip values and drum-surface conditions. Although the surface conditions change (i.e., from wet to dry or from dry to wet), the controller performs a reasonably good job of

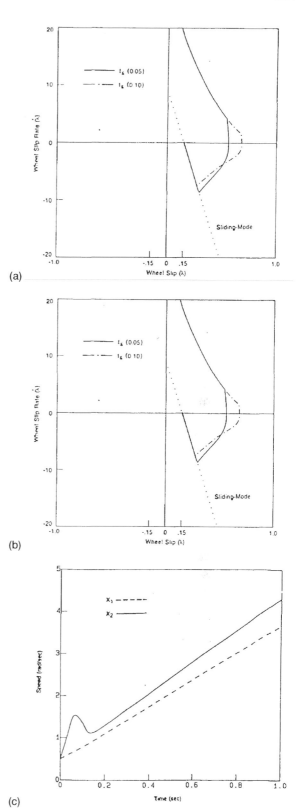

Figure 13.10. Simulated response in $\lambda - \dot{\lambda}$ plane: (a) trajectories for anti-spin control, (b) torque curves for anti-spin control, and (c) vehicle/wheel speeds for anti-spin control (Tan and Tomizuka 1990a).

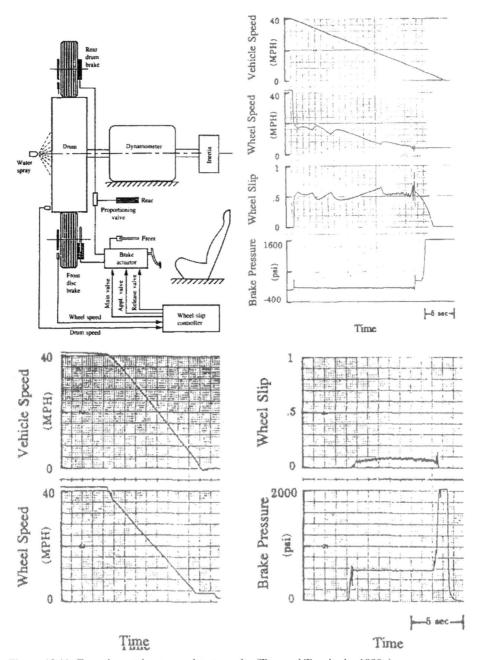

Figure 13.11. Experimental setup and test results (Tan and Tomizuka 1990a).

maintaining the desired slip values by estimating the actual $\mu - \lambda$ relationship during operation.

One of the most important variables in realizing the TCS/ABS control algorithms presented herein is the wheel slip ratio λ. Because the vehicle speed usually is not available and the estimation may be inaccurate, a reference vehicle speed may be generated. The reference speed is equal to the wheel speed at the beginning of the ABS application and decreases at a constant rate, which is determined by the

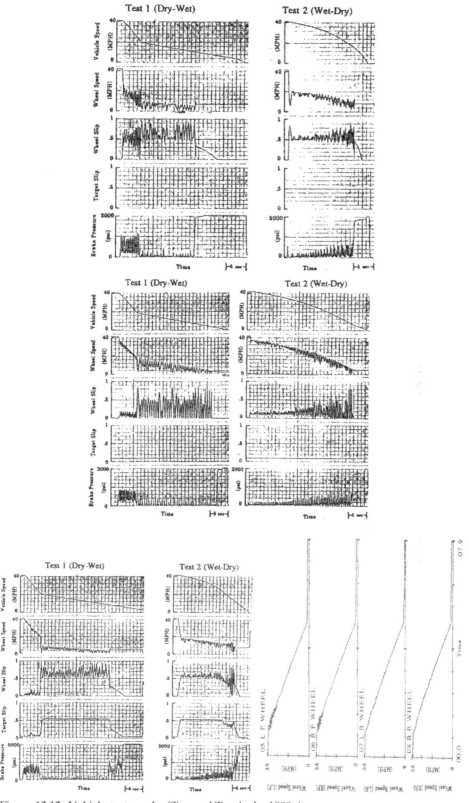

Figure 13.12. Vehicle test results (Tan and Tomizuka 1990a).

road-surface condition. ABS control algorithms implemented on production vehicles may have the following form:

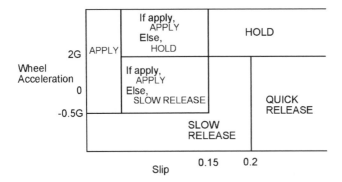

Significant tuning usually is necessary to determine a good set of rules. When implementing on vehicles, customer feedback can be as important a factor as engineering judgment.

PROBLEMS

1. The nonlinear simulation in Example 13.3 uses the tire model in Example 13.1. Redo the simulation in Example 13.3 using the Magic Formula tire model in Example 13.2.

2. Attempt to develop a "rule-based" ABS or traction controller similar to that suggested in Examples 13.5 and 13.6. Suggest some modification to the rules in those examples that you think will improve the performance of the controller. First, explain the rule or rules that you base the controller on and then try it out in simulations using a program such as those shown in Example 13.5 and 13.6. Compare your results to those in the examples.

REFERENCES

Anonymous, 1986, "Traction Aids at Three Levels," *Automotive Engineering*, January, pp. 39–41.

Anonymous, 1987, "Traction Control: An ABS Extension," *Automotive Engineering*, August, pp. 75–80.

Anonymous, 1988a, "Electronic Traction Control for Luxury Car," *Automotive Engineering*, February, p. 164.

Anonymous, 1988a, "Total Traction Four-Wheel Drive and Anti-Skid Brakes in a Mid-Size Car," *Automotive Engineering*, March, pp. 82–5.

Anonymous, 1988c, "Wheel Speed Sensing Developments," *Automotive Engineering*, June, pp. 67–73.

Anonymous, 1990a, "Traction Control for Front-Wheel Drive," *Automotive Engineering*, January, pp. 89–90.

Anonymous, 1990a, "Improved Electronics Aid ABS," *Automotive Engineering*, September, pp. 30–5.

Austin, L., and D. Morrey, 2000, "Recent Advances in Anti-Lock Braking Systems and Traction Control Systems," *Proceedings of the IMechE*, Vol. 214, Part D, pp. 625–38.

Bakker, E., H. B. Pacejka, and L. Lidner, 1989, "A New Tire Model with an Application in Vehicle Dynamics Studies," SAE Technical Paper 890087, 1989, doi:10.4271/890087.

Bakker, E., L. Nyborg and H. B. Pacejka, 1987, *Tyre modelling for use in vehicle dynamics studies*, SAE.

Borrelli, F., A. Bemporad, M. Fodor, and D. Hrovat, 2001, "A Hybrid Approach to Traction Control," *Lecture Notes in Computer Science*, Springer Publishing Co.

Brewer, H. K., 1983, "Design and Performance Aspects of Antilock Brake Control Systems," in W. E. Meyer and J. D. Walters (eds.), *Frictional Interaction of Tire and Pavement*, ASTM STP 793, American Society for Testing and Materials, pp. 79–115.

Canudas de Wit, C., and P. Tsiotras, 1999, "Dynamic Tire Friction Models for Vehicle Traction Control," *Proceedings of the 38th IEEE Conference on Decision and Control*, Vol. 4, pp. 3746–51.

Elgeskog, E., and S. Brodd, 1976, "The Influence of Wheel Slip Control Dynamics on Vehicle Stability During Braking and Steering," *Proceedings of the Institution of Mechanical Engineers*.

Forkenbrock, G. J., M. F. Flick and W. R. Garrott, 1999, *NHTSA Light Vehicle Antilock Brake System Research Program Task 4: A test Track Study of Light Vehicle ABS Performance Over a Broad Range of Surfaces and Maneuvers*, Final Report DOT HS 808 875, National Highway Traffic Safety Administration (NHTSA), January 1999.

Gohring, E., E. C. von Glasner, and C. Bremer, 1989, "The Impact of Different ABS-Philosophies on the Directional Behavior of Commercial Vehicles," *Vehicle Dynamics Related to Braking and Steering*, SAE SP-801, SAE Paper 892500.

Herb, E., K. Krusche, E. Schwartz, and H. Wallentowitz, 1988, "Stability Control and Traction Control at Four-Wheel Drive Cars," *Automotive Systems Technology, The Future*, Vol. 1.

Hori, Y., Y. Toyoda, and Y Tsuruoka, 1998, "Traction Control of Electric Vehicle," *IEEE Transactions on Industry Applications*.

Jurgen, R., 1995, *Automotive Electronics Handbook*, McGraw-Hill Publishers.

Kimbrough, S., 1991, "Coordinated Braking and Steering Control for Emergency Stops and Accelerations," in S. A. Velinsky, R. H. Fries, I. Haque, and D. Wang (eds.), *Advanced Automotive Technologies–1991*, ASME DE-Vol. 40, New York, pp. 229–44.

Lin, W.C., D. J. Dobner, and R. D. Fruechte, 1993, "Design and Analysis of an Antilock Brake Control System with Electric Brake Actuator," *International Journal of Vehicle Design*, Vol. 14, No. 1, pp. 13–43.

Maretzke, J., and B. Richter, 1987, *Traction and Direction Control of 4WD Passenger Cars*, Volkswagenwerk AG, Wolfsburg, Germany.

Mills, V., B. Samuels, and J. Wagner, 2002, "Modeling and Analysis of Automotive Antilock Brake Systems Subject to Vehicle Payload Shifting, *Vehicle System Dynamics*, Vol. 37, No. 4, April, pp. 283–310.

Rajamani, R., D. Piyabongkarn, J. Y. Lew, , K. Yi, and G. Phanomchoeng, 2010, "Tire-Road Friction-Coefficient Estimation," *IEEE Control Systems Magazine*, Vol. 30, No. 4, August, pp. 54–69.

Rittmannsberger, N., 1988, "Antilock Braking System and Traction Control," *Proceedings of the International Congress on Transportation Electronics*, October, pp. 195–202.

Robinson, B. J., and B. S. Riley, 1989, "Braking and Stability Performance of Cars Fitted with Various Types of Anti-Lock Braking Systems," *Proceedings of the 12th International Conference on Experimental Safety Vehicles*, NHTSA, Sweden, pp. 836–46.

Rompe, K., A. Schindler, and M. Wallrich, 1987, "Comparison of the Braking Performance Achieved by Average Drivers in Vehicles with Standard and Anti-Wheel Lock Brake Systems," *SAE International Congress and Exposition*. Detroit, MI, SAE Paper No. 870335.

Sado, H., S. Sakai, and Y. Hori, 1999, "Road Condition Estimation for Traction Control in Electric Vehicles," *Proceedings of the ISIE'99 – Industrial Electronics*.

SAE, 1989, *Vehicle Dynamics Related to Braking and Steering*, Society of Automotive Engineers.

Segel, L., 1975, "The Tire as a Vehicle Component," in B. Paul, K. Ullman, and H. Richardson (eds.), *Mechanics of Transportation Suspension Systems*, ASME, New York, AMD-Vol. 15.

Sigl, A., and H. Demel, 1990, "ASR-Traction Control, State of the Art and Some Prospects," ABS Traction Control and Brake Components, *SAE ICE*, SAE Paper No. 900204.

Taheri, S., and E. H. Law, 1991, "Slip Control Braking of an Automobile During Combined Braking and Steering Maneuvers," in S. A. Velinsky, R. H. Fries, I. Haque, and D. Wang (eds.), *Advanced Automotive Technologies–1991*, ASME DE-Vol. 40, New York, pp. 209–27.

Tan, H. S., and Y. K. Chin, 1991, "Vehicle Traction Control: Variable-Structure Control Approach," *ASME Journal of Dynamic Systems, Measures, and Control*, June, Vol. 113, pp. 223–30.

Tan, H. S., and M. Tomizuka, 1990a, "Discrete-Time Controller Design for Robust Vehicle Traction," *IEEE Control System Magazine*, April, pp. 107–13.

Tan, H. S., and M. Tomizuka, 1990b, "An Adaptive Sliding-Mode Vehicle Traction Controller Design," *Proceedings of the American Control Conference*, San Diego, CA, May, pp. 1856–61.

Watanabe, M., and N. Noguchi, 1990, "A New Algorithm for ABS to Compensate for Road-Disturbance," *ABS Traction Control and Brake Components*, SAE Paper 900205.

Yeh, E.C., and G. C. Day, 1992, "A Parametric Study of Anti-Skid Brake Systems Using Poincare Map Concept," *International Journal of Vehicle Design*, Vol. 13, No. 3, pp. 210–32.

14 Vehicle Stability Control

The concept of vehicle stability control (VSC), variously known as vehicle dynamics control (VDC) and electronic stability program (ESP), was first introduced in 1995 and has been studied extensively as an active safety device to improve vehicle stability and handling. The concepts are natural extensions of the ABS and TCS discussed in Chapter 13. When "differential braking" is applied – that is, braking forces of the left- and right-hand-side tires are different – a yaw moment is generated. This yaw moment then slows down the vehicle and influences the vehicle lateral/yaw/roll motion. By taking advantage of the widely available and mature ABS hardware, this differential braking function is added on with minimal additional cost. Therefore, this vehicle control system has enjoyed rapid market acceptance.

The vehicle yaw moment also can be generated through the manipulation of tire-traction forces – for example, through the control of differentials or power-split devices in AWD vehicles (Piyabongkarn et al. 2010). In such systems, the objectives of VSC can be achieved without reducing the longitudinal velocity – but at the cost of additional hardware. Vehicle-handling performance can be influenced by many different actuations. In addition to differential braking and traction, all-wheel steering and active/semi-active suspensions (including antiroll systems) can be used. Hac and Bodie (2002) discussed methods for improving vehicle stability and emergency handling using electronically controlled chassis systems. Small changes in the balance of tire forces between the front and rear axles may affect vehicle yaw moment and stability. They discuss methods of affecting vehicle-yaw dynamics using controllable brakes, steering, and suspension. Brake-steering techniques are discussed in detail in Pilutti et al. (1998).

Matsumoto et al. (1992) developed a braking-force-distribution control strategy using a steering-wheel-angle feed-forward and yaw-rate feedback design. A target yaw rate was calculated based on steering-wheel angle and actual yaw rate, and left–right brake-force-distribution control provided a corrective yaw moment as required. The vehicle model had 11 DOF, and the tire model generated forces as a function of side-slip angle, normal load, longitudinal-slip ratio, and coefficient of friction of the road surface. Simulations showed significant reduction in yaw-rate error for braking during a lane change. This work is one of the earliest known publications on the concept later known as VSC or VDC.

Bosch developed a VDC system (Zanten et al. 1995) that derives the desired vehicle motion from the steering angle, accelerator-pedal position, and brake pressure. Actual vehicle motion is determined from measurement of the yaw rate and lateral acceleration. Differences between actual and desired motion are minimized by regulating engine torque and brake pressures using TCS components.

Many automobile manufacturers have implemented VDC systems or have plans to make VDC systems available in future models. Inputs to the General Motors system-control strategy are vehicle speed, steering-wheel angle, yaw rate, brake pressure, and lateral acceleration. The system has been shown to reduce yaw-velocity rise time, overshoot, and settling time (Hoffman and Rizzo 1998).

14.1 Introduction

The VSC system is a relatively new, active safety concept introduced to control vehicle lateral and yaw motions in emergency conditions. The control system that achieves this function is known in the literature as direct yaw-moment control (Shibahata et al. 1993), VDC (Zanten et al. 1995), brake-force-distribution control (Matsumoto et al. 1992), differential braking (Kraft and Leffler 1990), ESP (Zanten 2000), and VSC (Koibuchi et al. 1996). In the previous decade, this vehicle-control concept has been studied by almost all major automobile manufacturers, and many models already offer VSC (or one of the equivalent alternative names) as an option.

The VSC function may be slightly different among designs but, essentially, it involves the following tasks. First, the desired vehicle motion is inferred from driver action (i.e., steering angle, acceleration-pedal position, or brake pressure). Second, the actual vehicle motion, measured by a set of sensors (i.e., yaw rate, lateral acceleration, and vehicle speed), is compared to the desired motion. Third, control action (i.e., wheel-brake pressures) is applied to regulate vehicle motion to follow the desired motion under certain stability constraints (e.g., vehicle side-slip angle and roll motion).

The motivation for VSC is understood best by examining the difficulty for a human driver (or control system) to control vehicle lateral dynamics under extreme conditions. Briefly, when the vehicle side-slip angle is large, the authority of the vehicle steering angle in generating a yaw moment becomes significantly reduced due to tire-force saturation. This fact was first illustrated by the "β-method" proposed in Shibahata et al. (1993). This method envisioned a vehicle driving straight (i.e., $r_d = 0$ in Eq. (4.55)). Using the bicycle model (see Eqs. (4.45) and (4.46)), we have:

$$
\begin{aligned}
F_{yf} + F_{yr} &= m(\dot{v} + ur) \\
F_{yf} \cdot a - F_{yr} \cdot b &= I_z \cdot \dot{r}
\end{aligned}
\tag{14.1}
$$

where all the variables are as defined in Chapter 4. The lateral tire forces are functions of tire slip angles, where the front- and rear-axle slip angles are calculated from:

$$
\begin{aligned}
\alpha_f &= \delta_f - \frac{v + ar}{u} \\
\alpha_r &= -\frac{v - br}{u}
\end{aligned}
\tag{14.2}
$$

The tire forces then are calculated from the Magic Formula model (see Chapter 13). The key idea of the β-method is that when the vehicle side-slip angle $\beta \approx \frac{v}{u}$ is large,

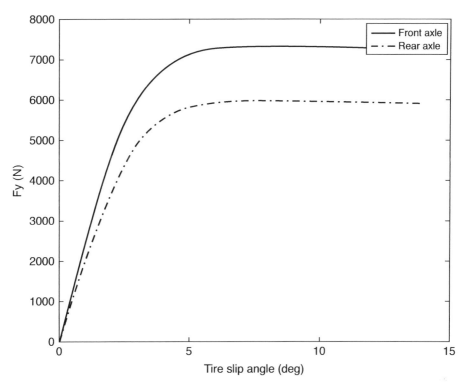

Figure 14.1. Tire lateral-force diagram (F_y versus α).

the yaw moment $aF_{yf} - bF_{yr}$ and total lateral force $F_{yf} + F_{yr}$ become almost constant, regardless of the steering angle, δ_f – that is, a driver's authority in controlling the vehicle lateral/yaw motion is reduced significantly. This fact is shown in the following example.

EXAMPLE 14.1. In this MATLAB example, three plots were generated. First, the tire lateral-force diagram (Figure 14.1) shows the relationship between tire slip angle α and lateral force. Notice that the tire forces are almost constant when the slip angle is larger than 5 degrees.

Second, this example program involves the calculation of vehicle control authority. Imagine a situation in which the vehicle side-slip angle varies between 0 and 15 degrees; under zero vehicle yaw rate ($r = 0$), the yaw moment $aF_{yf} - bF_{yr}$ and total lateral force $F_{yf} + F_{yr}$ can be calculated using the Magic Formula tire model. It can be seen in Figure 14.2 that the authority of front-axle steering is reduced greatly when the vehicle side-slip angle is large. In other words, regardless of whether the human driver (or automatic controller) is steering the front wheels to the left or to the right, similar amounts of lateral force and yaw moment are generated due to tire saturation.

Example 14.1 illustrates the problem of a large vehicle side-slip angle, which has been the key argument for regulating it. In early designs, many VSC systems were designed to limit vehicle side slip to be less than a larger threshold (e.g., 5 degrees). Recently, with the ability to estimate more accurately the side-slip angle, some companies have pushed for a lower threshold value ($\sim$3 degrees), especially

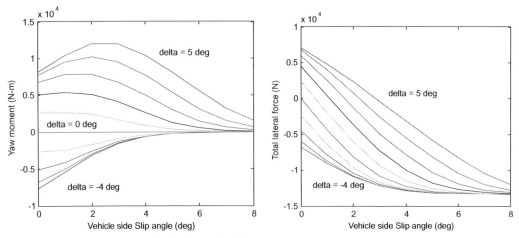

Figure 14.2. Control authority reduction.

on slippery (i.e., icy) road surfaces. Figures 14.1 and 14.2 are produced from tire data obtained on high-μ surfaces and they should not be used to generalize conclusions for low-μ cases. However, reduced steering authority also is observed for low-μ cases.

Another reason that it is a good idea to limit the vehicle side-slip angle is the fact that human drivers usually do not have much experience driving under limiting (i.e., emergency) situations. It is believed that they typically drive in situations when the vehicle side-slip angle is less than 2 degrees (Zanten 2000). Therefore, drivers usually are taken by surprise by the nonlinear behavior such as that shown in Example 14.1, and they commonly overreact to a situation. To avoid this problem, the proper control of vehicle dynamics so that the vehicle never enters the highly nonlinear region is an effective preemptive measure.

```
% Ex14_1.m
clear
% The Magic Formula for Fz = 3225N
By = 0.27;  Cy = 1.2; Dy = 2921;
Ey = -1.6; Shy = 0; Svy = 0;
% Vehicle parameters
a = 1.14;  L = 2.54;   b = L - a;
m = 1500; Iz = 2420.0; g = 9.81;
Fz_f = m*g*b/L; Fz_r = m*g*a/L;
Dy_f = Dy*Fz_f/3225; Dy_r = Dy*Fz_r/3225;
for i=1:101,
    alpha(i) = (i-1)*0.14;    % slip angle 0 to 14 deg
    phi_y = (1-Ey)*(alpha(i)+Shy) + Ey/By* ...
                atan(By*(alpha(i)+Shy));
    fy_f(i) = Dy_f*sin(Cy*atan(By*phi_y))+Svy;
    fy_r(i) = Dy_r*sin(Cy*atan(By*phi_y))+Svy;
end
plot(alpha, fy_f, '-r', alpha, fy_r,'-.b')
xlabel('Tire slip angle (deg)'); ylabel('Fy (N)')
```

```
legend('Front axle', 'Rear axle'); pause
% Part II, calculate Yaw moment & lateral force
r_over_u = 0;              % yaw rate = 0
for i = 1:10,
    delta = i-5;           % steering angle -4 to 5 deg
    for j = 1:9,
    beta(j) = j-1;         % side slip angle 0 to 8 deg
    alpha_f = delta - (beta(j) + 180/pi*a*r_over_u);
    alpha_r = - (beta(j) - 180/pi*b*r_over_u);
    phi_yf  = (1-Ey)*(alpha_f+Shy) + Ey/By ...
                *atan(By*(alpha_f+Shy));
    Fyf     = Dy_f*sin(Cy*atan(By*phi_yf))+Svy;
    phi_yr  = (1-Ey)*(alpha_r+Shy) + Ey/By ...
                *atan(By*(alpha_r+Shy));
    Fyr     = Dy_r*sin(Cy*atan(By*phi_yr))+Svy;
    yaw_moment(j, i)  = a * Fyf - b * Fyr;
    lateral_force(j,i) = Fyf + Fyr;
  end
end
plot(beta, yaw_moment)
xlabel('Vehicle side Slip angle (deg)')
ylabel('Yaw moment (N-m)'); pause
plot(beta, lateral_force)
xlabel('Vehicle side Slip angle (deg)')
ylabel('Total lateral force (N)')
```

14.2 Linear Vehicle Model

We first derive a vehicle model that is suitable for VSC simulations. Apparently, if we must simulate a vehicle roll motion (e.g., to study the effect of rollover reduction by the VSC system), the minimum number of vehicle DOFs needed is four (i.e., lateral, yaw, roll, and longitudinal). This model can be built easily on the basis of the three-DOF lateral/yaw/roll model introduced in Chapter 4 and described in Appendix B. Because the longitudinal dynamics of the vehicle is much slower than the yaw/roll dynamics, the simulation model does not need to be nonlinear. Rather, the yaw/roll model can be updated at each time step, assuming that the vehicle forward speed is constant. This separation of dynamics makes it possible to use the linear simulation command *lsim*() in MATLAB to perform the VSC simulation. Another factor to consider is actuator dynamics. We can assume that the control signal is individual brake pressure; however, this obviously adds complexity to the simulations. Assuming that the control signal is the yaw moment M_{VSC}, the vehicle lateral/yaw/roll model is then:

$$Y_\beta \beta + Y_r r + Y_\phi \phi + Y_\delta \delta_f = m(u\dot{\beta} + ur) + m_R h\dot{p}$$
$$N_\beta \beta + N_r r + N_\phi \phi + N_\delta \delta_f + M_{VSC} = I_z \dot{r} + I_{xz}\dot{p} \qquad (14.3)$$
$$L_\phi \phi + L_p p = m_R h(u\dot{\beta} + ur) + I_x \dot{p} + I_{xz}\dot{r}$$

and the longitudinal dynamics model is:

$$-\left|\frac{M_{VSC}}{T/2}\right| = m\dot{u} \tag{14.4}$$

where the vehicle-stability derivatives are defined as:

$$
\begin{aligned}
&Y_\beta \equiv -(C_{\alpha f} + C_{\alpha r}) && Y_r \equiv \frac{bC_{\alpha r} - aC_{\alpha f}}{u} \\
&Y_\phi \equiv C_{\alpha r}\frac{\partial \delta_r}{\partial \phi} + C_{\gamma f}\frac{\partial \gamma_f}{\partial \phi} && Y_\delta \equiv C_{\alpha f} \\
&N_\beta \equiv bC_{\alpha r} - aC_{\alpha f} && N_r \equiv -\frac{a^2 C_{\alpha f} + b^2 C_{\alpha r}}{u} \\
&N_\phi \equiv aC_{\gamma f}\frac{\partial \gamma_f}{\partial \phi} - bC_{\alpha r}\frac{\partial \delta_r}{\partial \phi} && N_\delta \equiv aC_{\alpha f}
\end{aligned} \tag{14.5}
$$

The control signal M_{VSC} is assumed to be generated by an electronic braking system; thus, it always slows down the vehicle regardless of the sign of M_{VSC}. The lateral/yaw/roll part of the vehicle dynamics (Eq. (14.3)) is not in the standard state-space form and therefore must be modified slightly. If we treat the front-wheel steering angle as a disturbance input and M_{VSC} as the control signal, Eq. (14.3) can be rearranged to:

$$
\begin{bmatrix} mu & 0 & m_R h & 0 \\ 0 & I_z & I_{xz} & 0 \\ m_R hu & I_{xz} & I_x & 0 \\ 0 & 0 & 0 & 1 \end{bmatrix}
\begin{bmatrix} \dot{\beta} \\ \dot{r} \\ \dot{p} \\ \dot{\phi} \end{bmatrix}
+
\begin{bmatrix} -Y_\beta & mu - Y_r & 0 & -Y_\phi \\ -N_\beta & -N_r & 0 & -N_\phi \\ 0 & m_R hu & -L_p & -L_\phi \\ 0 & 0 & -1 & 0 \end{bmatrix}
\begin{bmatrix} \beta \\ r \\ p \\ \phi \end{bmatrix}
$$

$$
=
\begin{bmatrix} 0 \\ 1 \\ 0 \\ 0 \end{bmatrix} M_{VSC}
+
\begin{bmatrix} Y_\delta \\ N_\delta \\ 0 \\ 0 \end{bmatrix} \delta_f
$$

$$\equiv E\dot{\underline{x}} + F\underline{x} = GM_{VSC} + H\delta_f \tag{14.6}$$

or:

$$\dot{x} = Ax + B_u M_{VSC} + B_w \delta_f \tag{14.7}$$

where $A = -E^{-1}F, B_u = E^{-1}G, B_w = E^{-1}H$.

EXAMPLE 14.2. In this example, the vehicle model described in Eqs. (14.4) and (14.6) is simulated. We can imagine a simple VSC algorithm: $M_{VSC} = Kx = [k_\beta \ 0 \ 0 \ -k_\phi]x$, where the state vector x is defined in Eq. (14.6) and the two control gains (i.e., k_β and k_ϕ) are both assumed to be positive. The previously mentioned control law is a simple proportional control law: A positive VSC yaw moment is requested when the vehicle side-slip angle (β) is positive or when the vehicle roll angle (ϕ) is negative.

Figure 14.3 demonstrates that when road friction is high and linear tire models are used, the need for VSC is not obvious. Notice that the simulation represents a severe "fishhook" maneuver, with the yaw rate as high as

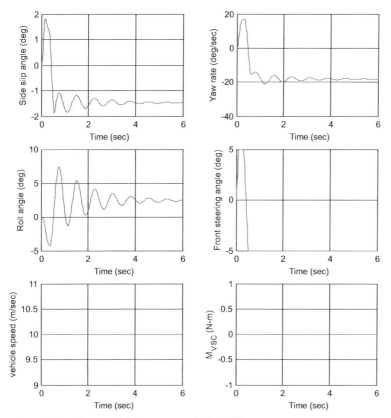

Figure 14.3. Linear vehicle response (VSC off).

20 degrees/second However, in this case, the vehicle slip angle is quite small; thus, there is no apparent need for VSC. This example demonstrates the need for a nonlinear simulation model in the simulation and validation of VSC designs.

14.3 Nonlinear Vehicle Model

Before we present the nonlinear vehicle model, it is important to point out that a combined-slip tire model must be used because of the possibility of simultaneous occurrence of high tire slip angle and large slip ratios. For combined-slip tire models, it is necessary to keep track of both tire slip angle and slip ratio to calculate tire forces (both lateral and longitudinal). To make it easier to perform the simulations, we assume that the slip ratios of the four tires are directly controllable or, equivalently, the wheel speeds are regulated by servo-loop controllers. This assumption makes it possible to simulate the vehicle response using a five-state model, which is similar to but somewhat more complicated than the one described in Section 14.2. This assumption is reasonable because the wheel dynamics is significantly faster than the vehicle dynamics. A good servo-loop, therefore, can regulate wheel-slip ratios around desired values fast enough. There is an added benefit of the assumption mentioned here: It is consistent with the evolution of VSC hardware. Most VSC systems were developed to reside on top of ABS/TCS; thus, setting the desired wheel-slip ratio is possible if the underlying ABS/TCS is known to work well in regulating wheel slips.

In the following discussion, we assume that the Magic Formula tire model with combined-slip correction is used to compute simultaneously the lateral and longitudinal tire forces. The calculated tire forces then become the inputs to the vehicle-dynamic equations. When the inputs to the vehicle yaw/roll model are lateral forces at the front and rear axles, the dynamic equations become:

$$\begin{aligned}
F_{yf} + F_{yr} &= m(ur + \dot{v}) + m_R h \dot{p} \\
F_{yf} \cdot a - F_{yr} \cdot b &= I_z \dot{r} + I_{xz} \dot{p} + M_{VSC} \\
L_\phi \phi + L_p p &= m_R h (\dot{v} + ur) + I_x \dot{p} + I_{xz} \dot{r}
\end{aligned} \tag{14.8}$$

or, in the state-space form:

$$\begin{bmatrix} m & 0 & m_R h & 0 \\ 0 & I_z & I_{xz} & 0 \\ m_R h & I_{xz} & I_x & 0 \\ 0 & 0 & 0 & 1 \end{bmatrix} \begin{bmatrix} \dot{v} \\ \dot{r} \\ \dot{p} \\ \dot{\phi} \end{bmatrix} + \begin{bmatrix} 0 & mu & 0 & 0 \\ 0 & 0 & 0 & 0 \\ 0 & m_R hu & -L_p & -L_\phi \\ 0 & 0 & -1 & 0 \end{bmatrix} \begin{bmatrix} v \\ r \\ p \\ \phi \end{bmatrix}$$

$$= \begin{bmatrix} 0 \\ 1 \\ 0 \\ 0 \end{bmatrix} M_{VSC} + \begin{bmatrix} 1 & 1 \\ a & -b \\ 0 & 0 \\ 0 & 0 \end{bmatrix} \begin{bmatrix} F_{yf} \\ F_{yr} \end{bmatrix} \tag{14.9}$$

$$\equiv E\underline{\dot{x}} + F\underline{x} = GM_{VSC} + H \begin{bmatrix} F_{yf} \\ F_{yr} \end{bmatrix}$$

or:

$$\dot{x} = Ax + B_u M_{VSC} + B_w \begin{bmatrix} F_{yf} \\ F_{yr} \end{bmatrix} \tag{14.10}$$

where $A = -E^{-1}F$, $B_u = E^{-1}G$, $B_w = E^{-1}H$. Eq. (14.9) illustrates that there are three inputs to the model: lateral forces at the front and rear axles and the yaw moment generated from the longitudinal tire forces. From a servo-loop perspective, there are five inputs: the front-axle steering angle (i.e., disturbance) and the slip ratios at the four tires. Given the steering angle and slip ratios, we can calculate all four tire slip angles and thereby compute the tire forces. Once lateral and longitudinal tire forces are known, the three inputs to the vehicle dynamics (i.e., M_{VSC}, F_{yf}, and F_{yr}) are calculated.

EXAMPLE 14.3. In this example, the vehicle model described in Eqs. (14.4) and (14.9) is simulated, in which the tire forces are calculated from the combined-slip Magic Formula tire model. A simple VSC algorithm again is used in this example – a proportional control law. The control action is turned on only if the magnitude of the vehicle side-slip angle is higher than a certain threshold. Figure 14.4 illustrates the case in which the VSC is not turned on. It shows that because of the nonlinear tire model used in this simulation, a very high vehicle slip angle is generated, whereas the vehicle yaw rate is much lower than that from the linear vehicle model. The vehicle gradually loses stability, with a slow increase of yaw rate and continuously increasing side-slip angle. This phenomenon is similar to what was observed in actual fishhook testing. When a simple proportional algorithm is applied, the vehicle side-slip angle is significantly reduced (Figure 14.5).

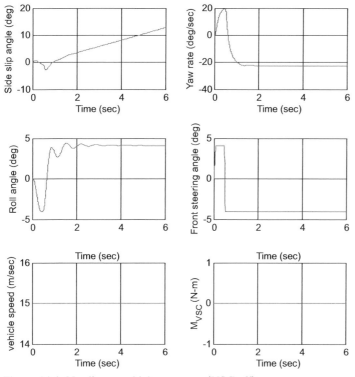

Figure 14.4. Nonlinear vehicle response (VSC off).

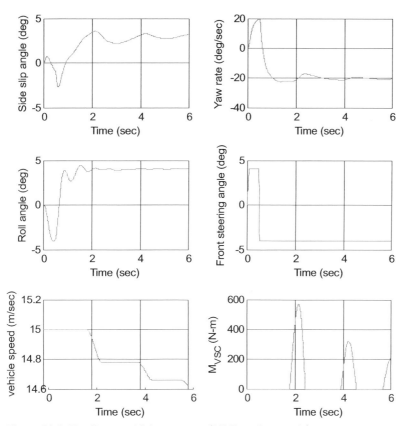

Figure 14.5. Nonlinear vehicle response (VSC on, front axle).

14.4 VSC Design Principles

In the previous section, we successfully develop a nonlinear vehicle model that is suitable for VSC studies. In this section, we discuss the design of VSC in more detail. At the most fundamental level, VSC aims to improve vehicle yaw and sometimes roll stability. Generally speaking, the design principle of the VSC control algorithm can be divided into three parts: (1) estimation of desired vehicle (i.e., yaw) motion and tracking; (2) vehicle side-slip estimation and regulation; and (3) (optional) rollover/wheel-liftoff prevention.

Yaw-Rate Following

The desired yaw motion of a vehicle usually is calculated from the human steering-wheel angle and vehicle states, the most important of which is the forward speed. In some cases, a human driver might steer suddenly to perform an obstacle-avoidance maneuver. In this situation, it is important to use differential braking to achieve the driver's "desired" yaw rate as quickly as possible. When the vehicle side slip increases, the controller then must examine the situation and achieve a good balance between yaw-rate following and limiting side-slip angle.

So, how can we determine a driver's desired yaw motion? Usually, the steady-state yaw rate corresponding to the current driver steering input is assumed to be a good approximation of what the driver wants to accomplish. If a linear (e.g., bicycle) model is used, this steady-state yaw rate can be computed from either the state-space equation or the transfer function from steering to yaw rate. Commonly, the bicycle model is used to figure out the desired yaw rate. Because the transfer function from the front steering angle to yaw rate is:

$$G_r = \frac{r}{\delta_f} = \frac{\frac{u}{R}}{\delta_f} = \frac{u}{L + K_{us}u^2} \qquad (14.11)$$

where u is vehicle forward speed (m/sec), L is the wheel base (m), and $K_{us} \equiv \frac{mb}{LC_{\alpha f}} - \frac{ma}{LC_{\alpha r}}$ is the vehicle understeer coefficient $rad/(m/\sec^2)$. It is common to measure the steering angle and then multiply by this speed-dependent yaw-rate gain to determine the desired vehicle yaw rate. A simple control algorithm then can be used to reduce the difference between the desired and the actual vehicle yaw rates. Frequently, dead-band thresholds are implemented; for example:

$$r_d = \dot{\psi}_d = G_r \delta_f \qquad (14.12)$$

If $\left| \dot{\psi} - \dot{\psi}_d \right| \geq \dot{\psi}_{thresold}$ and $\dfrac{\left| \dot{\psi} - \dot{\psi}_d \right|}{\dot{\psi}_d} \geq \dot{\psi}_{thresold\_perct} \qquad (14.13)$

Then $M_{VSC\_yaw} = K_{yaw}(\dot{\psi}_d - \dot{\psi})$

Else $M_{VSC\_yaw} = 0$

where two thresholds (i.e., absolute and percentage) are used and M_{VSC_yaw} is the VSC control torque due to the yaw-rate-following consideration.

Slip-Angle Estimation and Regulation

Yaw-rate following is an important consideration for vehicle-handling purposes. However, it is equally if not more important to consider vehicle stability. In other words, the vehicle slip angle and/or roll angle should not become excessively large. The accurate estimation of the vehicle side-slip angle is not a trivial task because of the lack of an affordable sensor for vehicle lateral-speed measurement. In an actual implementation, the vehicle side-slip angle usually is estimated from a bicycle model with the available measurements: steering angle, yaw rate, lateral acceleration, and forward speed (Üngören et al. 2002). If we assume that the vehicle side-slip angle can be either measured or estimated, a simple control law again can be used. A common algorithm uses a "sliding-surface" concept, which for this simple problem is equivalent to PD control:

$$\text{If} \quad |\beta| \geq \beta_{threshold} \quad \text{and} \quad \beta \cdot \Delta\beta > 0$$
$$\text{Then} \quad M_{VSC\_\beta} = K_{\beta p} \cdot \beta + K_{\beta d} \cdot \Delta\beta \qquad (14.14)$$
$$\text{Else} \quad M_{VSC\_\beta} = 0$$

where the constraint $\beta \cdot \Delta\beta > 0$ ensures that the control law is turned on only when the magnitude of the side-slip angle is increasing. M_{VSC_β} is the VSC control torque for slip-angle regulation.

Rollover or Wheel Liftoff Prevention

In addition to the yaw and handling performance considerations discussed previously, recent attention has turned to the possibility of using a VSC system for rollover-prevention purposes. This function can be achieved by modifying the desired yaw-rate calculation shown in Eqs. (14.11) and (14.12). For example:

$$\dot{\psi}_d = sat(G_r\delta_f, \dot{\psi}_{d\_threshold}) \qquad (14.15)$$

where $sat()$ is the saturation function and its magnitude limit can be computed according to the vehicle static stability factor (SSF). In other words, the desired yaw rate can be limited so that the corresponding lateral acceleration never exceeds a certain level that would cause rollover concerns. However, the idea shown herein is essentially "static-model–based" and therefore is vulnerable to external disturbances not considered in the model (e.g., a large road-bank angle). A feedback-based algorithm that uses vehicle response (i.e., roll angle, load transfer, or lateral acceleration) is more robust and could be implemented without or in addition to Eq. (14.15). An example is shown here:

$$\text{If} \quad |\phi| \geq \phi_{threshold}$$
$$\text{Then} \quad M_{VSC\_\phi} = -K_{ay} \cdot a_y \qquad (14.16)$$
$$\text{Else} \quad M_{VSC\_\phi} = 0$$

Table 14.1. *Parameter values used in Example 14.4 simulations*

Variable	Value	Unit
$\dot{\psi}_{threshold}$	0.5	deg/sec
$\dot{\psi}_{threshold\_perct}$	2	%
K_{us}	0.0 (assumed)	rad/(m/sec^2)
K_{yaw}	0.2 (or 0 when off)	slip per rad/sec
$\beta_{threshold}$	3	deg
$K_{\beta p}$	2.5 (or 0 when off)	slip per rad
$K_{\beta d}$	0.1 (or 0 when off)	slip per rad
$\phi_{threshold}$	4	deg
K_{ay}	0.0 (off)	slip per (m/sec^2)

where a_y is the vehicle lateral acceleration and M_{VSC_ϕ} is the VSC control torque for rollover prevention.

The overall VSC action then can be "combined." A simple way to combine these three separate control actions is to total them:

$$M_{VSC} = M_{VSC\_yaw} + M_{VSC\_\beta} + M_{VSC\_\phi} \tag{14.17}$$

which, of course, is not necessarily the best way to accomplish this. An alternative, for example, is to emphasize stability when the slip angle or roll angle is high. Only when there are no yaw/roll stability concerns is the yaw rate applied following the control action. This logic can be constructed easily and is an exercise for readers. In the following example, Eq. (14.17) is used as the combined VSC algorithm, which clearly shows the tradeoff among (sometimes) conflicting goals.

EXAMPLE 14.4. The vehicle model used in this example is identical to that shown in Example 14.3. The VSC algorithm, however, is based on Eqs. (14.13) through (14.17). There is another slight change: Rather than calculating desired yaw-control moments (i.e., $M_{VSC} = M_{VSC\_yaw} + M_{VSC\_\beta} + M_{VSC\_\phi}$), desired wheel-slip ratios (i.e., $s_{VSC} = s_{VSC\_yaw} + s_{VSC\_\beta} + s_{VSC\_\phi}$) are obtained. Depending on the sign of the final control action (i.e., $s_{VSC} = s_{VSC\_yaw} + s_{VSC\_\beta} + s_{VSC\_\phi}$), either the right- or left-front tire is actuated. This modification is necessary because tire forces are calculated from the combined-slip tire model. The tradeoff between yaw-rate following and slip-angle regulation is demonstrated in this example. Because the vehicle parameters used in this example correspond to a passenger vehicle, rollover is not a concern; therefore, rollover-prevention control is turned off to make it easier to see the tradeoff between the remaining two objectives.

For an SUV or light truck, the control logic must be more elaborate. When all three objectives are considered, careful examination of the behavior and timing of each control action is needed to ensure that they do not undermine one another.

The control parameters in the simulations are summarized in Table 14.1. All of the control gains have units for "desired slip" rather than "desired yaw moment," as explained previously. Four sets of data are shown here: no control, yaw only, slip angle, and yaw/slip.

No control:

Yaw only:

Slip only:

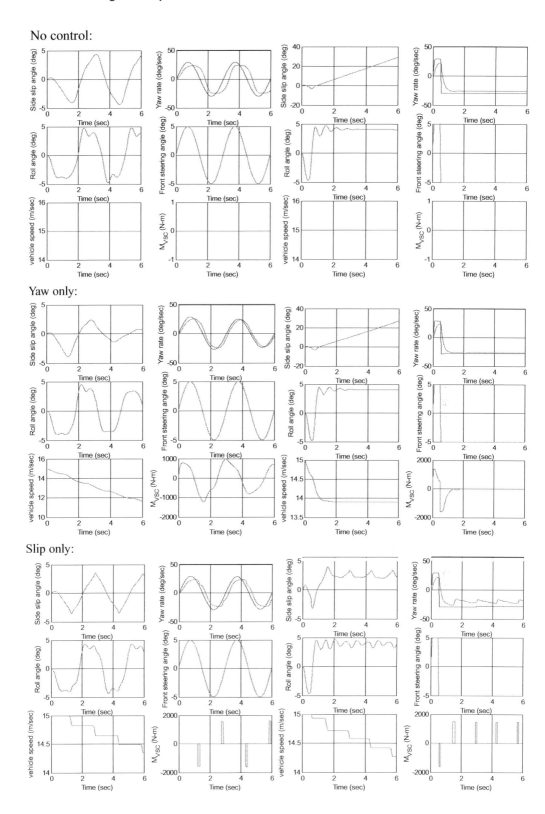

Yaw/Slip:

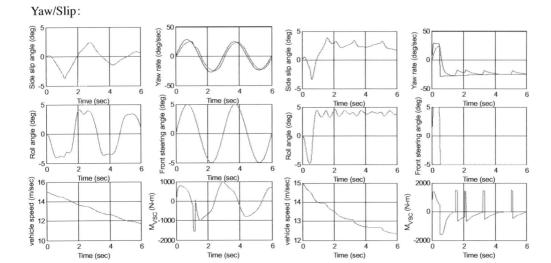

PROBLEMS

1. Use the parameter values given in Example 14.1 and change the Magic Formula parameters B_y, D_y, and E_y by ± 5 percent from their nominal values to show how these parameters affect the calculated forces. Plot the results.

2. Based on results obtained from Example 14.1, explain why one of the objectives of VSC is to maintain a small side-slip angle, β.

3. Equation 14.3 describes the effect of control yaw-moment M_{VSC} on the three-DOF vehicle lateral/yaw/roll model. Show the effect of the control yaw-moment M_{VSC} on the two-DOF vehicle lateral/yaw model developed in Equation 4.49 (see Chapter 4).

4. Repeat the simulation in Example 14.2 with non-zero values of the control gains k_b and k_ϕ. Initially select very small gains and then gradually increase them to show the effect of including the control.

5. Using the nonlinear model (see Example 14.3) and comparing your results to Figures 14.4 and 14.5, determine (by trial and error) appropriate control gains k_b and k_ϕ. The control action is turned on only if the magnitude of the vehicle side-slip angle is higher than a certain threshold; determine an appropriate threshold.

6. Repeat the simulation in Example 14.4 but also include – based on the simple strategy in Eq. (14.17) – the control action for preventing rollover.

7. Using simulations based on Example 14.4, propose and demonstrate an alternative to Eq. (14.17) for combining the various elements of a VSC logic.

REFERENCES

Allen, R. W., D. H. Klyde, T. J. Rosenthal and D. M. Smith 1988, "Analytical Modeling of Driver Response," in *Crash Avoidance Maneuvering, Vol. I: Technical Background*, DOT-HS-807-270, April.

Hac, A., and M. O. Bodie, 2002, "Improvements in Vehicle Handling Through Integrated Control of Chassis Systems," *International Journal of Vehicle Design*, Vol. 29, No. 1–2, pp. 23–50.

Hoffman, D. D., and M. D. Rizzo, 1998, "Chevrolet C5 Corvette Vehicle Dynamic Control System," SAE Paper No. 980233, Warrendale, PA: Society of Automotive Engineers.

Koibuchi, K., M. Yamamoto, Y. Fukada, and S. Inagaki, 1996, "Vehicle Stability Control in Limit Cornering By Active Brake," SAE Paper No. 960487, SAE Congress.

Kraft, H. J., and H. Leffler, 1990, "The Integrated Brake and Stability Control System of the New BMW 850i," SAE Paper No. 900209, SAE Congress.

Lie, Anders, C. Tingvall, M. Krafft, and A. Kullgren, 2006, "The Effectiveness of Electronic Stability Control (ESC) in Reducing Real-Life Crashes and Injuries," *Traffic Injury Prevention*, Vol. 7, No. 1, pp. 38–43.

Ma, W-H., 1998, *Worst-Case Evaluation Methods for Vehicles and Vehicle Control Systems*, Ph.D. Thesis (Mechanical Engineering), University of Michigan.

Matsumoto, S., H. Yamaguchi, H. Inoue, and Y. Yasuno, 1992, "Improvement of Vehicle Dynamics Through Braking Force Distribution Control," SAE Paper No. 920645, SAE Congress.

Mills, V., B. Samuels, and J. Wagner, 2002, "Modeling and Analysis of Automotive Antilock Brake Systems Subject to Vehicle Payload Shifting," *Vehicle System Dynamics*, Vol. 37, No. 4, April, pp. 283–310.

Nagai, M., M. Shino, and F. Gao, 2002, "Study on Integrated Control of Active Front Steer Angle and Direct Yaw Moment," *JSAE Review*, Vol. 23, No. 3, July, pp. 309–15.

Pilutti, T., A. G. Ulsoy, and D. Hrovat, 1998, "Vehicle Steering Intervention Through Differential Braking," *ASME Journal of Dynamic Systems, Measurement and Control*, Vol. 120, No. 3, September, pp. 314–21.

Piyabongkarn, D., J. Y. Lew, R. Rajamani, and J. A. Grogg, 2010, "Active Driveline Torque Management Systems," *IEEE Control Systems Magazine*, Vol. 30, No. 4, August, pp. 86–102.

Shibahata, Y., K. Shimada, and T. Tomari, 1993, "Improvement of Vehicle Maneuverability by Direct Yaw Moment Control," *Vehicle System Dynamics*, Vol. 22, pp. 465–81.

Tseng, H. E., B. Ashrafi, D. Madau, T. A. Brown, and D. Recker, 1999, "The Development of Vehicle Stability Control at Ford," *IEEE/ASME Transactions on Mechatronics*, Vol. 4, No. 3, September, pp. 223–34.

Üngören, A.Y., H. Peng, and H. Tseng, 2002, "Experimental Verification of Lateral Speed Estimation Methods," *Proceedings of the Advanced Vehicle Control Conference*, Hiroshima, Japan, September.

Zanten, A. V., R. Erhardt, and G. Pfaff, 1995, "VDC, the Vehicle Dynamics Control System of Bosch," SAE Paper No. 950759, SAE Congress.

Zanten, A. V., 2000, "Bosch ESP Systems: 5 Years of Experience," SAE Paper No. 2000-01-1633, SAE Congress.

15 Four-Wheel Steering

15.1 Basic Properties

In Chapter 4, we review the lateral dynamics of a front-wheel steering (FWS) vehicle. Following the same procedure, the dynamic equations for a four-wheel-steering (4WS) "bicycle" vehicle can be derived. The state-space model of the 4WS vehicle model is as follows:

$$
\begin{bmatrix} \dot{v} \\ \dot{r} \end{bmatrix} = \begin{bmatrix} \dfrac{-(C_{\alpha f} + C_{\alpha r})}{m u_o} & \dfrac{b C_{\alpha r} - a C_{\alpha f}}{m u_o} - u_o \\ \dfrac{b C_{\alpha r} - a C_{\alpha f}}{I_z u_o} & \dfrac{-(C_{\alpha f} a^2 + C_{\alpha r} b^2)}{I_z u_o} \end{bmatrix} \begin{bmatrix} v \\ r \end{bmatrix} + \begin{bmatrix} \dfrac{C_{\alpha f}}{m} & \dfrac{C_{\alpha r}}{m} \\ \dfrac{a C_{\alpha f}}{I_z} & \dfrac{-b C_{\alpha r}}{I_z} \end{bmatrix} \begin{bmatrix} \delta_f \\ \delta_r \end{bmatrix}
$$

$$(15.1)$$

If we include the lateral displacement and the yaw angle as the two extra state variables, then time-domain performance of 4WS vehicles (e.g., lane following) also can be simulated. The fourth-order state-space model is obtained by adding two trivial dynamic equations, $\dot{y} = v + u_o \psi$ and $\dot{\psi} = r$, to the two-state model:

$$
\frac{d}{dt} \begin{bmatrix} y \\ v \\ \psi \\ r \end{bmatrix} = \begin{bmatrix} 0 & 1 & u_o & 0 \\ 0 & \dfrac{-(C_{\alpha f} + C_{\alpha r})}{m u_o} & 0 & \dfrac{b C_{\alpha r} - a C_{\alpha f}}{m u_o} - u_o \\ 0 & 0 & 0 & 1 \\ 0 & \dfrac{b C_{\alpha r} - a C_{\alpha f}}{I_z u_o} & 0 & \dfrac{-(C_{\alpha f} a^2 + C_{\alpha r} b^2)}{I_z u_o} \end{bmatrix} \begin{bmatrix} y \\ v \\ \psi \\ r \end{bmatrix} + \begin{bmatrix} 0 & 0 \\ \dfrac{C_{\alpha f}}{m} & \dfrac{C_{\alpha r}}{m} \\ 0 & 0 \\ \dfrac{a C_{\alpha f}}{I_z} & \dfrac{-b C_{\alpha r}}{I_z} \end{bmatrix} \begin{bmatrix} \delta_f \\ \delta_r \end{bmatrix}
$$

$$(15.2)$$

Before introducing the 4WS control strategy, we examine basic characteristics of FWS and rear-wheel steering (RWS) vehicles, which helps us to understand what we gain by steering the rear wheels. In the following discussion, we first prove by using Eq. (15.1) that the rear wheels are as effective as the front wheels in terms of yaw-rate generation. This fact then is verified using the MATLAB program in Example 15.1. At steady-state, we obtain the following equations from Eq. (15.1):

$$
0 = -\frac{C_{\alpha f} + C_{\alpha r}}{u_o} v + \frac{b C_{\alpha r} - a C_{\alpha f} - m u_o^2}{u_o} r + C_{\alpha f} \delta_f + C_{\alpha r} \delta_r
$$

$$
0 = \frac{b C_{\alpha r} - a C_{\alpha f}}{u_o} v - \frac{a^2 C_{\alpha f} + b^2 C_{\alpha r}}{u_o} r + a C_{\alpha f} \delta_f - b C_{\alpha r} \delta_r
$$

Eliminating the terms containing the lateral speed, v, we obtain:

$$0 = \frac{-(bC_{\alpha r} - aC_{\alpha f})mu_o^2 - C_{\alpha f}C_{\alpha r}(a+b)^2}{u_o}r + (a+b)C_{\alpha f}C_{\alpha r}\delta_f - (a+b)C_{\alpha f}C_{\alpha r}\delta_r$$

which shows that steering angles of the same magnitude but opposite sign at the front and rear axles generate the same yaw-rate response (at steady-state). However, the following sketch clearly shows that the transient-force generation at the rear axle is in the opposite direction of steady-state force. In other words, there is a non-minimum-phase (NMP) response in the vehicle lateral acceleration for a RWS vehicle, which is not the case for a FWS vehicle. The NMP characteristic makes it difficult for human drivers to control a RWS vehicle at high vehicle speed, which is a primary reason why they are not as popular as FWS vehicles.

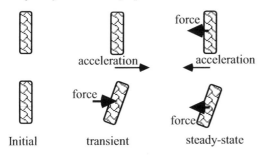

EXAMPLE 15.1: COMPARISON OF FWS AND RWS VEHICLES. The frequency responses of the FWS and RWS cases show that the steady-state yaw responses are indeed identical, shifted only by 180 degrees in phase. The acceleration responses, however, are very different. The 180-degree phase change of the lateral-acceleration response for a RWS vehicle also shows that it is NMP.

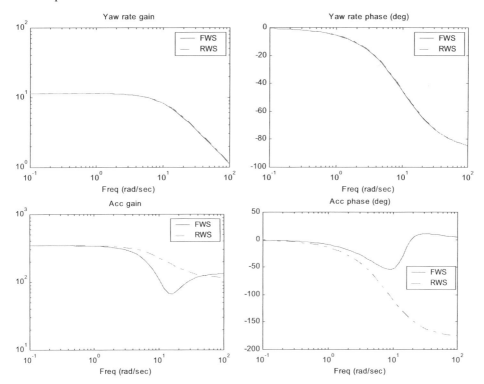

```
%Ex15_1.m
a=1.14; l=2.54; b=l-a;
g=9.81; u0=30.0;
m=1000; Iz=1400.0;
Caf=2400.0*57.2958; Car=2000*57.2958;
A=[ 0,    1,    u0,    0;
    0, -(Caf+Car)/(m*u0), 0, (b*Car-a*Caf)/(m*u0)-u0;
    0,    0,    0,    1;
    0, (b*Car-a*Caf)/(Iz*u0), 0,  -(a*a*Caf+b*b*Car)/(Iz*u0)];
B_2ws=[0; Caf/m; 0; a*Caf/Iz];
B_rws=[0; Car/m; 0; -b*Car/Iz];
C_r=[0,0,0,1];
C_acc=A(2,:)+u0*[0,0,0,1];
% Bode plots
w=logspace(-1,2,50);
[m_2ws_r, p_2ws_r]=bode(A,B_2ws,C_r,0,1,w);
[m_rws_r, p_rws_r]=bode(A,B_rws,C_r,0,1,w);
[m_2ws_acc, p_2ws_acc]=bode(A,B_2ws,C_acc,B_2ws(2),1,w);
[m_rws_acc, p_rws_acc]=bode(A,B_rws,C_acc,B_rws(2),1,w);
loglog(w,m_2ws_r, w, m_rws_r,'-.');
xlabel('Freq (rad/sec)'); title('Yaw rate gain');
legend('FWS','RWS'); pause
% reverse phase of RWS vehicle for comparison
semilogx(w,p_2ws_r, w, p_rws_r-180,'-.');
xlabel('Freq (rad/sec)'); title('Yaw rate phase (deg)');
legend('FWS','RWS'); pause
loglog(w,m_2ws_acc, w, m_rws_acc,'-.');
xlabel('Freq (rad/sec)'); title('Acc gain');
legend('FWS','RWS'); pause
semilogx(w,p_2ws_acc, w, p_rws_acc-180,'-.');
xlabel('Freq (rad/sec)'); title('Acc phase (deg)');
legend('FWS','RWS');
```

15.2 Goals of 4WS Algorithms

When the average operating speed of motor vehicles is not very high, the need for 4WS is not obvious. When higher performance at higher speeds is needed, the possibility of utilizing rear steering is explored.

The 4WS control strategies used in the past are mostly feed-forward in nature. In other words, the rear wheels usually are steered as a fixed function of the front-steering angle, regardless of vehicle yaw rate and lateral-acceleration response. The 4WS strategies were designed to achieve one of several of the following desirable characteristics:

- reduced vehicle side-slip angle
- reduced yaw rate and lateral-acceleration phase difference
- reduced low-speed turning radius
- consistent steering response (model matching)
- increased tire-force reserve (especially on a low-friction surface)

The most common goal of 4WS systems proposed in the past was to reduce the steady-state side-slip angle, β (or, equivalently, lateral speed v) to zero by adjusting the ratio of the RWS angle to the FWS angle. This has both ergonomic and dynamic aspects that must be considered. In terms of ergonomics, the vehicle becomes easier to steer because it proceeds in the forward direction without any side slip. In terms of dynamics, less energy is required when a vehicle enters a turn from a straight-line course because with 4WS, it is necessary to turn only the tires, which have low inertia. In two-wheel steering, in which a side-slip angle develops, the vehicle must be turned and it has a high inertia. If we set $v = 0$ in Eq. (15.1), eliminate r from the resulting equations, and solve for the ratio of the RWS angle to the FWS angle, we obtain the following condition for zero side-slip angle at steady-state:

$$K_r = \frac{\delta_r}{\delta_f} = \frac{mu_o^2 a - b(a+b)C_{\alpha r}}{mu_o^2 b + a(a+b)C_{\alpha f}} \cdot \frac{C_{\alpha f}}{C_{\alpha r}} \tag{15.3}$$

Clearly, the steering ratio K_r to achieve the zero side-slip angle is a function of longitudinal speed (Figure 15.1a). Equation (15.3) shows that at high speeds, the rear wheel should be turned in the same direction but somewhat less than the front wheel. At low speeds, the rear wheel should be turned in the direction opposite to that of the front wheel. Steering the wheels according to Eq. (15.3) achieves good stability in the intermediate to high speed ranges. However, it does not always provide a favorable steering sensation for the driver because it can produce a strong understeer characteristic (from the driver's perspective), which causes the yaw response to deteriorate. Steering the rear wheels in the opposite direction (i.e., phase) from the front wheels at low speed improves vehicle maneuverability. From this basic steady-state analysis, we observe the desirability of steering the rear wheels in proportion to the front wheels as a function of vehicle forward velocity. Development of an appropriate control strategy, however, is not as straightforward. Some authors have shown disadvantages of 4WS (Nalecz and Bidermann 1988) and argued against the desirability of a zero side-slip angle as the basis for 4WS control (Abe 1990). Examples 15.2 and 15.3 show simulation results for various 4WS control strategies.

EXAMPLE 15.2: TIRE SLIP-ANGLE COMPARISON. If the fixed-gain 4WS algorithm described in Eq. (15.3) is applied, we can see that the tire slip angle at both front and rear axles increases compared to 2WS vehicles in the transient (i.e., at steady-state, they are the same). In other words, direct implementation of the zero-vehicle-slip algorithm may cause undesirable side effects.

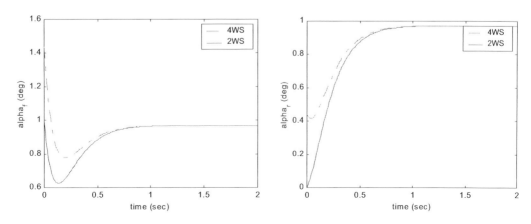

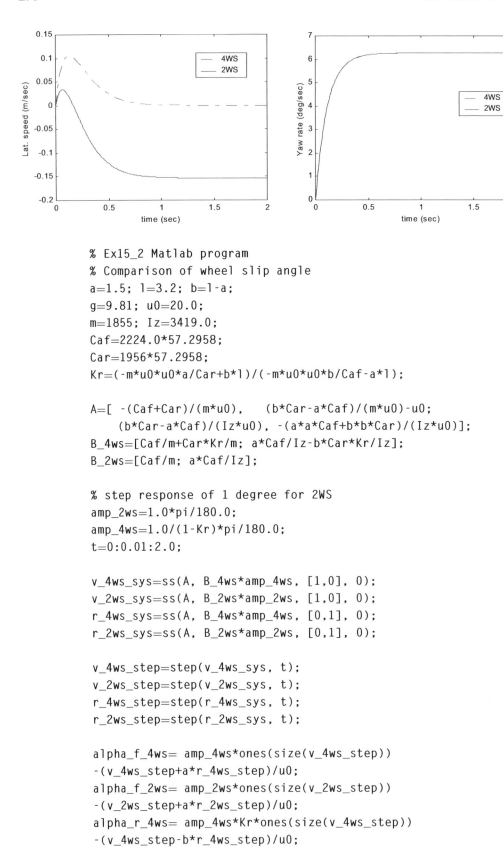

```
% Ex15_2 Matlab program
% Comparison of wheel slip angle
a=1.5; l=3.2; b=l-a;
g=9.81; u0=20.0;
m=1855; Iz=3419.0;
Caf=2224.0*57.2958;
Car=1956*57.2958;
Kr=(-m*u0*u0*a/Car+b*l)/(-m*u0*u0*b/Caf-a*l);

A=[ -(Caf+Car)/(m*u0),    (b*Car-a*Caf)/(m*u0)-u0;
    (b*Car-a*Caf)/(Iz*u0), -(a*a*Caf+b*b*Car)/(Iz*u0)];
B_4ws=[Caf/m+Car*Kr/m; a*Caf/Iz-b*Car*Kr/Iz];
B_2ws=[Caf/m; a*Caf/Iz];

% step response of 1 degree for 2WS
amp_2ws=1.0*pi/180.0;
amp_4ws=1.0/(1-Kr)*pi/180.0;
t=0:0.01:2.0;

v_4ws_sys=ss(A, B_4ws*amp_4ws, [1,0], 0);
v_2ws_sys=ss(A, B_2ws*amp_2ws, [1,0], 0);
r_4ws_sys=ss(A, B_4ws*amp_4ws, [0,1], 0);
r_2ws_sys=ss(A, B_2ws*amp_2ws, [0,1], 0);

v_4ws_step=step(v_4ws_sys, t);
v_2ws_step=step(v_2ws_sys, t);
r_4ws_step=step(r_4ws_sys, t);
r_2ws_step=step(r_2ws_sys, t);

alpha_f_4ws= amp_4ws*ones(size(v_4ws_step))
-(v_4ws_step+a*r_4ws_step)/u0;
alpha_f_2ws= amp_2ws*ones(size(v_2ws_step))
-(v_2ws_step+a*r_2ws_step)/u0;
alpha_r_4ws= amp_4ws*Kr*ones(size(v_4ws_step))
-(v_4ws_step-b*r_4ws_step)/u0;
```

```
alpha_r_2ws=-(v_2ws_step-b*r_2ws_step)/u0;

plot(t,alpha_f_4ws*180/pi,'-.r', t,alpha_f_2ws*180/pi,'b')
xlabel('time (sec)')
ylabel('alpha_f (deg)'); legend('4WS','2WS')
pause

plot(t,alpha_r_4ws*180/pi,'-.r', t,alpha_r_2ws*180/pi,'b')
xlabel('time (sec)'),
ylabel('alpha_r (deg)'); legend('4WS','2WS')
pause

plot(t,v_4ws_step,'-.r', t,v_2ws_step,'b')
xlabel('time (sec)')
ylabel('Lat. speed (m/sec)'); legend('4WS','2WS')
pause
plot(t,r_4ws_step*180/pi,'-.r', t,r_2ws_step*180/pi,'b')
xlabel('time (sec)')
ylabel('Yaw rate (deg/sec)')
legend('4WS','2WS')
```

EXAMPLE 15.3: A PROTOTYPE 4WS SYSTEM. Shown in Figure 15.1 is a block diagram of a simulation model for 4WS. Here, A_f and A_r represent candidate control algorithms, as shown in Figure 15.1c. The transfer functions H_f, H_r, G_f, and G_r represent the two-DOF vehicle dynamics in transfer-function form. Simulation and experimental results, comparing the candidate control strategies, are provided in Figure 15.2. Controller A is the FWS vehicle that is considered here as a basis for comparison. Controller B is simply a proportional controller with gains $A_r = K_r$ (calculated from Eq. (15.3)) and $A_f = 0$. Controllers C and D are similar to Controller B but include a mechanism to generate phase lag. Controller E includes a lead action for the front wheels and a lag for the rear wheels. The modification of the FWS angle has the effect of more quickly steering the wheels to a certain side-slip angle. These results show that both response and stability can be improved substantially by steering both wheels. However, reducing the side-slip angle to zero (as in Controller B) causes the yaw-rate gain to drop, resulting in the deterioration of cornering performance. Also, the side-slip angle, although zero at steady-state, becomes negative during the transient, which can be disconcerting to a driver. Thus, setting the side-slip angle to zero causes problems for the practical use of the vehicle. The introduction of lead and/or lag actions can improve these characteristics.

The results of Example 15.3 also show a dilemma for 4WS system designs. On the one hand, consumers (and engineers) are reluctant to accept steer-by-wire vehicles; on the other hand, modifying only the RWS angle reduces the yaw and lateral-acceleration characteristics of the vehicle. The latter fact has drawn some critics toward 4WS vehicles, based on the human-factors perspective. The major obstacle to the wide implementation of 4WS vehicles, however, is actually the cost.

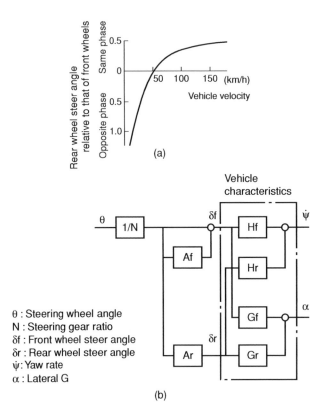

Figure 15.1. (a) Rear Wheel Steer angle relative to that of front wheels for zero sideslip angle. (b) Block diagram of simulation model. (c) Control functions.

EXAMPLE 15.4: YAW RATE AND LATERAL-ACCELERATION RESPONSE. In this example, we compare the frequency-response plots of the yaw rate and lateral-acceleration signals of 2WS and 4WS vehicles. By simply applying the proportional-gain 4WS strategy, we see that the phase difference becomes greatly reduced. This is somewhat expected because lateral acceleration determines how quickly the vehicle lateral speed builds up, and yaw rate determines how quickly a vehicle is reoriented. If the phase difference between these two signals is reduced, the side-slip angle also reduces (and vice versa).

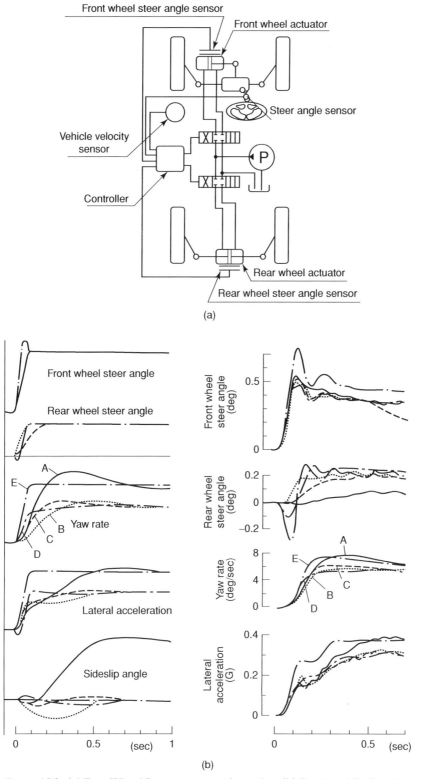

Figure 15.2. (a) Four Wheel Steer system configuration. (b) Simulated (*left*) and experimental (*right*) step response results for 4WS.

Phase diff. (deg)

```
% Ex15_4.m
a=1.14; l=2.54; b=l-a;
g=9.81; u0=30.0;
m=1000; Iz=1400.0;
Caf=2400.0*57.2958; Car=2000.0*57.2958;
Kr=(-m*u0*u0*a/Car+b*l)/(-m*u0*u0*b/Caf-a*l);

A=[ 0,     1,      u0,     0;
    0, -(Caf+Car)/(m*u0), 0, (b*Car-a*Caf)/(m*u0)-u0;
    0,     0,      0,      1;
    0, (b*Car-a*Caf)/(Iz*u0), 0, -(a*a*Caf+b*b*Car)/(Iz*u0)];
B_4ws=[0; Caf/m+Car*Kr/m; 0; a*Caf/Iz-b*Car*Kr/Iz];
B_2ws=[0; Caf/m; 0; a*Caf/Iz];
B_rws=[0; Car/m; 0; -b*Car/Iz];
C_r=[0,0,0,1];
C_acc=A(2,:)+u0*[0,0,0,1];
% Bode plots
w=logspace(-1,2,50);
[m_2ws_r, p_2ws_r]=bode(A,B_2ws,C_r,0,1,w);
[m_rws_r, p_rws_r]=bode(A,B_rws,C_r,0,1,w);
[m_4ws_r, p_4ws_r]=bode(A,B_4ws,C_r,0,1,w);
[m_2ws_acc, p_2ws_acc]=bode(A,B_2ws,C_acc,B_2ws(2),1,w);
[m_rws_acc, p_rws_acc]=bode(A,B_rws,C_acc,B_rws(2),1,w);
[m_4ws_acc, p_4ws_acc]=bode(A,B_4ws,C_acc,B_4ws(2),1,w);

loglog(w,m_2ws_r, w, m_rws_r,'-.');
xlabel('Freq (rad/sec)'); title('Yaw rate gain');
legend('FWS','RWS'); pause

% reverse phase of RWS vehicle for comparison
semilogx(w,p_2ws_r, w, p_rws_r-180,'-.');
```

```
xlabel('Freq (rad/sec)'); title('Yaw rate phase (deg)');
legend('FWS','RWS'); pause
loglog(w,m_2ws_acc, w, m_rws_acc,'-.');
xlabel('Freq (rad/sec)'); title('Acc gain');
legend('FWS','RWS'); pause
semilogx(w,p_2ws_acc, w, p_rws_acc-180,'-.');
xlabel('Freq (rad/sec)'); title('Acc phase (deg)');
legend('FWS','RWS'); pause
semilogx(w,p_4ws_r-p_4ws_acc,w,p_2ws_r-p_2ws_acc,'-.')
xlabel('Freq (rad/sec)'); title('Phase diff. (deg)');
legend('4WS','FWS')
```

EXAMPLE 15.5: LANE-CHANGE MANEUVER: 2WS VERSUS 4WS. In this example, we modify the simulation program discussed in Chapter 5 (i.e., driver modeling) to compare the 2WS and 4WS vehicles under the same double-lane–change maneuver. It is observed that the 4WS vehicle achieves a smaller tracking error (except at the first peak, when the vehicle is instantaneously changing orientation) and a smaller yaw rate and acceleration. Moreover, the improved performance is achieved with a smaller steering command from the driver. Therefore, it is fair to say that this 4WS vehicle is easier to handle than a 2WS vehicle.

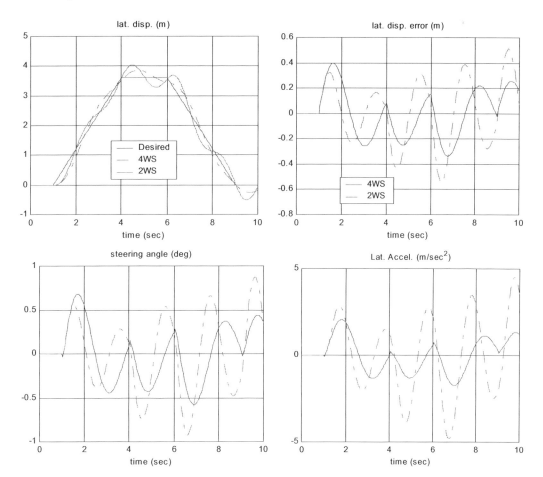

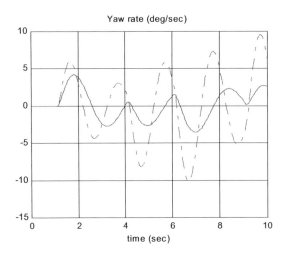

```
% Ex15_5.m
a=1.14; l=2.54; b=l-a;
g=9.81; u0=30.0;
m=1000; Iz=1400.0;
Caf=2400.0*57.2958; Car=2000*57.2958;
Kr=(-m*u0*u0*a/Car+b*l)/(-m*u0*u0*b/Caf-a*l);
A=[ 0,    1,    u0,    0;
    0, -(Caf+Car)/(m*u0), 0, (b*Car-a*Caf)/(m*u0)-u0;
    0,    0,    0,    1;
    0, (b*Car-a*Caf)/(Iz*u0), 0, -(a*a*Caf+b*b*Car)/(Iz*u0)];
B_4ws=[0; Caf/m+Car*Kr/m; 0; a*Caf/Iz-b*Car*Kr/Iz];
B_2ws=[0; Caf/m; 0; a*Caf/Iz];
C=[1,0,0,0]; D=0;
[num_4ws,Gvden]=ss2tf(A,B_4ws,C,D,1);
[num_2ws,Gvden]=ss2tf(A,B_2ws,C,D,1);
Gvnum_4ws=num_4ws(3:5);
Gvnum_2ws=num_2ws(3:5);
Kd = 0.03;
T = 0.1; Tp=0.3;
Gvnum_4ws=conv(Gvnum_4ws, [Tp^2/2 Tp 1]);
Gvnum_2ws=conv(Gvnum_2ws, [Tp^2/2 Tp 1]);
Gdnum=Kd*[-T/2 1];
Gdden=[T/2 1];
Gcnum_4ws=conv(Gdnum,Gvnum_4ws);
Gcnum_2ws=conv(Gdnum,Gvnum_2ws);
Gcden_4ws=conv(Gdden,Gvden) + Gcnum_4ws;
Gcden_2ws=conv(Gdden,Gvden) + Gcnum_2ws;
t=[0:0.05:10];
yd=[zeros(1,20), 0:0.06:3.6, 3.6*ones(1,39), 3.6:-
0.06:0, zeros(1,20)];
y_4ws=lsim(Gcnum_4ws,Gcden_4ws,yd,t);
y_2ws=lsim(Gcnum_2ws,Gcden_2ws,yd,t);
```

```
e_4ws=yd-y_4ws';
e_2ws=yd-y_2ws';
plot(t,yd,t,y_4ws','-.',t,y_2ws,':'); grid
xlabel('time (sec)'), legend('Desired','4WS','2WS')
title('lat. disp. (m)'), pause
plot(t,yd-y_4ws',t,yd-y_2ws','-.'); grid
xlabel('time (sec)'), legend('4WS','2WS')
title('lat. disp. error (m)'), pause
% Steering angle is the output of the driver block Gd
% under the input signal e=yd-y
steer_4ws=lsim(Gdnum,Gdden,e_4ws,t);
steer_2ws=lsim(Gdnum,Gdden,e_2ws,t);
plot(t,steer_4ws*180/pi, t,steer_2ws*180/pi,'-.'); grid
xlabel('time (sec)')
title('steering angle (deg)'), pause
% Lateral acceleration = v_dot + u0*r
C_acc=A(2,:)+u0*[0 0 0 1];
acc_4ws=lsim(A,B_4ws,C_acc,B_4ws(2),steer_4ws,t);
acc_2ws=lsim(A,B_2ws,C_acc,B_2ws(2),steer_2ws,t);
plot(t,acc_4ws,t,acc_2ws,'-.'); grid
xlabel('time (sec)')
title('Lat. Accel. (m/sec^2)')
pause
% Yaw rate
C_r=[0,0,0,1];
r_4ws=lsim(A,B_4ws,C_r,0,steer_4ws,t);
r_2ws=lsim(A,B_2ws,C_r,0,steer_2ws,t);
plot(t,r_4ws*180/pi,t,r_2ws*180/pi,'-.'); grid
xlabel('time (sec)')
title('Yaw rate (deg/sec)')
```

PROBLEMS

1. The state equation for the 4WS vehicle under curved-road following was found to be:

$$\frac{d}{dt}\begin{bmatrix} y \\ v \\ \psi - \psi_d \\ r \end{bmatrix} = \begin{bmatrix} 0 & 1 & u_o & 0 \\ 0 & -\dfrac{C_{\alpha f} + C_{\alpha r}}{m u_o} & 0 & \dfrac{b C_{\alpha r} - a C_{\alpha f}}{m u_o} - u_o \\ 0 & 0 & 0 & 1 \\ 0 & \dfrac{b C_{\alpha r} - a C_{\alpha f}}{I_z u_o} & 0 & -\dfrac{a^2 C_{\alpha f} + b^2 C_{\alpha r}}{I_z u_o} \end{bmatrix} \begin{bmatrix} y \\ v \\ \psi - \psi_d \\ r \end{bmatrix}$$

$$+ \begin{bmatrix} 0 & 0 \\ \dfrac{C_{\alpha f}}{m} & \dfrac{C_{\alpha r}}{m} \\ 0 & 0 \\ \dfrac{a C_{\alpha f}}{I_z} & \dfrac{-b C_{\alpha r}}{I_z} \end{bmatrix} \begin{bmatrix} \delta_f \\ \delta_r \end{bmatrix} + \begin{bmatrix} 0 \\ 0 \\ -1 \\ 0 \end{bmatrix} r_d$$

For a vehicle with the parameters:

$$l = 2.54 \text{ m} \qquad a = 1.14 \text{ m} \qquad b = l - a = 1.40 \text{ m}$$
$$g = 9.81 \text{ m/s}^2 \qquad u_0 = 25.0 \text{ m/s} \qquad m = 1,400 \text{ kg}$$
$$I_z = 2,500.0 \text{ kg-m}^2 \qquad C_{af} = 2,400.0 \text{ N/deg} \qquad C_{ar} = 2,000.0 \text{ N/deg}$$

plot the steering gain K_r as a function of vehicle longitudinal velocity, u_0.

2. For the same vehicle model and parameters given in Problem 1, use the preview driver program in Example 15.4 (using the parameters listed previously) to simulate the vehicle and driver. Simulate 2WS (FWS), 4WS (i.e., with zero-slip proportional RWS $\delta_r = K_r \delta_f$, the FWS angle is controlled by the human driver), and RWS vehicles under the same road-curvature conditions. The goal is to compare simulation results and assess the handling of the three vehicles (e.g., 4WS is easier to handle than 2WS). Adjust the preview driver gain, preview time, and neuromuscular time delay for the three cases separately, if necessary (i.e., to stabilize the closed-loop system, if possible, and generate responses that aid in the comparison). Submit simulation plots for:

- lateral displacement (i.e., tracking error)
- FWS angle (i.e., human control effort)
- lateral acceleration
- yaw rate

Also, if using different driver gains and time delays, make sure that these numbers are shown clearly on the graphs.

Hint: The input Matrix B for the 4WS case should be in MATLAB format, as follows:

B_4ws=[0; Caf/m+Car*Kr/m; 0; a*Caf/Iz-b*Car*Kr/Iz];

3. You work as an engineer for an automotive supplier that produces 4WS systems, and your customer asks you to compare the performance of a proposed 4WS system to a traditional 2WS system through simulations that include both a driver and a vehicle model. For the 2WS-vehicle model, you utilize the two-DOF vehicle lateral-yaw-dynamics model in Chapter 4 to obtain the following vehicle-dynamics transfer function:

$$G_v(s) = \frac{y(s)}{\delta(s)} = \frac{(b_2 s^2 + b_1 s + b_0)}{s^2(s^2 + a_1 s + a_0)}$$

where δ is the FWS angle input and y is the vehicle lateral position. Additionally, you develop a preview-based driver model as follows: (1) the basic driver model is assumed to be proportional plus delay, $G_d(s) = K_d \exp(-sT)$; (2) the error, e, is calculated as $e(t + T_p) = r(t + T_p) - y(t + T_p)$; and (3) it is assumed that the desired lateral position at some preview time, T_p, in the future, $r(t + T_p)$, is known and that the actual position at that same time, $y(t + T_p)$, is estimated from the simple rectilinear motion model:

$$y_{Tp} = y(t + T_p) = y(t) + T_p \dot{y}(t) + \left(T_p^{2/2}\right)\ddot{y}(t)$$

The parameter values to be used for the models are $a = 1.14$ m, $b = 1.40$ m, $m = 1,500$ kg, $I_z = 2,420$ (N.m.s^2)/rad, $C_{af} = (2,050)(57.3)$ N/rad, $C_{ar} = (1,675)(57.3)$ N/rad,

$T = 0.05$ sec, and $T_p = 0.5$ sec. Also, select the parameter value K_d to tune the response and specify which value was used.

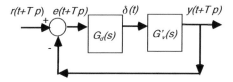

Based on the previous vehicle-dynamics transfer function and preview model, complete the following tasks:

(a) Simulate the 2WS driver–vehicle system for the reference input trajectory, $r(t)$, shown in the following figure and use three values of longitudinal vehicle speed u_0: low, medium, and high.

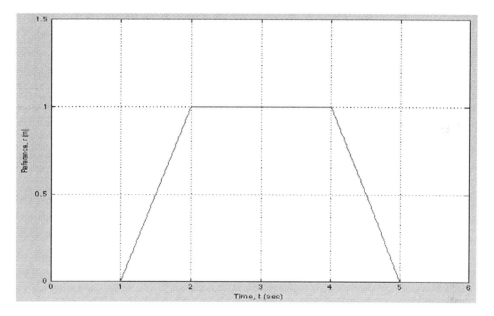

(b) Repeat the simulation in (a) but for the same vehicle with 4WS. Assume that the 4WS is based on the adjustment of the RWS angle in proportion to the FWS angle generated by the driver. That is,

$$\delta_r = K_r \delta = \left\{ \left(\frac{mu_0^2 a - b(a+b)C_{\alpha r}}{mu_0^2 b + a(a+b)C_{\alpha f}} \right) \left(\frac{C_{\alpha f}}{C_{\alpha r}} \right) \right\} \delta$$

Compare the results obtained in (a) and (b) at three different speeds (i.e., three values of u_0) designated as low, medium, and high. Justify how the three speeds used in these comparisons were selected.

REFERENCES

Abe, M., 1990, "Analysis on Free Control Stability of a Four-Wheel-Active-Steer Vehicle," *JSAE Review*, Vol. 11, No. 3, p. 28.

Fukunaga, Y., N. Irie, J. Kuroki, and F. Sugasawa, 1987, "Improved Handling and Stability Using Four-Wheel Steering," *11th International Technical Conference on Experimental Safety Vehicles*, Washington, DC, Section 4, pp. 415–25.

Furukawa, Y., N. Yuhara, S. Sano, H. Takeda, and Y. Matsushita, 1989, "A Review of Four-Wheel Steering Studies from the Viewpoint of Vehicle Dynamics and Control," *Vehicle System Dynamics: International Journal of Vehicle Mechanics and Mobility*, Vol. 18, No. 1, pp. 151–86.

Nalecz, A. G., and A. C. Bidermann, 1988, "Investigation into the Stability of Four-Wheel-Steering Vehicles," *International Journal of Vehicle Design*, Vol. 9, No. 2, pp. 159–78.

16 Active Suspensions

Automotive suspensions are discussed in Chapter 4 in connection with the vertical motion and ride properties of vehicles. A two-DOF quarter-car model was used, which is simple but sufficiently detailed to capture many of the key suspension-performance tradeoffs, such as ride quality (represented by sprung-mass acceleration); handling (represented by tire deflection); and packaging (represented by suspension stroke, also known as the rattle space). The performance index (see Chapter 4, Example 4.9) combines these three performance measures by assigning adjustable weights to the three performance terms.

Studies show that passive suspensions frequently are tuned to achieve good tradeoffs. Any improvement in one aspect of performance always is achieved at the expense of the deteriorated performance in another. The extra DOF offered by an active suspension could provide improved performance compared with a strictly passive suspension. The optimal design of a suspension for a quarter-car one-DOF model, as shown in Figure 16.1a (i.e., no unsprung-mass [wheel] dynamics), and the performance index, $J_1 = x_{1rms}^2 + r\, u_{rms}^2$, has the structure shown in Figure 16.1b. Clearly, this structure, which includes a so-called skyhook damper, cannot be realized by the passive-suspension configuration shown in Figure 16.1c. Note that x_1 in this one-DOF model represents the suspension stroke, r is a weight on control signal, and u is the control force, which also is directly proportional to sprung-mass acceleration. Clearly, an active suspension can provide performance benefits that cannot be achieved by using a strictly passive design (Figure 16.2). Furthermore, an active design can allow the performance to be user-selectable. For example, if a softer or a firmer ride characteristic is preferred by a user, the weights in the performance index used in the controller design can be changed (Hrovat 1988), leading to different controller gains and, consequently, different performance characteristics.

Several types of "active" suspensions have been developed (Bastow 1988; Sharp and Crolla 1987): (1) semi-active suspensions (Hac and Youn 1991; Redfield 1991); (2) high-bandwidth, fully active suspensions (Chalasani 1986; Chalasani and Alexandridis 1986); and (3) low-bandwidth, fully active suspensions (Sharp and Hassan 1987). Although they have been available for decades, active suspensions have not found widespread commercial application. This is due mainly to their large power requirements (e.g., 3 times the power required by the air-conditioning compressor and 2.5 times the peak power required by the starter), especially the high-bandwidth

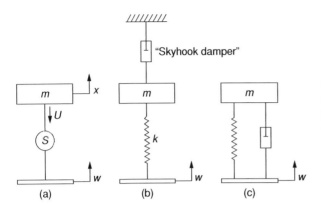

Figure 16.1. (a) One-DOF vehicle model with active suspension; (b) corresponding LQG-optimal structure; and (c) passive one-DOF model (Hrovat 1988).

versions. Thus, commercial implementations are primarily for low-bandwidth active suspensions to emphasize attitude-holding performance during maneuvers and for semi-active suspensions. Active stabilizer bars for roll control also have been implemented. In semi-active suspensions, the damping forces can be adjusted by control of the damping coefficient, for example, by using electro-rheological fluids.

In the following sections, the design of optimal, fully active, high-bandwidth suspensions is described first based on a single DOF model, then on a two-DOF model, and finally the optimal active suspension for the two-DOF model with state estimation.

16.1 Optimal Active Suspension for Single-DOF Model

Consider the design of an active suspension based on the single-DOF model, as in Figures 16.1a–c. As shown in Figure 16.3, we can define the two states x_1 = suspension stroke (positive in extension) and x_2 = sprung-mass velocity (positive downwards).

The sprung mass is denoted by m_s, the suspension force by $u(t)$, and the ground-velocity input by $w(t)$. It is assumed that the ground-velocity input can be well modeled as a zero-mean white-noise input, w, with variance W.

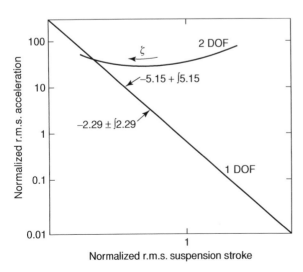

Figure 16.2. Comparison between performance of conventional passive suspension for two-DOF vehicle model and optimal one-DOF active suspension (with representative eigenvalues) (Hrovat 1988).

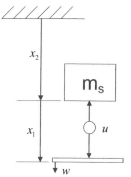

Figure 16.3. Single-DOF quarter-car model for active-suspension design.

The equations of motion then are written as:

$$\dot{x}_1 = w - x_2$$
$$\dot{x}_2 = -\frac{u}{m_s}$$

In standard state equation form, these can be written as:

$$\frac{d}{dt}\begin{bmatrix} x_1 \\ x_2 \end{bmatrix} = \begin{bmatrix} 0 & -1 \\ 0 & 0 \end{bmatrix}\begin{bmatrix} x_1 \\ x_2 \end{bmatrix} + \begin{bmatrix} 0 \\ -\dfrac{1}{m_s} \end{bmatrix} u + \begin{bmatrix} 1 \\ 0 \end{bmatrix} w$$

or

$$\dot{\mathbf{x}} = \mathbf{A}\mathbf{x} + \mathbf{b}u + \mathbf{g}w$$

As discussed in Chapter 4, the control-design objective can be represented as a quadratic form in the states and control input. For example, if we consider a weighted sum with weight r of the suspension stroke and control effort, we can write:

$$J = E\left[x_1^2(t) + ru^2(t)\right] = E\left[\mathbf{x}(t)^T \mathbf{R}_{\mathbf{xx}}\mathbf{x}(t) + ru^2(t)\right]$$

From the Certainty Equivalence Principle, the optimal control gains are known to be the same as for the corresponding deterministic LQ problem:

$$J = \int_0^\infty \left[\mathbf{x}(t)^T \mathbf{R}_{\mathbf{xx}}\mathbf{x}(t) + ru^2(t)\right] dt$$

The optimal control is given by:

$$u^*(t) = -r^{-1}\mathbf{b}^T \mathbf{P}\mathbf{x}(t) = -\mathbf{k}_r^T \mathbf{x}(t)$$

where $\mathbf{P}$ is the symmetric, positive-definite solution of the following algebraic Riccati equation:

$$\mathbf{A}^T\mathbf{P} + \mathbf{P}\mathbf{A} - r^{-1}\mathbf{P}\mathbf{b}\mathbf{b}^T\mathbf{P} + \mathbf{R}_{\mathbf{xx}} = 0$$

For our problem, we solve the previous equation with:

$$A = \begin{bmatrix} 0 & -1 \\ 0 & 0 \end{bmatrix}; \quad b = \begin{Bmatrix} 0 \\ -\dfrac{1}{m_s} \end{Bmatrix}; \quad P = \begin{bmatrix} P_1 & P_2 \\ P_2 & P_3 \end{bmatrix}$$

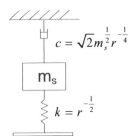

$$c = \sqrt{2} m_s^{\frac{1}{2}} r^{-\frac{1}{4}}$$

m_s

$$k = r^{-\frac{1}{2}}$$

Figure 16.4. Optimal "skyhook damper" active suspension.

which yields:

$$P_1 = \sqrt{2 m_s^{\frac{3}{2}} r^{\frac{1}{4}}}; \quad P_2 = -m_s r^{\frac{1}{2}}; \quad P_3 = \sqrt{2} m_s^{\frac{3}{2}} r^{\frac{4}{3}}$$

and we then can obtain the optimal control:

$$u^*(t) = -r^{-1} \mathbf{b}^T \mathbf{P} \mathbf{x}(t) = -\mathbf{k}_r^T \mathbf{x}(t)$$

$$= -r^{-1} \left[0 - \frac{1}{m_s} \right] \begin{bmatrix} \sqrt{2} m_s^{\frac{1}{2}} r^{\frac{1}{4}} & -m_s r^{\frac{1}{2}} \\ -m_s r^{\frac{1}{2}} & \sqrt{2} m_s^{\frac{3}{2}} r^{\frac{3}{4}} \end{bmatrix} \mathbf{x}(t)$$

$$= \left[-r^{-\frac{1}{2}} \sqrt{2} m_s^{\frac{1}{2}} r^{-\frac{1}{4}} \right] \mathbf{x}(t) = \begin{bmatrix} k_1 & k_2 \end{bmatrix} \mathbf{x}(t)$$

Thus, the optimal control is a state-feedback controller:

$$\mathbf{u}^*(t) = \left[-r^{-\frac{1}{2}} \quad \sqrt{2} m_s^{\frac{1}{2}} r^{\frac{1}{4}} \right] \cdot \begin{bmatrix} x_1 \\ x_2 \end{bmatrix}$$

which feeds back – with optimal gains that depend on r and m_s – the suspension stroke x_1 and the sprung-mass velocity x_2. The physical interpretation of this control, as shown in Figure 16.1b, is that of a passive suspension with a skyhook damper, which is illustrated in Figure 16.4.

Consequently, the optimal active suspension cannot be implemented through strictly passive means. In other words, it is not only the gains of the optimal controller but also its structure, which is different from the passive suspension. As shown in Figure 16.1c, the passive-suspension structure feeds back not x_1 and x_2 as in the active suspension but rather x_1 and $\dot{x}_1$.

16.2 Optimal Active Suspension for Two-DOF Model

Active suspension systems for automobiles also can be designed based on the two-DOF quarter-car model for vertical motion (see Chapter 4) by using optimal control methods to achieve the desired tradeoffs among passenger comfort, packaging requirements, and vehicle-handling requirements. This typically is accomplished by using a quadratic performance criterion (similar to the one discussed in Example 4.9) and a linear model of the vertical vehicle dynamics. The linear model was derived previously in Chapter 4 and is given in Eqs. (4.63) and (4.64).

The controller design is based on the minimization of a quadratic performance index including weighted combinations of the squares of the rms values of sprung-mass acceleration, wheel hop, rattle space, and applied-force terms:

$$
\begin{aligned}
J &= E\left\{\dot{x}_4^2 + r_1 x_1^2 + r_2 x_3^2 + r_3 u^2\right\} \\
&= \dot{x}_{4rms}^2 + r_1 x_{1rms}^2 + r_2 x_{3rms}^2 + r_3 u_{rms}^2\} \\
&= \mathbf{x}_{rms}^T \mathbf{R}_{xx} \mathbf{x}_{rms} + 2u\mathbf{x}_{rms}^T \mathbf{R}_{xu} + \mathbf{R}_{uu} u_{rms}^2
\end{aligned}
\tag{16.1}
$$

where $R_{uu} = (1 + r_3)$:

$$
R_{xx} = \begin{bmatrix}
r_1 & 0 & 0 & 0 \\
0 & (2\zeta_2\omega_2)^2 & -2\zeta_2\omega_2^3 & -(2\zeta_2\omega_2)^2 \\
0 & -2\zeta_2\omega_2^3 & (r_2 + \omega_2^4) & 2\zeta_2\omega_2^3 \\
0 & -(2\zeta_2\omega_2)^2 & 2\zeta_2\omega_2^3 & (2\zeta_2\omega_2)^2
\end{bmatrix}
\qquad
R_{xu} = \begin{bmatrix}
0 \\
-2\zeta_2\omega_2 \\
\omega_2^2 \\
2\zeta_2\omega_2
\end{bmatrix}
\tag{16.2}
$$

and the expectation operator, E, in Eq. (16.1) is defined by:

$$
E\{x(t)\} = \lim_{T \to \infty} \frac{1}{T} \int_0^T x(\tau)d\tau
\tag{16.3}
$$

Other formulations of the performance index have been used – for example, including jerk (i.e., time rate of change of the acceleration) as part of the passenger-comfort criterion (Hrovat and Hubbard 1987).

Consider the design of an LQ optimal active suspension based on the model shown in Chapter 4 and minimization of the performance index defined by Eqs. (16.1) and (16.2) with respect to the control variable $u(t)$. Initially, it is assumed that all of the states, $\mathbf{x}(t)$, are measurable and can be used directly in the controller implementation. Later, the implementation of the controller with measurement of only some state variables is discussed. The LQ optimal controller is given by the state-feedback law:

$$
u(t) = -\mathbf{K_r}\mathbf{x}(t)
\tag{16.4}
$$

where the optimal controller gain $\mathbf{K_r}$ is given by:

$$
\mathbf{K_r} = \mathbf{R}_{uu}^{-1}[\mathbf{B}'\mathbf{P} + \mathbf{R}_{xu}']
\tag{16.5}
$$

and $\mathbf{P}$ is obtained from the solution to the algebraic Riccati equation:

$$
\mathbf{P}(\mathbf{A} - \mathbf{BR}_{uu}^{-1}\mathbf{R}_{xu}') + (\mathbf{A}' - \mathbf{R}_{xu}\mathbf{R}_{uu}^{-1}\mathbf{B}')\mathbf{P} - \mathbf{PBR}_{uu}^{-1}\mathbf{B}'\mathbf{P} + \mathbf{R}_{xx} - \mathbf{R}_{xu}\mathbf{R}_{uu}^{-1}\mathbf{R}_{xu}' = \mathbf{0}
\tag{16.6}
$$

Equation (16.6) can be solved numerically using computer-aided-design software, as illustrated in Example 16.1.

EXAMPLE 16.1: LQ ACTIVE-SUSPENSION DESIGN. An LQ active-suspension design is illustrated in this example. The LQ controller design consists of selecting weights for use in the performance index, then determining the controller gain $\mathbf{K_r}$. The MATLAB program also generates Bode Plots showing the response of the system to a white-noise ground-velocity input, $w(t)$. The Bode Plot includes

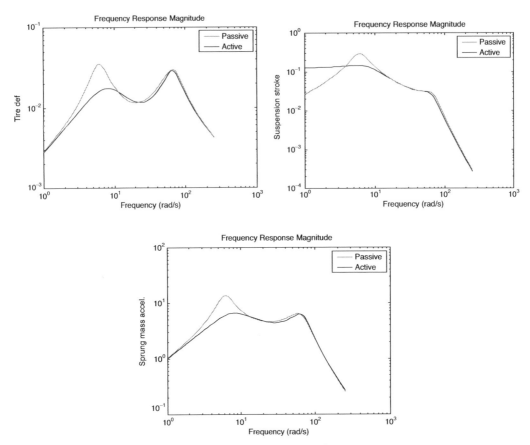

Figure 16.5. Bode Plots of passive and active suspension systems.

frequency responses for the tire displacement, $x_1(t)$; the suspension stroke, $x_3(t)$; and the sprung-mass acceleration, $\dot{x}_4(t)$. These results are in Figure 16.5, where it can be seen that the active-suspension system achieves lower tire deflection and sprung-mass acceleration in the low-frequency region, and its suspension deflection is somewhat worse. The active suspension was found to improve overall performance by about 17 percent.

```
% Ex16_1.m
clear
% Normalized vehicle parameters
w1 = 20*pi; % w1 = sqrt(kus/mus)
w2 = 2.0*pi; % w2 = sqrt(ks/ms)
z1 = 0.0; % z1 = cus/(2*ms*w1)
z2 = 0.3; % z2 = cs/(2*ms*w2)
rho = 10.0; % rho = ms/mus

% Passive system equations:
A = [0 1 0 0
     -w1^2 -2*(z2*w2*rho+z1*w1) rho*w2^2 2*z2*w2*rho
```

```
     0 -1 0 1
     0 2*z2*w2 -w2^2 -2*z2*w2];
B = [0 rho 0 -1]'; G = [-1 2*z1*w1 0 0]';
% Define outputs for plotting results:
C= [1 0 0 0;  % tire displacement
    0 0 1 0;  % suspension stroke
    A(4,:)];  % sprung mass acceleration
Du = [0.0; 0.0; B(4)]; Dw = [0.0; 0.0; 0.0];

% Select weights for use in performance index:
% r1=1.1E3; r2=100.; r3=0.0; % Soft (S) ride case
  r1=5.0e4; r2=5.0E3; r3=0.0; % Typical (T) ride case
% r1=1.0E6; r2=1.0E5; r3=0.0; % Harsh (H) ride case
Rxx = [r1 0 0 0
       0 (2*z2*w2)^2 -2*z2*w2^3 -(2*z2*w2)^2
       0 -2*z2*w2^3 (r2+w2^4) 2*z2*w2^3
       0 -(2*z2*w2)^2 2*z2*w2^3 (2*z2*w2)^2];
Rxu = [0 -2*z2*w2 w2^2 2*z2*w2]'; Ruu = (1+r3);
% Calculate the LQ optimal gain Kr:
[Kr,S] = lqr(A,B,Rxx,Ruu,Rxu);
Ac=(A-B*Kr); Cc=(C-Du*Kr);

% Frequency response curves for the closed-loop and
% open-loop (passive) systems:
w=logspace(0,2.4,100);
[mag_p_tire, phase_p_tire] = bode(A,G,C(1,:),Dw(1),1,w);
[mag_a_tire, phase_a_tire] = bode(Ac,G,Cc(1,:),Dw(1),1,w);
[mag_p_susp, phase_p_susp] = bode(A,G,C(2,:),Dw(2),1,w);
[mag_a_susp, phase_a_susp] = bode(Ac,G,Cc(2,:),Dw(2),1,w);
[mag_p_ride, phase_p_ride] = bode(A,G,C(3,:),Dw(3),1,w);
[mag_a_ride, phase_a_ride] = bode(Ac,G,Cc(3,:),Dw(3),1,w);

loglog(w,mag_p_tire,'r',w,mag_a_tire,'b-.');
title('Frequency Response Magnitude');
xlabel('Frequency (rad/s)');
ylabel('Tire def');
legend('Passive', 'Active'); pause

loglog(w,mag_p_susp,'r',w,mag_a_susp,'b-.');
title('Frequency Response Magnitude');
xlabel('Frequency (rad/s)');
ylabel('Suspension stroke');
legend('Passive', 'Active'); pause

loglog(w,mag_p_ride,'r',w,mag_a_ride,'b-.');
title('Frequency Response Magnitude');
xlabel('Frequency (rad/s)');
```

```
ylabel('Sprung mass accel.');
legend('Passive', 'Active')

% calculate the performance index
% of the system with and without control
Xss=lyap(A,G*G');
x3barrms=sqrt(Xss(3,3));
x1barrms=sqrt(Xss(1,1));
x4dotbarrms=sqrt([A(4,:)]*Xss*[A(4,:)]'+ [G(4)]*[G(4)]');
Pindex=x4dotbarrms^2+r1*(x1barrms^2)+ r2*(x3barrms^2);

Xss_act=lyap(Ac,G*G');
x3barrms=sqrt(Xss_act(3,3));
x1barrms=sqrt(Xss_act(1,1));
x4dotbarrms=sqrt([Ac(4,:)]*Xss_act*[Ac(4,:)]'+ [G(4)]*[G(4)]');
ubarrms=sqrt(Kr*Xss_act*Kr');                    % control signal
Pindex_act=x4dotbarrms^2+r1*(x1barrms^2)+
r2*(x3barrms^2)+r3*(ubarrms^2);

% Ratio of active performance/passive performance
Pindex_act/Pindex
```

16.3 Optimal Active Suspension with State Estimation

A challenge in implementing state-feedback control algorithms, including the LQ optimal-control approach, is that all states of the system must be measurable. From a cost perspective, it is desirable to minimize the measurements needed for active-suspension implementation. In such cases, some of the states required for the feedback-control strategy are estimated from available measurements. This leads to a so-called Linear Quadratic Gaussian (LQG) optimal control problem. The unmeasured states are estimated using an optimal filter, known as the Kalman filter. The LQG optimal active-suspension design also is based on the quarter-car model (i.e., Eqs. (4.63) and (4.64)); the performance index given in Eqs. (16.1) and (16.2); and an output equation, which defines the measurable outputs of the system:

$$\mathbf{y}(t) = \mathbf{C}\mathbf{x}(t) + \mathbf{D}u(t) \tag{16.7}$$

where the coefficients $\mathbf{C}$ and $\mathbf{D}$ must be selected to define the measurable signals that can be used in the controller. For example, if the suspension stroke, $x_3(t)$, is the only measurable variable, then $\mathbf{C}$ and $\mathbf{D}$ in Eq. (16.7) become $\mathbf{C} = [0 \ 0 \ 1 \ 0]$, and $\mathbf{D} = 0$. Similarly, if the measured variables are the suspension stroke, $x_3(t)$, and the sprung-mass acceleration, $\dot{x}_4(t)$, then $\mathbf{C}$ and $\mathbf{D}$ in Eq. (16.7) become:

$$\mathbf{C} = \begin{bmatrix} 0 & 0 & 1 & 0 \\ 0 & 2\zeta_2\omega_2 & -\omega_2^2 & -2\zeta_2\omega_2 \end{bmatrix}; \quad \mathbf{D} = \begin{Bmatrix} 0 \\ 0 \end{Bmatrix} \tag{16.8}$$

where the second entry of the **D** matrix is obtained from the last entry of the input matrix corresponding to the **G** matrix, which is the matrix for the road-velocity input. The optimal LQG controller has the form:

$$u(t) = -\mathbf{K_r}\hat{\mathbf{x}}(t) \tag{16.9}$$

and **K_r** is calculated exactly from the same procedure as in the LQ control case. The state estimates, $\hat{\mathbf{x}}(t)$, are calculated from the equations:

$$\frac{d}{dt}\hat{\mathbf{x}}(t) = \mathbf{A}\hat{\mathbf{x}}(t) + \mathbf{B}\mathbf{u}(t) + \mathbf{K_e}(\mathbf{y}(t) - \hat{\mathbf{y}}(t)) \tag{16.10}$$

and

$$\hat{\mathbf{y}}(t) = \mathbf{C}\hat{\mathbf{x}}(t) + \mathbf{D}\mathbf{u}(t) \tag{16.11}$$

The optimal estimator gain, **K_e**, is calculated from:

$$\mathbf{K_e} = -\mathbf{P_e}\mathbf{C}\mathbf{V}^{-1}$$

and **P_e** is computed from the solution of another algebraic Riccati equation:

$$\mathbf{P_e}\mathbf{A}^\mathbf{T} + \mathbf{A}\mathbf{P}_e - \mathbf{P_e}\mathbf{C}\mathbf{V}^{-1}\mathbf{C}^\mathbf{T}\mathbf{P_e} + \mathbf{W} = \mathbf{0} \tag{16.12}$$

where **V** is the measurement-noise covariance matrix and **W** is the process-disturbance covariance matrix. Like the weights $\mathbf{R_{xx}}$, $\mathbf{R_{uu}}$, and $\mathbf{R_{xu}}$, these matrices must be selected before the optimal estimator can be designed. The optimal estimator relies more on the measurement by producing a large gain, **K_e**, when the confidence in the measurement is high relative to the model (i.e., when **V** is small compared to **W**). Similarly, when **V** is large compared to **W**, the estimator gain is small and the estimates rely more on the model than the measurement.

For the closed-loop system, the state-space model has twice the number of state variables as the open-loop system. When the plant model is completely known and measurement noise is small, the state-space model of the augmented system is:

$$\frac{d}{dt}\begin{Bmatrix} \mathbf{x} \\ \hat{\mathbf{x}} \end{Bmatrix} = \begin{bmatrix} \mathbf{A} & -\mathbf{B}\mathbf{K_r} \\ \mathbf{K_e}\mathbf{C} & \mathbf{A} - \mathbf{B}\mathbf{K_r} - \mathbf{K_e}\mathbf{C} \end{bmatrix}\begin{Bmatrix} \mathbf{x} \\ \hat{\mathbf{x}} \end{Bmatrix} + \begin{Bmatrix} \mathbf{G} \\ \mathbf{0} \end{Bmatrix}w \tag{16.13}$$

When the "perceived" plant state and input matrices are different from those of the true plant and measurement noise is included, the state-space model is:

$$\begin{Bmatrix} \dot{\mathbf{x}} \\ \dot{\hat{\mathbf{x}}} \end{Bmatrix} = \begin{bmatrix} \mathbf{A} & -\mathbf{B}\mathbf{K_r} \\ \mathbf{K_e}\mathbf{C} & \hat{\mathbf{A}} - \hat{\mathbf{B}}\mathbf{K_r} - \mathbf{K_e}\hat{\mathbf{C}} - \mathbf{K_e}(\mathbf{D} - \hat{\mathbf{D}})\mathbf{K_r} \end{bmatrix}\begin{Bmatrix} \mathbf{x} \\ \hat{\mathbf{x}} \end{Bmatrix} + \begin{bmatrix} \mathbf{G} & \mathbf{0} \\ \mathbf{0} & \mathbf{K_e} \end{bmatrix}\begin{Bmatrix} w \\ v \end{Bmatrix} \tag{16.14}$$

where $\hat{\mathbf{A}}$, $\hat{\mathbf{B}}$, $\hat{\mathbf{C}}$, and $\hat{\mathbf{D}}$ are the perceived model matrices. Typically, the **C** and **D** matrices are the same as the actual ones. In a case in which the feedback signal contains acceleration terms, however, they might be different from the actual matrices. It is important to note that because the augmented system has twice the number of state variables as the original model, the output matrices all must be adjusted accordingly. For example, when we want to obtain the actual tire-deflection signal, the corresponding **C** matrix is $\mathbf{C} = [1\ 0\ 0\ 0; \text{zeros}(1,4)]$. The estimated tire deflection, conversely, is obtained from a **C** matrix that is $\mathbf{C} = [\text{zeros}(1,4); 1\ 0\ 0\ 0]$.

Yue, Butsuen, and Hedrick (1989) consider the LQG design of active suspensions using only the suspension-stroke measurement. This LQG design approach

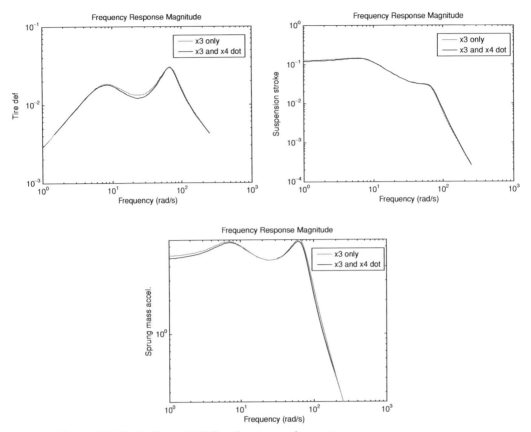

Figure 16.6. Bode Plots of LQG active-suspension systems.

is illustrated in Example 16.2. Several possible measurement sets are considered in Ulsoy et al. (1994. They also investigate the robustness of LQ and LQG controllers with respect to unmodeled sensor and actuator dynamics and their sensitivity to variations in the parameters of a parallel passive suspension.

EXAMPLE 16.2: LQG ACTIVE-SUSPENSION DESIGN. An LQG active-suspension design, based only on suspension-stroke measurement, is illustrated in this example. The LQG controller design consists of selecting the weights for use in the performance index, selecting the covariance matrices for use in the Kalman filter, determining the controller gain $\mathbf{K_r}$, and determining the estimator gain $\mathbf{K_e}$. The program also generates the Bode Plots showing the response of the closed-loop system to a white-noise ground-velocity input, $w(t)$. The results shown in Figure 16.6 can be compared to those in Figure 16.5.

```
% Ex16_2.m
clear
% Specify model parameter values:
w1 = 20*pi; % w1 = sqrt(kus/mus); w2 = 2.0*pi;
      % w2 = sqrt(ks/ms)
z1 = 0.0; % z1 = cus/(2*ms*w1); z2 = 0.3; % z2 = cs/(2*ms*w2)
rho = 10.; % rho = ms/mus
```

```
% Open loop system equations:
A = [0 1 0 0
  -w1^2 -2*(z2*w2*rho+z1*w1) rho*w2^2 2*z2*w2*rho
   0 -1 0 1
   0 2*z2*w2 -w2^2 -2*z2*w2];
B = [0 rho 0 -1]'; G = [-1 2*z1*w1 0 0]';
% Select weights for use in performance index:
% r1=1.1E3; r2=100.;  r3=0.0; % Soft (S) ride case
  r1=5.0e4; r2=5.0E3; r3=0.0; % Typical (T) ride case
% r1=1.0E6; r2=1.0E5; r3=0.0; % Harsh (H) ride case
Rxx = [r1 0 0 0
      0 (2*z2*w2)^2 -2*z2*w2^3 -(2*z2*w2)^2
      0 -2*z2*w2^3 (r2+w2^4) 2*z2*w2^3
      0 -(2*z2*w2)^2 2*z2*w2^3 (2*z2*w2)^2];
Rxu = [0 -2*z2*w2 w2^2 2*z2*w2]'; Ruu = (1+r3);
% Calculate the LQ optimal gain Kr:
[Kr,S] = lqr(A,B,Rxx,Ruu,Rxu);
% Define C1 and D1 for suspension stroke
% Define C2 and D2 for sprung mass acceleration
C1=[0 0 1 0]; D1=0;
C2=[0 0 1 0;A(4,:)]; D2=[0;G(4)];
% parameters of the noise model:
Amp=1.65E-5; Vel=80; p=0.01;
% calculation of the covariances used in the KF design
Xss=lyap(A,G*G');
x3barrms=sqrt(Xss(3,3));
x1barrms=sqrt(Xss(1,1));
x4dotbarrms=sqrt([A(4,:)]*Xss*[A(4,:)]'+ [G(4)]*[G(4)]');
Pindex=x4dotbarrms^2+r1*(x1barrms^2)+ r2*(x3barrms^2);
W=(2.0*pi*Amp*Vel);
V1=(p^2)*(2.0*pi*Amp*Vel)*(x3barrms^2);
V2=(p^2)*(2.0*pi*Amp*Vel)*[(x3barrms^2) 0;  0 (x4dotbarrms^2)];
% calculation of the steady state KF gains
Ke1=lqe(A,G,C1,W,V1);
Ke2=lqe(A,G,C2,W,V2);
% Compute the state matrices for LQG systems
Ac1 = [A, -B*Kr; Ke1*C1, A-B*Kr-Ke1*C1];
Ac2 = [A, -B*Kr; Ke2*C2, A-B*Kr-Ke2*C2];
% Define various outputs for plotting results:
Cc_lqg = [1 0 0 0 0 0 0 0; % tire displacement
        0 0 1 0 0 0 0 0; % suspension stroke
        A(4,:) 0 0 0 0]; % sprung mass acceleration
Dw = [0.0; 0.0; G(4)]; Du = [0.0;0.0;B(4)]; Gc = [G;0;0;0;0];
% Frequency response
w=logspace(0,2.4,100);
[mag_p_tire, phase_p_tire] = bode(A,G,[1 0 0 0],0.0,1,w);
[mag_a1_tire, phase_a1_tire] = bode(Ac1,Gc,Cc_lqg(1,:),
   Dw(1),1,w);
```

```
[mag_a2_tire, phase_a2_tire] = bode(Ac2,Gc,Cc_lqg(1,:),
   Dw(1),1,w);
[mag_p_susp, phase_p_susp] = bode(A,G,[0 0 1 0],0,1,w);
[mag_a1_susp, phase_a1_susp] = bode(Ac1,Gc,Cc_lqg(2,:),
   Dw(2),1,w);
[mag_a2_susp, phase_a2_susp] = bode(Ac2,Gc,Cc_lqg(2,:),
   Dw(2),1,w);
[mag_p_ride, phase_p_ride] = bode(A,G, A(4,:), G(4),1,w);
[mag_a1_ride, phase_a1_ride] = bode(Ac1,Gc,Cc_lqg(3,:),
   Dw(3),1,w);
[mag_a2_ride, phase_a2_ride] = bode(Ac2,Gc,Cc_lqg(3,:),
   Dw(3),1,w);
loglog(w,mag_a1_tire,'r',w,mag_a2_tire,'b-.');
title('Frequency Response Magnitude');
xlabel('Frequency (rad/s)');
ylabel('Tire def');
legend('x3 only', 'x3 and x4 dot'); pause
loglog(w,mag_a1_susp,'r',w,mag_a2_susp,'b-.');
title('Frequency Response Magnitude');
xlabel('Frequency (rad/s)');
ylabel('Suspension stroke');
legend('x3 only', 'x3 and x4 dot'); pause
loglog(w,mag_a1_ride,'r',w,mag_a2_ride,'b-.');
title('Frequency Response Magnitude');
xlabel('Frequency (rad/s)');
ylabel('Sprung mass accel.');
legend('x3 only', 'x3 and x4 dot');
% calculate the rms response to a unit variance white
% noise input of the system with LQG control
Xss_a1=lyap(Ac1,[G;0;0;0;0]*[G;0;0;0;0]');
x3barrms1=sqrt([0 0 1 0 0 0 0 0]*Xss_a1*[0 0 1 0 0 0 0 0]');
x4dotbarrms1=sqrt([A(4,:) -B(4)*Kr]*Xss_a1*[A(4,:) -
B(4)*Kr]'+[G(4)]*[G(4)]');
x1barrms1=sqrt([1 0 0 0 0 0 0 0]*Xss_a1*[1 0 0 0 0 0 0 0]');
ubarrms1=sqrt([0 0 0 0 Kr]*Xss_a1*[0 0 0 0 Kr]');
Pindex_a1=x4dotbarrms1^2+r1*(x1barrms1^2)+r2*(x3barrms1^2)
   +r3*(ubarrms1^2);
Xss_a2=lyap(Ac2,[G;0;0;0;0]*[G;0;0;0;0]');
x3barrms2=sqrt([C1 0 0 0 0]*Xss_a2*[C1 0 0 0 0]');
x4dotbarrms2=sqrt([A(4,:) -B(4)*Kr]*Xss_a2*[A(4,:) -
B(4)*Kr]'+[G(4)]*[G(4)]');
x1barrms2=sqrt([1 0 0 0 0 0 0 0]*Xss_a2*[1 0 0 0 0 0 0 0]');
ubarrms2=sqrt([0 0 0 0 Kr]*Xss_a2*[0 0 0 0 Kr]');
Pindex_a2=x4dotbarrms2^2+r1*(x1barrms2^2)+r2*(x3barrms2^2)
   +r3*(ubarrms2^2);
Pindex_a1/Pindex
Pindex_a2/Pindex
```

EXAMPLE 16.3: COMPARE PASSIVE, ACTIVE LQ, AND ACTIVE LQG SUSPENSIONS. The responses of three suspension systems under the excitation of the same ground-velocity input are compared in this example. The three suspensions compared are the passive suspension, the active LQ suspension designed in Example 16.1, and the active LQG suspension designed in Example 16.2. The responses for tire deflection are compared in Figure 16.7a, for suspension stroke in Figure 16.7b, and for sprung-mass acceleration in Figure 16.7c. The LQ and LQG results are virtually indistinguishable. The active suspensions show some improvement over the passive suspensions. These results also show the importance of working with the rms values in the performance index because the time responses are difficult to interpret in terms of performance.

```
% Ex16_3.m
clear;      % Specify model parameter values:
w1 = 20*pi; % w1 = sqrt(kus/mus); w2 = 2.0*pi;
   % w2 = sqrt(ks/ms)
z1 = 0.0; % z1 = cus/(2*ms*w1); z2 = 0.3; % z2 = cs/(2*ms*w2)
rho = 10.; % rho = ms/mus
% Open loop system equations:
A = [0 1 0 0
   -w1^2 -2*(z2*w2*rho+z1*w1) rho*w2^2 2*z2*w2*rho
   0 -1 0 1
   0 2*z2*w2 -w2^2 -2*z2*w2];
B = [0 rho 0 -1]'; G = [-1 2*z1*w1 0 0]';
C= [1 0 0 0; % tire displacement
     0 0 1 0; % suspension stroke
     A(4,:)]; % sprung mass acceleration
Dw = [0.0; 0.0; G(4)]; Du = [0.0;0.0;B(4)];
% Select weights for use in performance index:
% r1=1.1E3; r2=100.; r3=0.0;  % Soft (S) ride case
   r1=5.0e4; r2=5.0E3; r3=0.0; % Typical (T) ride case
% r1=1.0E6; r2=1.0E5; r3=0.0; % Harsh (H) ride case
Rxx = [r1 0 0 0
       0 (2*z2*w2)^2 -2*z2*w2^3 -(2*z2*w2)^2
       0 -2*z2*w2^3 (r2+w2^4) 2*z2*w2^3
       0 -(2*z2*w2)^2 2*z2*w2^3 (2*z2*w2)^2];
Rxu = [0 -2*z2*w2 w2^2 2*z2*w2]'; Ruu = (1+r3);
% Calculate the LQ optimal gain Kr:
[Kr,S] = lqr(A,B,Rxx,Ruu,Rxu);
% LQ results
Ac=(A-B*Kr); Cc=(C-Du*Kr);
% Define C1 and D1 for suspension stroke
% Define C2 and D2 for sprung mass acceleration
C1=[0 0 1 0]; D1=0;
C2=[0 0 1 0;A(4,:)]; D2=[0;G(4)];
% parameters of the noise model:
```

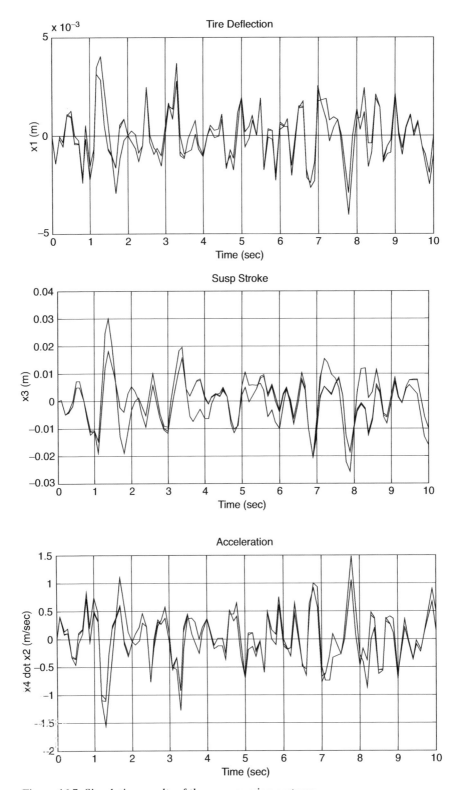

Figure 16.7. Simulation results of three suspension systems.

```
Amp=1.65E-5; Vel=80; p=0.01;
% calculation of the covariances used in the KF design
Xss=lyap(A,G*G');
x3barrms=sqrt(Xss(3,3));
x1barrms=sqrt(Xss(1,1));
x4dotbarrms=sqrt([A(4,:)]*Xss*[A(4,:)]'+ [G(4)]*[G(4)]');
Pindex=x4dotbarrms^2+r1*(x1barrms^2)+ r2*(x3barrms^2);
W=(2.0*pi*Amp*Vel);
V1=(p^2)*(2.0*pi*Amp*Vel)*(x3barrms^2);
V2=(p^2)*(2.0*pi*Amp*Vel)*[(x3barrms^2) 0;  0 (x4dotbarrms^2)];
% calculation of the steady state KF gains
Ke1=lqe(A,G,C1,W,V1);
Ke2=lqe(A,G,C2,W,V2);
% Compute the state matrices for LQG systems
Ac1 = [A, -B*Kr; Ke1*C1, A-B*Kr-Ke1*C1];
Ac2 = [A, -B*Kr; Ke2*C2, A-B*Kr-Ke2*C2];
% Define various outputs for plotting results:
Cc_lqg = [1 0 0 0 0 0 0 0;   % tire displacement
        0 0 1 0 0 0 0 0;   % suspension stroke
        A(4,:) 0 0 0 0];   % sprung mass acceleration
Gc = [G;0;0;0;0];
t=[0:0.1:10];
w=sqrt(2*pi*Amp*Vel)*randn(size(t));
yp=lsim(A,G,C,Dw,w,t);
ylq=lsim(Ac,G,Cc,Dw,w,t);
ylqg=lsim(Ac1,Gc,Cc_lqg-Du*[0 0 0 0 Kr],Dw,w,t);
clf; plot(t,[yp(:,1) ylq(:,1) ylqg(:,1)]);
title('Tire Deflection');
xlabel('Time (sec)'); ylabel('x1 (m)'); grid; pause;
plot(t,[yp(:,2) ylq(:,2) ylqg(:,2)]);
title('Susp Stroke');
xlabel('Time (sec)'); ylabel('x3 (m)'); grid; pause;
plot(t,[yp(:,3) ylq(:,3) ylqg(:,3)]);
title('Acceleration');
xlabel('Time (sec)'); ylabel('x4 dot (m/sec^2)'); grid;
```

PROBLEMS

1. Rerun the MATLAB program in Example 16.1 to generate frequency-response plots for the LQ active suspensions with soft, typical, and harsh ride characteristics. Compare these plots and discuss why these particular combinations of the performance index weights r_1 and r_2 are referred to as "soft," "typical," and "harsh." For these three LQ active suspensions and the passive-only suspension, create a table that contains the following information: (a) the value of the performance index; (b) the rms values of the individual terms in the performance index (e.g., suspension stroke, tire deflection, and sprung-mass acceleration); and (c) the poles (or eigen-values). Comment on how the various active designs change the values of these quantities as compared to the passive case.

2. Use a MATLAB program similar to the one in Example 16.2 to design an LQG active suspension that uses the "typical" ride weights, suspension-stroke measurement, and both sprung-mass and unsprung-mass accelerations. Provide (a) frequency response plots, (b) values of the performance index, and (c) closed-loop eigenvalues.

3. Your supervisor asks you to use a simple one-DOF quarter-car model to develop an adaptive version of an active suspension based on RLS estimation. She wants you to compare the performance of the adaptive active suspension to a nonadaptive version in simulations in which the vehicle sprung mass can vary by ±50 percent around its nominal value and to recommend whether adaptation is needed. The model is given by:

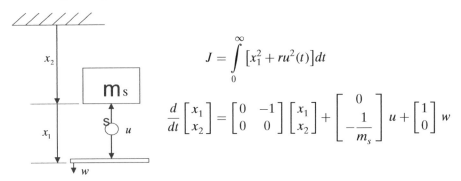

$$J = \int_0^\infty \left[x_1^2 + ru^2(t) \right] dt$$

$$\frac{d}{dt}\begin{bmatrix} x_1 \\ x_2 \end{bmatrix} = \begin{bmatrix} 0 & -1 \\ 0 & 0 \end{bmatrix}\begin{bmatrix} x_1 \\ x_2 \end{bmatrix} + \begin{bmatrix} 0 \\ -\dfrac{1}{m_s} \end{bmatrix} u + \begin{bmatrix} 1 \\ 0 \end{bmatrix} w$$

where x_1 is the suspension stroke, x_2 is the sprung-mass velocity, and the output $y =$ the suspension stroke. Thus:

$$y = [1 \quad 0] \begin{Bmatrix} x_1 \\ x_2 \end{Bmatrix} = x_1$$

The nonadaptive controller is an optimal controller that minimizes J, given by $u = k_1 x_1 + k_2 x_2$, where $k_1 = -(1/\sqrt{r})$ and $k_2 = \sqrt{(2m_s/\sqrt{r})}$. The model parameters have nominal values of $r = 0.0001$ and $m_s = 1,500$ kg. Thus, $k_1 = -100$ and $k_2 = 547.723$, and the closed-loop eigenvalues are at $s_{1,2} = -0.1826 \pm 0.1826j$ for the nominal conditions. However, because the sprung mass can vary, the actual closed-loop performance also varies. Consequently, an adaptive, RLS-based, active suspension based on this one-DOF quarter-car model also can be designed.

Note that for the plant-transfer function:

$$G(s) = \frac{Y(s)}{U(s)} = \frac{1}{m_s s^2}$$

with a sampling period of h and a zero-order hold, the equivalent pulse-transfer function is:

$$H(z) = \left(\frac{h^2}{2m_s} \right)\left(\frac{z+1}{(z-1)^2} \right) = \alpha \left(\frac{z+1}{z^2 - 2z + 1} \right)$$

Consequently, the discrete-time plant can be represented as:

$$y(k) = [\,2 \quad -1 \quad \alpha \quad \alpha\,] \begin{Bmatrix} y(k-1) \\ y(k-2) \\ u(k-1) \\ u(k-2) \end{Bmatrix}$$

Formulate this as $z(k) = \theta^T f(k)$ in terms of unknown parameters θ. Then, the RLS algorithm can be used to estimate the unknown parameters θ online from measurements of y and u. The control gains can be updated accordingly. Assume a sampling period of $h = 1$ second.

(a) Simulate the nonadaptive version of the active suspension for a unit-step input $w(t)$. Display the results for both suspension stroke and sprung-mass acceleration for the nominal value of mass (i.e., typical vehicle load), the maximum value of mass (i.e., fully loaded vehicle), and the minimum value of mass (i.e., empty vehicle).

(b) Repeat the simulations in (a) but with an adaptive version of the active-suspension design; compare the results to the nonadaptive version in (a). What will you recommend to your supervisor?

REFERENCES

Aburaya, T., M. Kawanishi, H. Kondo, T. Hamada, 1990, "Development of an Electronic Control System for Active Suspension," *Proceedings of the Conference on Decision and Control*, Honolulu, HI, pp. 2220–5.

Alexandridis, A. A., and T. R. Weber, 1984, "Active Vibration Isolation of Truck Cabs," *Proceedings of the American Control Conference*, San Diego, CA, pp. 1199–208.

Alleyne, A., and J. K. Hedrick, 1995, "Nonlinear Adaptive Control of Active Suspensions," *IEEE Transactions on Control Systems Technology*, Vol. 3, No. 1, March.

Bastow, D., 1988, *Car Suspension and Handling*, second edition, Pentech Press, London.

Ben Mrad, R., J. A. Levitt, and S. D. Fassois, 1991, "A Nonlinear Model of an Automobile Hydraulic Active Suspension System," in S. A. Velinsky, R. H. Fries, I. Haque, and D. Wang (eds.), *Advanced Automotive Technologies–1991*, ASME DE-Vol. 40, New York, pp. 347–59.

Bender, E. K., 1967, "Optimization of the Random Vibration Characteristics of Vehicle Suspensions Using Random Process Theory," Sc.D. Thesis, MIT, Cambridge, MA.

Berman, A., and A. J. Hannibal (eds.), 1975, *Passenger Vibration in Transportation Vehicles*, ASME AMD-Vol. 24, New York.

Chalasani, R. M., 1986, "Ride Performance Potential of Active Suspension Systems – Part I," ASME Monograph, AMD-Vol. 80, New York, December.

Chalasani, R. M., and A. A. Alexandridis, 1986, "Ride Performance Potential of Active Suspension Systems – Part II," ASME Monograph, AMD-Vol. 80, New York, December.

DeBenito, C., and S. J. Eckert, 1988, "Control of an Active Suspension System Subject to Random Component Failures," *presented at the ASME Winter Annual Meeting*, Chicago, IL, November, Paper No. 88-WA/DSC-33.

Fabien, B. C., 1991, "Controller Gain Selection for an Electromagnetic Suspension Under Random Excitation," *Proceedings of the American Control Conference*, Boston, MA, June.

Hac, A., 1985, "Suspension Optimization of a 2-DOF Vehicle Model Using a Stochastic Optimal Control Technique," *Journal of Sound and Vibration*, Vol. 100, No. 3, pp. 343–57.

Hac, A., and I. Youn, 1991, "Optimal Semi-Active Suspension with Preview Based on a Quarter-Car Model," *Proceedings of the American Control Conference*, Pittsburg, PA, June.

Hammond, J. K., and R. F. Harrison, 1981, "Nonstationary Response of Vehicles on Rough Ground – A State Space Approach," *ASME Journal of Dynamic Systems, Measurement, and Control*, Vol. 103, September, pp. 245–50.

Hayakawa, K., K. Matsumoto, M. Yamashita, Y. Suzuki, K. Fujimori, and H. Kimura, 1999, "Robust H^∞-output Feedback Control of Decoupled Automobile Active Suspension Systems," *Automatic Control, IEEE Transactions on*, Vol. 44, No. 2, February, pp. 392–6.

Hedrick, J. K., and T. Butsuen, 1988, "Invariant Properties of Automotive Suspension," *Advanced Suspensions*, Proceedings of the Institute of Mechanical Engineers, Paper No. C423/88, October.

Hrovat, D., 1988, "Influence of Unsprung Weight on Vehicle Ride Quality," *Journal of Sound and Vibration*, Vol. 124, No. 3, pp. 497–516.

Hrovat, D., 1991, "Optimal Active Suspensions for 3D Vehicle Models," *Proceedings of the American Control Conference*, Boston, MA, June.

Hrovat, D., 1997, "Survey of Advanced Suspension Developments and Related Optimal Control Applications," *Automatica*, Vol. 33, No. 10, pp. 1781–817.

Hrovat, D., and M. Hubbard, 1987, "A Comparison Between Jerk Optimal and Acceleration Optimal Vibration Isolation," *Journal of Sound and Vibration*, Vol. 112, No. 2, pp. 201–10.

Ivers, D. E., and L. R. Miller, 1991, "Semi-Active Suspension Technology: An Evolutionary View," in S. A. Velinsky, R. H. Fries, I. Haque, and D. Wang, (eds.), *Advanced Automotive Technologies–1991*, ASME DE-Vol. 40, New York, pp. 327–46.

Karnopp, D., 1983, "Active Damping in Road Vehicle Suspension Systems," *Vehicle System Dynamics*, Vol. 12, No. 6, pp. 291–311.

Karnopp, D. C., 1985, "Two Contrasting Versions of the Optimal Active Vehicle Suspension," *ASME Monograph*, DSC-Vol. 1, pp. 341–6.

Kasprzak, J. L., 1991, "Research and Development Needs for Road Vehicle Suspension Systems," in S. A. Velinsky, R. H. Fries, I. Haque, and D. Wang (eds.), *Advanced Automotive Technologies–1991*, ASME DE-Vol. 40, New York, pp. 35–6.

Khulief, Y. A., and S. P. Sun, 1989, "Finite Element Modeling and Semi-Active Control of Vibrations in Road Vehicles," *ASME Journal of Dynamic Systems, Measurement, and Control*, Vol. 111, September, pp. 521–7.

Krtolica, R., and D. Hrovat, 1992, "Optimal Active Suspension Control Based on a Half-Car Model: An Analytical Solution," *IEEE Transactions on Automatic Control*, Vol. 37, No. 4, April, pp. 528–32.

Levitt, J. A., and N. G. Zorka, 1991, "The Influence of Tire Damping in Quarter-Car Active-Suspension Models," *ASME Journal of Dynamic Systems, Measurement, and Control*, Vol. 113, March, pp. 134–7.

Lieh, J., 1991, "Modeling and Simulation of an Elastic Vehicle with Semi-Active Suspensions," in S. A. Velinsky, R. H. Fries, I. Haque, and D. Wang (eds.), *Advanced Automotive Technologies–1991*, ASME DE-Vol. 40, New York, pp. 315–26.

Lin, J. S., and I. Kanellakopoulos, 1997, "Nonlinear Design of Active Suspensions," *IEEE Control Systems Magazine*, Vol. 17, No. 3, pp. 45–59.

Michelberger, P., L. Palkovics, and J. Bojkor, 1993, "Robust Design of Active-Suspension System," *International Journal of Vehicle Design*, Vol. 14, Nos. 2/3, pp. 145–65.

Patten, W. N., R. M. Chalasani, D. Allsup, and J. Blanks, 1990, "Analysis of Control Issues for a Flexible One-Half Car Suspension Model," *Proceedings of the American Control Conference*, San Diego, CA, May.

Patten, W. N., P. Kedar, and E. Abboud, 1991, "A Variable Damper Suspension Design for Phase-Related Road Inputs," in S. A. Velinsky, R. H. Fries, I. Haque, and D. Wang (eds.), *Advanced Automotive Technologies–1991*, ASME DE-Vol. 40, New York, pp. 361–74.

Rajamani, R., and J. K. Hedrick, 1991, "Semi-Active Suspensions – A Comparison between Theory and Experiments," *Proceedings of the 12th IAVSD Symposium*, August.

Rajamani, R., and J. K. Hedrick, 1995, "Adaptive Observers for Active Automotive Suspensions: Theory and Experiment," *IEEE Transactions on Control Systems Technology*, Vol. 3, No. 1, March.

Ray, L. R., 1991, "Robust Linear-Optimal Control Laws for Active Suspension Systems," in S. A. Velinsky, R. H. Fries, I. Haque, and D. Wang (eds.), *Advanced Automotive Technologies–1991*, ASME DE-Vol. 40, New York, pp. 291–302.

Ray, L. R., 1992, "Robust Linear Optimal Control Laws for Active Suspension Systems," *ASME Journal of Dynamic Systems, Measurement and Control*, Vol. 114, No. 4, December, pp. 592–8.

Redfield, R. C., 1990, "Low-Bandwidth Semi-Active Damping for Suspension Control," *Proceedings of the American Control Conference*, San Diego, CA, May, pp. 1357–62.

Redfield, R. C., 1991, "Performance of Low-Bandwidth, Semi-Active Damping Concepts for Suspension Control," *Vehicle System Dynamics*, Vol. 20, pp. 245–67.

Redfield, R. C., and D. C. Karnopp, 1988, "Optimal Performance of Variable Component Suspensions," *Vehicle System Dynamics*, Vol. 17, No. 5.

Satoh, M., N. Fukushima, Y. Akatsu, I. Fujimura and K. Fukuyama, 1990, "An Active Suspension Employing an Electrohydraulic Pressure Control System," *Proceedings of the Conference on Decision and Control*, Honolulu, HI, pp. 2226–31.

Sharp, R. S., and D. A. Crolla, 1987, "Road Vehicle Suspension Design – A Review," *Vehicle System Dynamics*, Vol. 16, pp. 167–92.

Sharp, R. S., and S. A. Hassan, 1987, "On the Performance Capabilities of Active Automobile Suspension Systems of Limited Bandwidth," *Vehicle System Dynamics*, Vol. 16, pp. 213–25.

Sunwoo, M., and K. C. Cheok, 1990, "An Application of Explicit Self-Tuning Controller to Vehicle Active Suspension Systems," *Proceedings of the Conference on Decision and Control*, Honolulu, HI, pp. 2251–7.

Sunwoo, M., and K. C. Cheok, 1991, "Investigation of Adaptive Control Approaches for Vehicle Active Suspension Systems," *Proceedings of the American Control Conference*, Boston, MA, June.

Sunwoo, M., K. C. Cheok, and N. J. Huang, 1990, "Application of Model Reference Adaptive Control to Active Suspension Systems," *Proceedings of the American Control Conference*, San Diego, CA, May, pp. 1340–6.

Thompson, A. G., 1976, "An Active Suspension with Optimal Linear State Feedback," *Vehicle System Dynamics*, Vol. 5, pp. 187–203.

Ullman, P. B., and H. Richardson (eds.), 1975, *Mechanics of Transportation Suspension Systems*, ASME AMD-Vol. 15, New York.

Ulsoy, A. G., D. Hrovat, and T. Tseng, 1994, "Stability Robustness of LQ and LQG Active Suspensions," *ASME Journal of Dynamic Systems, Measurement, and Control*, Vol. 116, No. 1, March, pp. 123–31.

Veillette, R. J., 1991, "Projective Controls for 2-DOF Quarter-Car Suspension," *Proceedings of the American Control Conference*, Boston, MA, June.

Williams, R. A., 1997a, "Automotive Active Suspensions – Part 1: Basic Principles," *Proceedings of the Institution of Mechanical Engineers*, Vol. 211 Part D, pp. 415–26.

Williams, R. A., 1997a, "Automotive Active Suspensions – Part 2: Practical Considerations," *Proceedings of the Institution of Mechanical Engineers*, Vol. 211 Part D, pp. 427–44.

Wilson, D. A., R. S. Sharp, and S. A. Hassan, 1986, "The Application of Linear Optimal Control Theory to the Design of Active Automotive Suspensions," *Vehicle System Dynamics*, Vol. 15, pp. 105–18.

Yamashita, M., K. Fujimori, C. Uhlik, P. Kawatani and H. Kimura, 1990, "H_∞ Control of an Automotive Active Suspension," *Proceedings of the Conference on Decision and Control*, Honolulu, HI, pp. 2244–50.

Yue, C., T. Butsuen, and J. K. Hedrick, 1989, "Alternative Control Laws for Automotive Active Suspensions," *ASME Journal of Dynamic Systems, Measurement, and Control*, Vol. 111, No. 2, pp. 286–91.

PART IV

INTELLIGENT TRANSPORTATION SYSTEMS

17 Overview of Intelligent Transportation Systems

Mobility is essential to the economic growth of any modern country and the well-being of its population. This is especially true for the United States because of its size and diffuse population. Without the efficient transport of people and goods, U.S. industries cannot compete effectively with overseas producers. However, the rapid growth in demand and the slower growth in the capacity of highway systems have led to congestion that is estimated to cost more than $40 billion annually. In 1970, motorists in the United States drove approximately 1 trillion vehicle-miles; by 1985, this had increased to 1.8 trillion vehicle-miles; and, by 2000, to 2.8 trillion vehicle-miles. These increases have led to serious congestion problems. For example, peak-hour traffic operating in congested conditions on urban Interstate highways increased from 40 percent in 1970 to nearly 70 percent in 1990. From 1982 to 2002, the vehicle-miles traveled increased by 79 percent, whereas highway-lane miles increased by only 3 percent. The number of roadways considered congested grew from 34 to 58 percent. However, construction of the more than 40,000 miles of the multilane, controlled-access Interstate Highway System essentially is completed. Major new construction, especially in dense urban areas, generally is not feasible and definitely cannot keep up with future traffic demand. Although some growth of the highway system is inevitable, the more efficient use of the existing system is essential.

In recent decades, there also have been tremendous changes in the areas of information technology, electronics, computers, and communications. The pace of these developments is simply astounding, and electronic devices have infiltrated every aspect of life, including vehicles (see Chapter 1). Intelligent Transportation Systems (ITS) (formerly known as Intelligent Vehicle Highway Systems [IVHS]), however, represent more than simply advances in automotive electronics. ITS incorporate a wide variety of electronic-based technologies, both on the vehicle and as part of the highway infrastructure, which collectively are moving the world into the next generation of highway operations. These technologies offer the promise of increased throughput on existing highways at reduced congestion levels as well as improved safety and convenience. The ITS vision encompasses smart (i.e., control, sensing, and communications) automobiles and highways collaborating for improved safety, mobility, trip quality, and productivity while also reducing congestion and environmental impact.

ITS represents a long-term vision and an evolving concept of how the personal automobile can continue to be a primary means of transportation despite the near saturation of current highway capacities. Some argue that this is a dangerous vision to meet transportation needs into the 21st century and that other high-speed public-transportation systems must become an increasingly important part of the mix. Nevertheless, it is clear that ITS technologies will have an important – even if not an exclusive or primary – role in meeting transportation needs in the next few decades.

In many respects, the traffic-congestion problems faced today in the United States are more severe in other parts of the world (e.g., Europe and Japan). Perhaps it is not surprising then that Europe and Japan have taken a leadership role in the development of ITS. A survey of ITS being tested worldwide is in Jurgen (1991), and the situation relative to ITS in Japan is summarized in Ervin (1991). At the Mobility 2000 Workshop (1990), ITS activities were grouped in the following four major areas:

- Advanced Traffic Management Systems (ATMS)
- Advanced Traveler Information Systems (ATIS)
- Commercial Vehicle Operation (CVO)
- Advanced Vehicle Control Systems (AVCS)

Each of these areas is discussed briefly in the sections that follow (Special Issue on IVHS 1991).

17.1 Advanced Traffic Management Systems

ATMS permit the real-time adjustment of traffic-control systems and variable-message signs for drivers. Their application in selected corridors has reduced delay, travel time, and accidents. Traffic-control systems have used control and communications technology for more than a half-century but have been slow to incorporate the latest developments (e.g., distributed-system architecture, fiber-optic communications, and microprocessor-based equipment). The first traffic-control system using analog technology to optimize a flow of traffic through a series of intersections was installed in Chicago in 1926. It provided a fixed signal-timing sequence and did not allow for variations in traffic-flow characteristics. The first digital, computer-based system was installed in 1963 in Toronto; this system provided for more flexible signal control by monitoring detectors as well as operation of the traffic-signal controllers and detectors. Stored signal-timing plans were selected based on detector data or a time-of-day basis. With the availability of reliable and inexpensive microprocessor equipment in the 1980s, a form of signal control known as a "closed-loop system" became available (Figure 17.1). A closed-loop system uses a central personal computer that communicates with an intermediate level of control known as "master controllers" or "local area masters." These, in turn, communicate with the intersection controllers. The master controllers reduce the processing load on the central computer as well as the need to communicate directly with local intersection controllers.

ATMS require sensors in traffic lanes to identify the presence of vehicles and to communicate that information to distributed or central processors. They depend on real-time analysis of traffic data and systems for the control of traffic- and ramp-metering signals. ATMS inform drivers about the status of traffic through

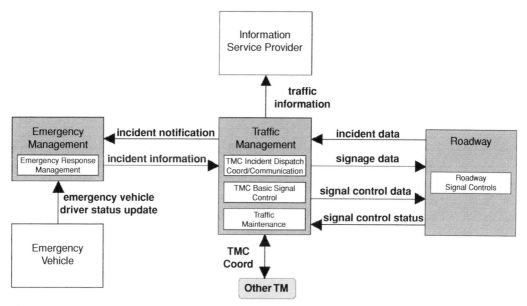

Figure 17.1. Information flow in closed-loop traffic-management systems.

variable-message signs. Ultimately, they indicate alternative routes to be followed in case of a traffic incident. ATMS incorporate incident-management procedures, which may reap the largest benefits. For example, an accident blocking one of three lanes reduces highway capacity by 50 percent and a 20-minute blockage wastes 2,100 vehicle-hours. This can lead to a queue that is almost 2 miles long and can take 2.5 hours to clear. During peak-traffic periods, the time wasted and delays associated with accidents are 50 times worse. To be effective, systems require cooperation among adjacent jurisdictions to permit switching traffic from throughways to arterials as required by incidents. Competent staff and maintenance crews are required to ensure high-reliability systems. ATMS are being introduced with current technology and will benefit from advanced technology. Wherever they have been installed, they are reducing congestion by improving traffic flow and reducing accidents and emissions (Table 17.1). The following lists the benefits of ramp-metering

Table 17.1. *Benefits realized from implementation of ATMS*

Location	Reduced incident-clearance time (%)	Reduced response time (%)	Accident reduction (%)	Secondary accident reduction (%)	Reduced accident rates	Cost savings/yr ($M)	Delay savings (hrs/yr)
Brooklyn, NY	66.0						
Philadelphia, PA		40.0					
San Antonio, TX		20.0	35.0	30.0	41.0%	1.65	255,500
Japan				50.0			
Houston, TX						8.40	572,095
Denver, CO						0.95	95,000
Atlanta, GA							2,000,000
Minnesota						1.40	

and incident-management systems (Proper 1999):

- *Portland, Oregon*: 58 ramp meters, 43 percent accident reduction, 39 percent travel-time reduction, 25 percent demand increase, 60 percent increase in speed
- *Minneapolis/St. Paul, MN*: 6 ramp meters, 8 km of freeway, 24 percent accident reduction, 38 percent accident-rate reduction, 16 percent increase in speed
- *Minneapolis, MN*: 39 ramp meters, 27 km of freeway, 27 percent accident reduction, 38 percent demand increase, 35 percent increase in speed
- *Seattle, WA*: 22 ramp meters, 52 percent travel-time reduction, 39 percent accident-rate reduction, 86 percent demand increase
- *Denver, CO*: 5 ramp meters, 50 percent accident reduction, 18.5 percent demand increase
- *Detroit, MI*: 28 ramp meters, 50 percent accident reduction, 8 percent increase in speed, 12.5 percent demand increase
- *Austin, TX*: 3 ramp meters, 4.2 km of freeway, 60 percent increase in speed, 7.9 percent demand increase
- *Long Island, NY*: 70 ramp meters, 207 km of freeway, 15 percent accident reduction, 9 percent increase in speed

Thus, ATMS represent the highway side of the cooperative ITS concept. They are being implemented in many major urban areas using existing technology that can evolve as advanced technologies mature. An important feature in the future development of ATMS is the ability to communicate with individual vehicles. This discussion continues in the next two sections.

17.2 Advanced Traveler Information Systems

ATIS allow drivers and passengers to know the current location of their and other vehicles and to find desired services. ATIS permit communication between the driver and ATMS for continuous advice regarding traffic conditions, alternate routes, and safety issues.

ATIS equipment can show vehicle location and movement on an electronic map. This enables route planning from origin to destination, identification of businesses or services that a driver may need, and the route to those sites. When ATIS and ATMS are in full communication, drivers are informed of incidents and alternate routes to avoid congestion. ATIS includes the following features:

- vehicle location (via GPS) and map-matching navigation system
- traffic-information receiver
- route planning for minimum distance of travel
- color video displays for maps, traffic information, and route guidance
- onboard database with detailed maps, business directory, specific location of services, hospitals, and tourist information
- information from traffic-management centers about congestion, incidents, and other problems
- electronic vehicle identification for toll debiting

Thus, ATIS represents the vehicle side of the cooperative ITS concept. Figure 17.2 shows information that was collected and used in ATIS services in the United States

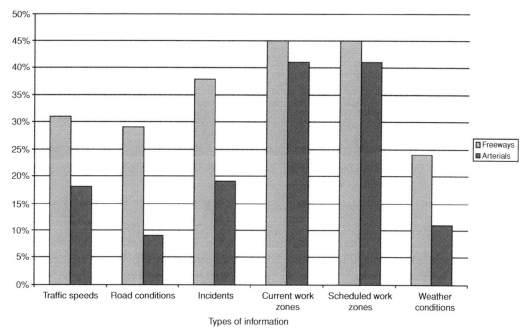

Figure 17.2. 1999 transfer of information in ATIS services in the United States (Radin et al. 2000).

in 1999. ATIS is envisioned in the following three general stages of development in North America:

- *Information Stage.* The primary emphasis is providing drivers with information to improve individual planning and decision making.
- *Advisory Stage.* The static onboard information available to drivers is supplemented with dynamic traffic information collected and transmitted by the infrastructure.
- *Coordination Stage.* Vehicles and infrastructure automatically exchange information to optimize the flow and safety of traffic throughout the entire network.

Vehicle-navigation systems use the following three main technologies to determine the location of a vehicle:

- GPS
- inertial navigation system (i.e., dead reckoning)
- map databases

Each has advantages and disadvantages that often are used in conjunction with one another for best results. These systems can be used for in-vehicle route guidance, route optimization, fleet management, and emergencies.

Various technologies are used to collect the data in ATIS, including video cameras, surveillance systems, loop detectors, cellular phones, police patrol, and GPS, as well as communications from transit authorities (e.g., road repairs) and updates from private-sector firms (e.g., hotels and restaurants). The information can be presented to a driver using various technologies such as a cellular phone, cable TV, the Internet, e-mail, and an in-vehicle personal digital assistant (PDA). Commercially available systems such as the GPS navigation systems, OnStar and Sync, are examples of various ATIS capabilities.

17.3 Commercial Vehicle Operations

CVO select from ATIS those features critical to commercial and emergency vehicles. They expedite deliveries, improve operational efficiency, and increase safety. CVO are designed to interact with ATMS as both become fully developed.

Global competition is changing the way that companies conduct business. Carriers are expected to provide faster, more reliable, and more cost-effective services. ITS technologies are emerging as a key to reducing costs and improving productivity. Commercial and emergency vehicles will adopt ATIS and link them to ATMS as soon as it is feasible to do so. Additional ITS technologies (some already have been developed) including weigh-in-motion sensors, automated vehicle-identification transponders, and automated vehicle-classification devices will reduce time spent in weigh stations, reduce labor costs to states, and minimize red tape for commercial operators. Commercial vehicles are leading the way in the development and implementation of ITS technologies. They already are using automatic vehicle location, tracking, and two-way communications; routing algorithms for dispatch; and in-vehicle text and map displays. ITS technologies being used in commercial vehicles include the following:

- automatic vehicle identification
- weigh-in-motion
- automatic vehicle classification
- onboard computer
- two-way, real-time communication
- automatic clearance sensing

Thus, CVO represents the vehicle side of the cooperative ITS concept for the special needs of commercial and emergency vehicles. Figure 17.3 is a list of current and near-future CVO devices on a commercial truck (Capps et al. 2001).

17.4 Advanced Vehicle-Control Systems

AVCS apply additional technology to vehicles to identify obstacles and adjacent vehicles, thereby assisting in the prevention of collisions and resulting in safer operation at higher speeds. AVCS also may interact with the fully developed ATMS to provide automated vehicle operations. A vision of the future of the automobile, in which they drive themselves while the drivers relax, was first introduced in the General Motors pavilion at the 1939 World's Fair. Since then, there have been relatively few efforts aimed at realizing this vision. In 1979–1981, General Motors conducted a systems study of highway automation (Bender 1991). In 1964–1980, studies on vehicle guidance and control were conducted at Ohio State University (Fenton and Mayhan 1991). In the same year, the University of California's PATH program conducted extensive studies on vehicle longitudinal and lateral control (Shladover et al. 1991).

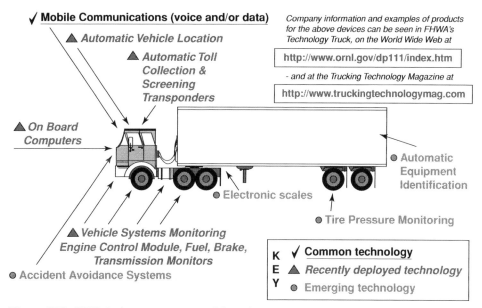

Figure 17.3. CVO devices on a commercial truck.

This section, an overview of AVCS technologies, identifies important classes of problems and studies the potential benefits and impacts. Chapters 18 through 20 provide more detail on the following AVCS technologies: collision detection and avoidance, longitudinal motion control and platooning, and lateral motion control and automated steering.

The goal of AVCS is to enhance vehicle control by facilitating and augmenting driver performance. Ultimately, AVCS aim to relieve drivers of most driving tasks. Three levels of development are foreseen for AVCS technologies, as follows:

- *ACVS-I:* Individual vehicle control includes only vehicle-based systems that detect the presence of obstacles or other vehicles. Studies show that half of all rear-end collisions and as many as a third of intersection accidents can be prevented if drivers have an additional half-second of warning. AVCS-I can provide the additional warning time by sensing the presence of vehicles and obstacles in blind spots. They also can warn drivers when their alertness starts to wane.
- *AVCS-II:* Cooperative driver/vehicle/highway systems implement lateral and longitudinal vehicle control functions in specific applications, such as high-occupancy vehicle lanes. Vehicles enter the lane voluntarily using manual control but then are under full or partial automatic control while in the lane. The advantages include increased speed and safety. "Platooning," the linking of cadre vehicles, also is possible. For private vehicles, AVCS-II provides vehicle-to-vehicle communication about travel paths, which will reduce collisions.
- *AVCS-III:* Automated vehicle-highway systems include complete automation of the driving function for vehicles operating on specially equipped freeway facilities. It builds on AVCS-I and AVCS-II technologies to provide "automated chauffeuring" of vehicles from on-ramp arrival to off-ramp departure.

Considerable research and development are needed before many AVCS technologies can be deployed commercially. Important areas for further research include the following:

- availability and reliability of devices (i.e., ultrasonic, infrared, radar, and vision) to detect spatial relationships of a vehicle to obstacles and other vehicles, and the use of this information in the automatic control of vehicles
- the change in speed of a vehicle under automatic control to be compatible with the limitations of human occupants and available equipment, and a facility for extensive full-scale testing, including human-factor considerations
- special traffic lanes for AVCS-equipped vehicles and automatic inspection procedures to ensure that the AVCS are functioning before equipped vehicles can enter them

It is anticipated that AVCS will reduce accidents and increase traffic flow; they are predicted to double traffic flow on current freeways. These goals may seem ambitious; however, development of the Interstate Highway System in the 1950s and 1960s doubled lane capacity and reduced accidents by 60 percent by grade-separating intersections and controlling access. AVCS technologies are described in further detail in subsequent chapters.

As discussed previously, several stages of the evolution of AVCS technologies are envisioned: beginning with in-vehicle control systems, which can detect the presence of obstacles or other vehicles and warn the driver (i.e., AVCS-I); evolving to partially automated driving systems on special high-occupancy vehicle lanes (i.e., AVCS-II); to fully cooperative vehicle–highway automation on major highways (i.e., AVCS-III). The human-factors aspect of the research is important to the successful development of these proposed AVCS systems. Each phase is discussed in more detail in the following subsections.

ACVS-I: Individual Vehicle Control

The goal of ACVS-I is to develop vehicle-based systems that detect the presence of obstacles or other vehicles and that provide a warning to drivers. AVCS-I includes only those advanced systems that are vehicle-based and that do not require inter-vehicle or vehicle–highway communications and coordination. The principal benefits of this technology are expected in the area of safety – that is, reductions in the annual toll of crashes, fatalities, and injuries as well as the resulting economic costs. However, AVCS-I technologies comprise the basis for the evolution to the AVCS-II and AVCS-III phases.

In recent years, considerable advances have been made in sensing-, warning-, and control-systems technologies with potential application to AVCS-I. The potential for improved safety will be realized as follows:

- reduction in driver exposure to high-risk environments
- reduction in the incidence of high-risk driver behavior
- facilitation of earlier driver response to an imminent crash by providing additional seconds of warning time

- improvement in the overall speed and quality of driver–vehicle response in a likely crash scenario

AVCS-I systems can be classified by the manner in which they aid a driver. Systems that provide an enhanced image of the driving scene are referred to as "perceptual enhancement" systems (e.g., night-vision systems). A driver is expected to interpret the enhanced images and to control the vehicle in a manner that improves mobility and safety. "Warnings" go one step further by providing an interpretation of sensor signals (e.g., lane-departure warning). AVCS-I systems that alter control actions to supplement those provided by a driver are referred to as "control enhancement" systems (e.g., headway control in ACC systems).

AVCS-II: Cooperative Driver/Vehicle/Highway Systems

The goal of AVCS-II is to implement lateral and longitudinal vehicle control functions in specific applications, such as high-occupancy vehicle lanes. Vehicles enter lanes voluntarily under manual control but then are under full or partial automatic control. The advantages are increased speed and safety; platooning also is possible. For private vehicles, AVCS-II offers vehicle-to-vehicle communication of travel paths, which will reduce collisions.

AVCS-II requires both vehicle- and highway-based equipment and utilizes vehicle-to-vehicle and roadway-to-vehicle communications systems developed in ATMS and/or ATIS. Vehicle lateral and longitudinal position is controlled when suitably equipped vehicles operate in dedicated instrumented lanes. Vehicles voluntarily enter and exit these lanes and under manual control but are under full or partial system control while in them. The benefits include enhanced safety and increased travel speed through bottleneck locations at a modest cost compared to achieving the same benefit by increasing the number of parallel lanes. The specific technological elements of AVCS-II, which provides a bridge between AVCS-I and AVCS-III, include the following:

- automatic lateral control
- automatic longitudinal control
- vehicle-to-vehicle communications (e.g., for merge or demerge)
- system integration of AVCS-II technologies (e.g., platooning)
- intersection-hazard warning
- electric propulsion

Platooning is a principal focus of AVCS-II system development. When the first three component technologies are combined, true driver/vehicle/highway platoon systems can be developed. In platooning, vehicles are linked electronically into "platoons" on one lane of a freeway. The first platooning facility likely consists of 20 to 25 miles of a two-lane freeway separated from other lanes by a barrier; a realistic operation involves 5,000 to 10,000 vehicles. Systems required include a vehicle-to-vehicle headway-control system, accurate vehicle-speed control, platoon-to-platoon spacing control, and automated entrance diagnostics. Intersection-hazard warning systems entail vehicle-to-vehicle communication about intended travel paths and extend beyond the capabilities of obstacle-detection systems developed in AVCS-I.

To reduce pollution in congested urban areas, roadway electrification and the use of roadway-powered electric vehicles also is envisioned as a part of AVCS-II systems. A short demonstration facility for an electrified roadway already operates, with a single bus being used to test the concept at low speeds. High-occupancy vehicle lanes that combine the use of electric vehicles and platooning are anticipated in congested urban areas.

AVCS-III: Automated Vehicle–Highway Systems

The goal of AVCS-III is to achieve complete automation of the driving task on limited-access highways. A limited demonstration of the concept has been successful in California. The benefits of this system are derived from the application of the following technologies as a complete package:

- drive-by-wire
- steer-by-wire
- automatic onboard diagnostics (which must be interrogated and found accept-able before entry to AVCS-II facilities is permitted)
- automatic lateral control
- automatic longitudinal control
- vehicle-to-vehicle and vehicle-wayside communication for control
- human interfaces for transitions to and from control
- integration of automated roadways with arterials and local streets
- automatic traffic-merging control
- automatic lane-changing control
- automatic trip routing and scheduling
- automatic obstacle detection and avoidance
- reliability and safety enhancement features for all functions (e.g., real-time condition monitoring, fault detection, and separate degraded performance and emergency operating modes)

For both technical and environmental reasons, it also would be beneficial for vehicles to be equipped with electric powertrains. Only a few elements of AVCS-III technology are commercially available today, such as drive-by-wire (which is available on only a few luxury vehicles).

There are no major technological barriers to the development of AVCS systems; however, public acceptance is a major concern. Who will benefit from the technology? How will the costs be allocated? Will AVCS systems be perceived as a threat to contemporary lifestyles, privacy, or individual autonomy? What are the education and training requirements for AVCS drivers? Safety is a major potential benefit of AVCS systems but also increases risk if the systems do not work as intended. Substantial effort in system design must be directed to minimizing the probability of failures, as well as the consequences when failures do occur. It is difficult to antic-ipate how the public will respond to AVCS, in which vehicles may be operating in closer proximity and higher speeds than today. Much attention must be given to the public's emotional responses to this form of travel and its perceptions of its safety.

PROBLEMS

1. The field of ITS is developing rapidly in terms of new technologies. Find a recent article about one of the new developments and summarize the main points.

2. Find a Web site (e.g., through the local or regional department of transportation) that provides information about how ITS technologies are being used in your region and summarize the main points.

REFERENCES

Anonymous, 1990, "Final Report of the Working Group on Advanced Vehicle Control Systems (AVCS)," Mobility 2000 Workshop on Intelligent Vehicles and Highway Systems, March.

Anonymous, 2001, "Digital Driving," *New York Times*, June 13.

Anonymous, 2002, "Autos: A New Industry," *Business Week*, July 15 (cover story).

Barfield, W., and T. A. Dingus, 1998, *Human Factors in Intelligent Transportation Systems*, Lawrence Erlbaum Associates.

Bender, J. G., 1991, "An Overview of Systems Studies of Automated Highway Systems," *IEEE Transactions on Vehicular Technology*, February 1991, Vol. 40, No. 1, pp. 82–99.

Betsold, R. J. and J. H. Rillings, 1990, *Final Report of the Working Group on Advanced Driver Information Systems (ADIS)*, Dallas, TX: Mobility 2000, March.

Burns, L. D., J. B. McCormick, and C. E. Borroni-Bird, 2002, "Vehicle of Change," *Scientific American*, October, pp. 64–73.

Buxton, J. L., S. K. Honey, W. E. Suchowerskyj, and A. Tempelhof, 1991, "The Travel Pilot: A Second-Generation Automotive Navigation System," *IEEE Transactions on Vehicular Technology*, February, Vol. 40, No. 1, pp. 41–4.

Capps, G. J., K. P. Gambrell, K. J. Johnson, 2001, *Demonstration Project 111 ITS/CVO Technology Truck*, Final Project Report ORNL/TM-2001/277.

Chang, K. S., W. Li, A. Shaikhbahai, F. Assaderaghi, and P. Varaiya, 1991, "A Preliminary Implementation for Vehicle Platoon Control System," *Proceedings of the American Control Conference*, Boston, MA, June, pp. 3078–83.

Chee, W., and M. Tomizuka, 1995, "Lane Change Maneuvers for Automated Highway Systems: Control, Estimation Algorithm and Preliminary Experimental Study," *Proceedings of the International Mechanical Engineering Congress and Exposition*, San Francisco, CA, November.

Davis, J. R., 1989, *Back Seat Driver: Voice-Assisted Automobile Navigation*, Ph.D. dissertation, Cambridge, Massachusetts Institute of Technology.

Davis, J. R., and C. M. Schmandt, 1989, "The Back Seat Driver: Real-Time Spoken Driving Instructions," pp. 146–50 in D. H. M. Reekie, E. R. Case, and J. Tsai (eds.), *First Vehicle Navigation and Information Systems Conference (VNIS'89)*, IEEE.

Dudek, C. L., R. D. Huchingson, W. R. Stockton, R. J. Koppa, S. H. Richards, and T. M. Mast, 1978, *Human Factors Requirements for Real-Time Motorist Information Displays, Volume I – Design Guide* (Technical Report FHWA-RD-78-5), Washington, DC, U.S. Department of Transportation, Federal Highway Administration.

Eby, D. W., and L. P. Kostyniuk, 1999, "An On-The-Road Comparison of In-Vehicle Navigation Assistance Systems," *Human Factors*, Vol. 41, pp. 295–311.

Ervin, R. D., 1991, *An American Observation of IVHS in Japan*, University of Michigan IVHS Program Report.

Esterberg, M. A., E. D. Sussman, and R. A. Walter, 1986, *Automotive Displays and Controls – Existing Technology and Future Trends* (Report PM-45-U-NHT-86-11), Washington, DC, U.S. Department of Transportation, Federal Highway Administration, National Highway Traffic Safety Administration.

Fenton, R. E., and R. J. Mayhan, 1991, "Automated Highway Studies at the Ohio State University," *IEEE Transactions on Vehicular Technology*, Vol. 40, No. 1, pp. 100–113.

Fontaine, H., G. Malaterre, and P. Van Elslande, 1989, "Evaluation of the Potential Efficiency of Driving Aids," *Vehicle Navigation and Information Systems Conference Proceedings-VNIS'89*, pp. 454–9, IEEE.

Ford Motor Company, 1986, "Automotive Electronics in the Year 2000: A Ford Motor Company Perspective," *Proceedings of the CONVERGENCE '86 Conference*, Dearborn, MI, October.

Furukawa, Y., 1992, "The Direction of the Future Automotive Safety Technology," *Proceedings of the International Congress on Transportation Electronics*, SAE Paper No. 92C013.

Harris, W. J., and G. S. Bridges, 1989, *Proceedings of the Workshop on Intelligent Vehicle/Highway Systems by Mobility 2000*, College Station, Texas Transportation Institute, Texas A&M University.

Haver, D. A., and P. J. Tarnoff, 1991, "Future Directions for Traffic Management Systems," *IEEE Transactions on Vehicular Technology*, February, Vol. 40, No. 1, pp. 4–10.

Hedrick, J. K., M. Tomizuka, and P. Varaiya, 1994, "Control Issues in Automated Highway Systems," *IEEE Control Systems Magazine*, December.

Himmelspach, T., and W. Ribbens, 1989, "Radar Sensor Issues for Automotive Headway Control Applications," *Fall 1989 Summary Report for the Special Topics Course in IVHS*, Appendix G, University of Michigan.

Jacobs, I. M., A. Salmasi, and T. J. Bernard, 1991, "The Application of a Novel Two-Way Mobile Satellite Communications and Vehicle Tracking System to the Transportation Industry," *IEEE Transactions on Vehicular Technology*, February, Vol. 40, No. 1, pp. 57–63.

Jurgen, R. K., 1991, "Smart Cars and Highways Go Global," *IEEE Spectrum*, May, pp. 26–36.

Labiale, G., 1989, *Influence of in Car Navigation Map Displays on Drivers Performances* [sic], SAE Paper No. 891683, Warrendale, PA: Society of Automotive Engineers.

LeBlanc, D. J., R. D. Ervin, G. E. Johnson, P. J. Th. Venhovens, G. Gerber, R. DeSonia, C.-F. Lin, T. Pilutti, and A. G. Ulsoy, 1996, "CAPC: An Implementation of a Road-Departure Warning System," *IEEE Control Systems Magazine*, Vol. 16, No. 6, December, pp. 61–71.

LeBlanc, D. J., P. J. Th. Venhovens, C.-F. Lin, T. Pilutti, R. D. Ervin, A. G. Ulsoy, C. MacAdam, and G. E. Johnson, 1996, "Warning and Intervention System to Prevent Road-Departure Accidents," *Vehicle System Dynamics*, Vol. 25, Suppl., pp. 383–96.

McMahon, D. H., J. K. Hedrick, and S. E. Shladover, 1990, "Vehicle Modeling for Automated Highway Systems," *Proceedings of the American Control Conference*, San Diego, CA, May, pp. 297–303.

Michigan Department of Transportation, 1981, *Scandi Project*, Freeway Operations Division, November.

Okuno, A., A. Kutami, and K. Fujita, 1990, "Towards Autonomous Cruising on Highways," SAE Paper No. 901484.

Oshizawa, H., and C. Collier, 1990, "Description and Performance of NAVMATE, and In-Vehicle Route Guidance System," *Proceedings of the American Control Conference*, San Diego, CA, May, pp. 782–7.

Proper, A. T. 1999, *Intelligent Transportation Systems Benefits: 1999 Update*, Federal Highway Administration Report FHWA-OP-99-012.

Radin, S., B. Sen, J. Lappin, 2000, "Advanced Traveler Information Service (ATIS): Private Sector Perceptions and Public Sector Activities," Volpe Center, Boston, MA, January.

Rajamani, R., 2005, *Vehicle Dynamics and Control*, Springer-Verlag.

Rajamani, R., H. S. Tan, B. K. Law, and W. B. Zhang, 2000, "Demonstration of Integrated Longitudinal and Lateral Control for the Operation of Automated Vehicles in Platoons," *IEEE Transactions on Control Systems Technology*, Vol. 8, No. 4, July, pp. 695–708.

Shladover, S., 1991, "Research and Development Issues in Intelligent Vehicle/Highway Systems (IVHS)," in S. A. Velinsky, R. H. Fries, I. Haque, and D. Wang (eds.), *Advanced Automotive Technologies-1991*, ASME DE-Vol. 40, New York, pp. 5–9.

Shladover, S., C. A. Desoer, J. K. Hedrick, M. Tomizuka, J. Walrand, W. B. Zhang, D. H. McMahon, H. Peng, S. Sheikholeslam, and N. McKeown, 1991, "Automatic Vehicle

Control Developments in the PATH Program," *IEEE Transactions on Vehicular Technology*, February, Vol. 40, No. 1, pp. 114–30.

Special Issue on IVHS, 1991, *IEEE Transactions on Vehicular Technology*, February, Vol. 40, No. 1.

Stehr, R. A., 1990, "Motorist Information Systems in Minnesota," *Proceedings of the American Control Conference*, San Diego, CA, May, pp. 287–90.

Trabold, W. G., and T. A. Prewitt, 1969, "A Design for an Experimental Route Guidance System," *Highway Research Record*, No. 265 (Route Guidance), pp. 50–61, Washington, DC, Highway Research Board, National Research Council (also released as GM Research Publication GMR-828, January 1968).

Walker, J., C. Sedney, E. Alicandri, and K. Roberts, 1989, *In-Vehicle Navigation Devices: Effects on Driver Performance* (Draft Technical Report FHWA/RD-89-xxx), Washington, DC, U.S. Department of Transportation, Federal Highway Administration.

Zwahlen, H. T., and D. P. DeBald, 1986, "Safety Aspects of Sophisticated In-Vehicle Information Displays and Controls," *Proceedings of the Human Factors Society – 30th Annual Meeting*, pp. 256–60, Santa Monica, CA: The Human Factors Society.

18 Preventing Collisions

18.1 Active Safety Technologies

An important motivation for AVCS technologies is safety, and a key safety technology is collision detection and avoidance systems. This type of safety enhancement is termed "active safety," which is different from the traditional passive-safety concept (i.e., crashworthiness) (Sun and Chen 2010). The goal is to prevent collisions, not simply mitigate their effects. There are two major driving forces behind recent progress in the active-safety area, as follows:

1. The continuous progress in passive-safety systems has pushed the technology into a return/cost plateau. For example, a recent study shows that 42 percent of fatal-crash occupants can be saved by safety belts, and 47 percent can be saved with safety belts plus an air bag (Figure 18.1). For the remainder of accidents, the impact energy level is simply too high to be managed by reasonable engineering means using current technology. Most of these high-impact energy impacts, however, can be avoided altogether by active safety technologies (ASTs).
2. Recent changes in the standards for Corporate Average Fuel Economy (CAFE) continue to move toward reduced petroleum consumption in the United States. An important engineering approach for higher fuel efficiency is to lower vehicle weight; however, this solution is likely to raise safety concerns. It has been verified consistently that vehicle weight is the third-most important safety attribute for automobiles (i.e., after safety belts and air bags). Again, a possible solution to this safety concern is to apply ASTs.

Many enabling technologies and subsystems, which are useful for AST, have been widely available on passenger vehicles since 2005 (Table 18.1). Therefore, the add-on complexity and cost of introducing AST are greatly reduced. This fact, together with the obvious diminishing returns from passive-safety devices, has made active-safety systems increasingly attractive.

18.2 Collision Detection and Avoidance

Human drivers do not always pay close attention to traffic conditions because of distractions (e.g., looking at the mirror, adjusting the radio, or using a cellular phone) or

Table 18.1. *Vehicle control functions and active safety technology (AST) enablement*

Control function	Installation base	AST function
Cruise control	90%	Throttle actuator
ABS	85%	Automatic braking and yaw
TCS	20%	Small deceleration and yaw
Electronically controlled transmissions	85%	Down-shifting
Multiplex	35%	Information and control
ACC	10%	Range sensor

cannot react in a timely manner because of weather, lighting, or physical disabilities. The value of collision detection and warning systems is in detecting hazardous situations and providing warning or intervention actions to avoid a collision or mitigate damage. These systems require the sensing of objects in a vehicle's near field and the appropriate interpretation of the sensor signals for purposes of warning or control. Also, this function must be carried out while a vehicle is traveling at a high speed. To date, this difficult and important area has not been researched sufficiently. However, there is considerable related research in areas such as obstacle avoidance in robot trajectory planning and mobile robots. The problems, although clearly related, are not identical because robots may not operate at high speeds and are not likely to encounter other obstacles that may be moving at high speeds. Therefore, the vehicle dynamics is important, whereas the robotic problems usually can be treated as kinematic or even static problems. Also, robots must act autonomously on the information obtained about obstacles. AVCS systems – at least AVCS-I and AVCS-II systems – are expected to provide information on obstacles to a driver, or perhaps warnings, but not to take over control of the vehicle. Studies show that many accidents can be avoided if a driver has an additional 0.5 to 1 second in which to take evasive action. Thus, the solution to the obstacle-detection and warning problem is important and is expected to provide significant safety benefits.

One of the key enabling technologies for a collision detection and avoidance system is the development of sensors. Ultrasonic, infrared, laser, video imaging, and

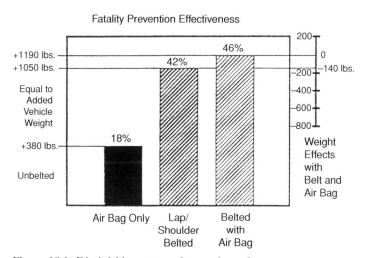

Figure 18.1. Diminishing returns for passive safety systems.

radar have been suggested as the most plausible sensing devices, some of which have been demonstrated in applications The collision-detection functions are divided into the following two general groups:

1. *Near-obstacle detection*: The application scenarios are low speed and/or low range (i.e., rear and blind-spot detection). The detecting signals are usually wide beam.

2. *Forward-looking sensors and detection*: For high-speed applications, the range needs to be much higher than those for near-obstacle (e.g., approximately 300 versus 30 feet). The beam width is usually much narrower so that vehicles in neighboring lanes are not constantly detected. For control-design purposes, it is desirable to have both range and range-rate information. The same sensing requirement can be used for ACC, which is one reason that the development of forward-looking sensors has received significant attention.

Figure 18.2 illustrates a typical scenario for obstacle detection of a vehicle in a highway environment. A long-range forward-looking detection system (e.g., radar) could be coupled with a shorter range forward- and rear-obstacle detection system (e.g., ultrasonic) or even a short-range side-detection system. Some blind spots may remain depending on sensor placement and range. The presence of many stationary roadside objects (e.g., trees and shrubs, guardrails, and signposts) increases the complexity of the problem of processing sensor information. The use of computer vision also has been considered but can be limited by computation times necessary for image processing. In an ITS setting, some of these sensors may be on the highway infrastructure – for example, induction coils or other vehicle detectors (Michigan Department of Transportation 1981) – as well as on vehicles. The information available on a vehicle may be from its own near-field sensors and/or transmitted to it from other vehicles and the roadway. For example, for merging maneuvers onto a freeway, installing overhead vision systems at the on-ramp site used to signal vehicles when it is safe to merge has been proposed (Pilutti et al. 1990). Such an overhead vision system is being used in a SMART CRUISE demonstration project in Tsukuba City, Japan, to monitor traffic flows for use in vehicle-navigation systems.

Ultrasonic sensors, already used in many applications such as autofocus cameras and mobile robots, typically are limited to an approximate 10-meter range and are not suitable for many highway applications. However, they may be useful (and have been used) in vehicles for near-field obstacle detection at low speeds – for example, to detect the curb in parallel parking and obstacles in a driveway when a driver is backing out. The sensors can operate in conditions such as fog, smoke, and darkness when visibility is poor. They also can be used in nonhighway vehicles, such as the autonomous guided vehicles (AGVs) used in manufacturing plants and warehouses.

Radar is considered a leading obstacle-sensing technology for possible use in automotive applications but there are important issues, including the appropriate frequency band to be used, possible interference between similarly equipped vehicles, and reflected radiation from road surfaces. Infrared sensors can be used to supply range information. Alternatively, radar can be used to enhance a driver's perception – allowing for better visibility and longer reaction times – in night driving and other conditions when visibility is poor. Prototype systems have been developed and demonstrated for use with vehicles.

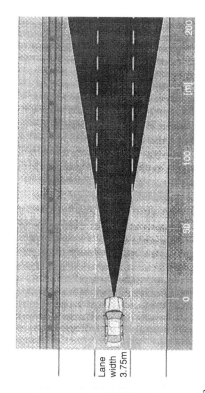

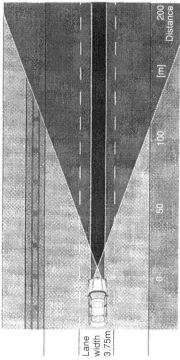

Figure 18.2. Representative forward-looking sensor azimuth coverage.

When the range and possibly the range-rate information are obtained, the collision-warning algorithm typically is triggered based on the following two types of indexes, both of which are popular and used widely:

Time to Collision (TTC): The simple use of range over range rate gives a reasonable index for collision warning. Depending on engineering judgment, this index usually is selected to be between 5 and 8 seconds.

Safe Distance: Based on physics, a "safe following distance" d can be constructed, which has the form:

$$d = V_c t_d + \frac{V_c^2}{2a_c} - \frac{V_l^2}{2a_l}$$

where V_c is the controlled (i.e., host) vehicle speed and t_d is the delay time, which consists of human neuromuscular delay plus judgment time. V_l is the lead vehicle speed and a_c and a_l are deceleration limits, which usually are assumed to be the same but can change with the environment (e.g., 4.5 $m/\sec^2$ for a dry road and 3.3 $m/\sec^2$ for a wet road).

Most collision-warning systems have multiple warning levels: the first level usually leaves judgment time for a driver to take action; the second level is triggered if a threat becomes imminent. The second level may involve intervention – for example, automated braking or differential braking (Pilutti et al. 1998) – as well as warnings to the driver. Nevertheless, it is important that the driver maintain ultimate control and be able to override any automated intervention.

Collision-avoidance systems consider and are developed for various scenarios, such as longitudinal collisions, road-departure accidents, vehicle rollover, and jackknifing of articulated vehicles. The TTC and safe-distance metrics described previously are suitable for longitudinal (i.e., accelerating or braking) collision scenarios. For single-vehicle road-departure (SVRD) accidents, which account for nearly a fourth of all highway accidents and a third of all fatalities, other metrics such as the time to lane crossing (TLC) are described in LeBlanc et al. (1996) and Lin and Ulsoy (1996). Such systems also may require additional sensors – for example, to measure yaw rate (Sivashankar and Ulsoy 1998) – and may include systems to assess the driver state (Pilutti and Ulsoy 1999). Other active-safety systems address serious scenarios such as vehicle rollover and jackknifing of trucks (Chen et al. 2010; Chen and Peng 1999; Ma and Peng 1999b). The next generation of active-safety systems for vehicles also is expected to include inter-vehicle communications, allowing for the sharing of information among vehicles to achieve cooperative active-safety systems (Caveney 2010).

EXAMPLE 18.1: PREVENTING RUN-OFF-ROAD ACCIDENTS. In this example (revisited in Chapter 20), we consider an active-safety system to prevent SVRD accidents (LeBlanc et al. 1996). The system is based on the TLC metric, which is computed from two key elements (Lin and Ulsoy 1996): (1) a projection of the vehicle path; and (2) an estimate of the upcoming lane geometry (Lin et al. 1999). A vision system is used to look head and provide information about the upcoming roadway geometry. On-vehicle sensors also are used to provide information about vehicle motion, which then is used with a vehicle model and disturbance estimation (Lin et al. 2000), such as in Eq. (4.49), to determine the path projection. Both

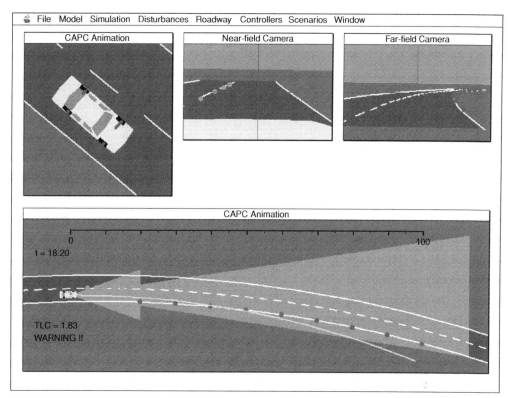

Figure 18.3. Simulation environment for design and evaluation of the SVRD active safety system.

the path projection and the roadway-geometry estimation use Kalman filters to best combine the sensor information with model information and to reduce uncertainty in the TLC calculations.

When the TLC is below a certain threshold value, the system issues a warning to the driver. This can be thought of as an "electronic rumble strip," similar to a physical rumble strip located along the edge of a highway that generates a sound audible to the driver when the wheels are outside the lane edge (Pilutti and Ulsoy 2003. These electronic rumble-strip systems have been implemented in some long-range trucking vehicles, in which SVRD due to driver drowsiness is a serious concern.

The system has the capability to not only warn the driver but, in some cases, also to intervene or even fully control the vehicle when the driver is not responsive. Therefore, the system includes the capability to assess driver status (i.e., alert, drowsy, or unresponsive), which is accomplished by utilizing a driver model. Input to the driver is assumed to be the view captured by the onboard vision system and the output of the driver is the measured steering-wheel angle (Pilutti and Ulsoy 1999). For example, if the camera detects a steady right turn ahead and the driver does not respond by steering to the right, this indicates that the driver is inattentive and may be drowsy or even asleep. In this case, the system uses differential braking to generate a yaw moment (i.e., to steer the vehicle to the right) (Pilutti et al. 1998).

As shown in Figure 18.3, a detailed simulation environment was used to develop and evaluate this active safety system for preventing SVRD accidents. In addition, driving simulators and road tests were used to evaluate and refine the proposed system.

PROBLEMS

1. A dead-reckoning system for vehicle navigation relies on integration of a velocity signal to determine vehicle position:

$$x(t) = \int_0^t v(\tau)d\tau + x(0)$$

If discrete-time measurements are made at time intervals, T, this can be approximated by:

$$x(t) = \sum_{i=0}^{t=NT} v(i)T + x(0)$$

The dead-reckoning approach can lead to position errors when there are errors in the measured velocity values. Assume that the vehicle-measured velocity signal is $v(t) = \sin(2\pi t) + e(t)$ and that $T = 0.02$. Consider the following cases and, for each one, determine $x(t = 0.5)$ given $x(0) = 0$:

(a) $e(t) = 0$ (i.e., no measurement error)
(b) $e(t)$ = normal random error with zero mean and a standard deviation of 1
(c) $e(t) = 1.0$ is a constant

Use MATLAB to compare and discuss the differences in the value of $x(0.5)$ in each case.

2. Consider a vehicle with velocity $v_f = 31$ m/s that follows a lead vehicle with velocity $v_l = 30$ m/s at a distance of 10 m. The total delay time, $t_d = 1$ sec, and acceleration/deceleration limits for both vehicles are $a_f = a_l = 3$ m. What are the TTC and the safe distance? How do you interpret these metrics in the context of this scenario?

3. Consider a simple scenario in which a vehicle is driving at constant longitudinal velocity, u, along the centerline of a lane in a straight section of roadway (i.e., no curvature) with a constant heading angle, ψ. Assuming that u and ψ are both measured exactly, show that the TLC can be calculated as:

$$\text{TLC} = (\Delta/\sin\psi)/u$$

where Δx is the distance along the lane, as shown in the following figure:

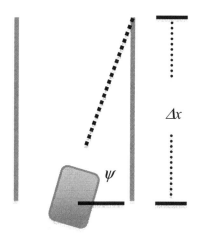

REFERENCES

Caveney, D., 2010, "Cooperative Vehicular Safety Applications," *IEEE Control Systems Magazine*, Vol. 30, No. 4, August, pp. 38–53.

Chen, B-C., and H. Peng, 1999, "Rollover Warning of Articulated Vehicles Based on a Time-to-Rollover Metric," American Society of Mechanical Engineers, Dynamic Systems and Control Division (Publication) DSC, Vol. 67, pp. 247–54, *Proceedings of the ASME International Mechanical Engineering Congress and Exposition*, Nashville, TN.

Chen, L. K., and A. G. Ulsoy, 2002, "Design of a Vehicle Steering Assist Controller Using Driver Model Uncertainty," *International Journal of Vehicle Autonomous Systems*, Vol. 1, No. 1, pp. 111–32.

Chen, L. K., and A. G. Ulsoy, 2006, "Experimental Evaluation of Steering Assist Controllers on a Driving Simulator," *Vehicle System Dynamics*, Vol. 44, No. 3, March, pp. 223–45.

Chen, S., T. B. Sheridan, H. Kusunoki, and N. Komoda, 1995, "Car Following Measurements, Simulations, and a Proposed Procedure for Evaluating Safety," *Proceedings of the IFAC Symposium on Analysis, Design and Evaluation of Man-Machine Systems*, Pergamon Press, pp. 603–8.

Chen, S. K., N. Moshchuk, F. Nardi, and J. Ryu, 2010, "Vehicle Rollover Avoidance," *IEEE Control Systems Magazine*, Vol. 30, No. 4, August, pp. 70–85.

Davis, J. R., 1989, *Back Seat Driver: Voice Assisted Automobile Navigation*, Ph.D. Dissertation, Cambridge, Massachusetts Institute of Technology.

Davis, J. R., and C. M. Schmandt, 1989, "The Back Seat Driver: Real Time Spoken Driving Instructions," pp. 146–50 in D. H. M. Reekie, E. R. Case, and J. Tsai (eds.), *Proceedings of the IEEE Vehicle Navigation and Information Systems Conference (VNIS'89)*.

Fontaine, H., G. Malaterre, and P. Van Elslande, 1989, "Evaluation of the Potential Efficiency of Driving Aids," *Proceedings of the IEEE Vehicle Navigation and Information Systems Conference (VNIS'89)*.

Hatsopoulos, N., and J. A. Anderson, 1992, "Collision-Avoidance System Based on Optical Flow," *Proceedings of the Intelligent Vehicles '92 Symposium*, Detroit, June–July.

Himmelspach, T., and W. Ribbens, 1989, "Radar Sensor Issues for Automotive Headway Control Applications," *Fall 1989 Summary Report for the Special Topics Course in IVHS*, Appendix G, University of Michigan.

Hirano, M., 1993, "Development of Vehicle-Following Distance Warning System for Trucks and Buses," *Proceedings of the IEEE-IEE Vehicle Navigation and Information Systems Conference*, Ottawa, October.

Jones, W. D., 2002, "Building Safer Cars," *IEEE Spectrum*, January, pp. 82–5.

Kaempchen, N., K. C. Fuerstenberg, A. G. Skibicki, and K. C. J. Dietmayer, 2004, "Sensor Fusion for Multiple Automotive and Safety Comfort Applications," in *Advanced Microsystems for Automotive Applications*, pp. 137–63, Springer, Berlin.

Kehtarnavaz, N., J. S. Lee, and N. C. Griswold, 1990, "Vision-Based Convoy Following by Recursive Filtering," *Proceedings of the American Control Conference*, San Diego, CA, May, pp. 268–72.

Kehtarnavaz, N., and W. Sohn, 1991, "Steering Control of Autonomous Vehicle by Neural Networks," *Proceedings of the American Control Conference*, Boston, MA, June, pp. 3096–101.

Kimbrough, S., 1991, "Coordinated Braking and Steering Control for Emergency Stops and Accelerations," in S. A. Velinsky, R. H. Fries, I. Haque, and D. Wang (eds.), *Advanced Automotive Technologies–1991*, ASME DE-Vol. 40, New York, pp. 229–44.

Kimbrough, S., and C. Chiu, 1990, "Automatic Steering System for Utility Trailers to Enhance Stability and Maneuverability," *Proceedings of the American Control Conference*, San Diego, CA, May, pp. 2924–9.

Kimbrough, S., and C. Chiu, 1991, "A Brake Control Algorithm for Emergency Stops (Which May Involve Steering) of Tow-Vehicle/Trailer Combinations," *Proceedings of the American Control Conference*, Boston, MA, June, pp. 409–14.

LeBlanc, D. J., R. D. Ervin, G. E. Johnson, P. J. Th. Venhovens, G. Gerber, R. DeSonia, C.-F. Lin, T. Pilutti, and A. G. Ulsoy, 1996, "CAPC: An Implementation of a Road-Departure Warning System," *IEEE Control Systems Magazine*, Vol. 16, No. 6, December, pp. 61–71.

LeBlanc, D. J., P. J. Th. Venhovens, C.-F. Lin, T. Pilutti, R. D. Ervin, A. G. Ulsoy, C. MacAdam, and G. E. Johnson, 1996, "Warning and Intervention System to Prevent Road-Departure Accidents," *Vehicle System Dynamics*, Vol. 25, Suppl., pp. 383–96.

Lin, C. F., and A. G. Ulsoy, 1996, "Time to Lane Crossing Calculation and Characterization of Its Associated Uncertainty," *ITS Journal*, Vol. 3, No. 2, pp. 85–98.

Lin, C. F., A. G. Ulsoy, and D. J. LeBlanc, 1999, "Lane Geometry Perception and the Characterization of Its Associated Uncertainty," *ASME Journal of Dynamic Systems, Measurement and Control*, Vol. 121, No. 1, March, pp. 1–9.

Lin, C.F., A. G. Ulsoy, and D. J. LeBlanc, 2000, "Vehicle Dynamics and External Disturbance Estimation for Vehicle Path Prediction," *IEEE Transactions on Control System Technology*, Vol. 8, No. 3, May, pp. 508–18.

Ma, W.-H., and H. Peng, 1999a, "Worst-Case Evaluation Method for Dynamic Systems," *ASME Journal of Dynamic Systems, Measurement and Control*, Vol. 121, No. 2, pp. 191–9.

Ma, W.-H., and H. Peng, 1999b, "Worst-Case Vehicle Evaluation Methodology – Examples on Truck Rollover/Jackknifing and Active Yaw Control Systems," *Vehicle System Dynamics*, Vol. 32, No. 4, pp. 389–408.

Michigan Department of Transportation, 1981, *Scandi Project*, Freeway Operations Division, November.

Nguyen, H. G., and J. Y. Laisne, 1992, "Obstacle Detection Using Bi-Spectrum CCD Camera and Image Processing," *Proceedings of the Intelligent Vehicles '92 Symposium*, Detroit, June–July.

Niehaus, A., and R. F. Stengel, 1991, "An Expert System for Automated Highway Driving," *IEEE Control Systems Magazine*, Vol. 11, No. 3, April, pp. 53–61.

Pilutti, T., U. Raschke, and Y. Koren, 1990, "Computerized Defensive Driving Rules for Highway Maneuvers," *Proceedings of the American Control Conference*, San Diego, CA, May.

Pilutti, T., and A.G. Ulsoy, 1999, "Identification of Driver State for Lane Keeping Tasks," *IEEE Transactions on Systems, Man and Cybernetics*, Vol. 29, No. 5, September, pp. 486–502.

Pilutti, T., and A. G. Ulsoy, 2003, "Fuzzy Logic Based Virtual Rumble Strip for Road Departure Warning Systems," *IEEE Transactions on Intelligent Transportation Systems*, Vol. 4, No. 1, March, pp. 1–12.

Pilutti, T., A. G . Ulsoy, and D. Hrovat, 1998, "Vehicle Steering Intervention Through Differential Braking," *ASME Journal of Dynamic Systems, Measurement and Control*, Vol. 120, No. 3, September, pp. 314–21.

Schneider, M., 2005, "Automotive Radar – Status and Trends," *Proceedings of the German Microwave Conference*, Ulm, Germany, April, pp. 144–7.

Sivashankar, N., and A. G. Ulsoy, 1998, "Yaw Rate Estimation for Vehicle Control Applications," *ASME Journal of Dynamics, Measurement, and Control*, Vol. 120, No. 2, June, pp. 267–74.

Sun, Z., and S. K. Chen, 2010, "Automotive Active Safety Systems," *IEEE Control Systems Magazine*, Vol. 30, No. 4, August, pp. 36–7.

Thorpe, C., M. H. Hebert, T. Kanade, S. A. Shafer, 1988, "Vision and Navigation for the Carnegie-Mellon NAVLAB," *IEEE Transactions on Pattern Analysis and Machine Intelligence*, Vol. 10, No. 3, May, pp. 362–73.

Truett, R., 1992, "The Rides of Their Lives," *Orlando Sentinel*, March 25, p. B-8.

Ulke, W., R. Adomat, K. Butscher, and W. Lauer, 1994, "Radar-Based Automotive Obstacle Detection System," SAE Paper No. 940904.

Yasui, Y., and D. L Margolis, 1992, "Lateral Control of Automobiles Using a Looking-Ahead Sensor," *AVEC '92*, pp. 292–7.

Zwahlen, H. T., and D. P. DeBald, 1986, "Safety Aspects of Sophisticated In-Vehicle Information Displays and Controls," *Proceedings of the Human Factors Society Annual Meeting*, pp. 256–60, Santa Monica, CA: The Human Factors Society.

19 Longitudinal Motion Control and Platoons

This chapter discusses the longitudinal control of vehicle motion in the context of AVCS for ITS. It begins with a slightly refined version of a cruise-control system with preview, which uses site-specific information available through the highway infrastructure. It then builds on the ACC topic of Chapter 12 as an intermediate step to the control of vehicles operating in platoons.

19.1 Site-Specific Information

It is assumed that highway infrastructure can provide information to drivers and vehicles, which can be useful in numerous ways; for example, drivers can be warned of accidents, roadwork, inclement weather or road-surface conditions, congestion, and roadway characteristics such as grade and curvature. This already is being implemented in many areas using programmable road signs on busy urban Interstate highways (see Chapter 17). Here, we consider specifically the use of site-specific information to improve the performance of vehicle control systems discussed in previous chapters. The manner in which such site-specific information can be provided to a vehicle is discussed first and then the use of such information by vehicle control systems.

Site-specific information can be provided to drivers and vehicles in a variety of ways. For example, the Prometheus project in Europe uses transmitters on signposts located along certain roadways in the Autocruise project to provide references to the cruise-control system. A driver can choose to set the cruise-control system to pick up speed-limit information transmitted from roadside signposts and then use this information as the reference speed for the cruise-control system. Another concept for providing site-specific roadside information, illustrated in Figure 19.1, involves the use of message chips embedded in pavement and placed along a roadside or on roadway bridges, overpasses, and signposts. The chips are similar to those already in use in manufacturing plants and warehouses for inventory control and parts identification: erasable programmable read only memory (EPROM) devices with a transmitter, which can be reprogrammed remotely by authorized vehicles to update the information they contain and transmit. These systems have been tested in under-the-pavement and roadside use and demonstrated to be feasible. In both systems, there are issues such as placement, range, durability, tamper-resistance, and power sources.

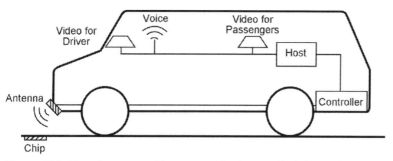

Figure 19.1. Use of message chips to provide site-specific information.

Other key issues include: What kind of information should they contain and transmit? How frequently should they be spaced? How "accurate" does the information they transmit need to be? How often does the information need to be updated? These issues require an understanding of the way in which the information may be used and the potential benefits to be derived from its use.

Here, we consider use of site-specific information by the vehicle control systems. If it is available, can this information be used to improve the capabilities and performance of in-vehicle control systems? The Autocruise system mentioned previously indicates that the answer may be affirmative. In that system, speed-limit information is provided to the vehicle as a reference for the cruise-control system and potentially relieves drivers of the need to reset the desired speed with changing road conditions. Similarly, if road-roughness information is available, it may be used by an active-suspension system to change automatically ride characteristics (e.g., soft versus typical versus harsh) rather than requiring drivers to make the selection. Additional possibilities are as follows:

- Lane and road-curvature information could be provided for automated vehicle lateral control, supplementing or alleviating the need for lane marking with magnets or an onboard lane-sensing capability.
- Information on road-surface conditions (e.g., an icy bridge) could be provided for use by TCS for antilock braking and anti-spin acceleration.
- Road-roughness information could be provided for use by the active-suspension system not only to select ride characteristics but also to improve the control algorithm by providing an estimate of the previously unknown road input.
- Altitude information could be useful for engine-control functions such as the air–fuel ratio control.
- In addition to speed-limit information, road-grade information could be useful for improving cruise-control-system performance, which is discussed in Example 19.1.

EXAMPLE 19.1: CRUISE CONTROL WITH PREVIEW BASED ON SITE-SPECIFIC INFOR- MATION. We first consider the use of site-specific information to improve the performance of a simple cruise-control system (see Chapter 12). This assumes that the highway infrastructure provides information to vehicles, specifically road-grade information. Thus, if a vehicle can be warned of an upcoming grade, the cruise control should be able to take advantage of this information to improve speed-regulation performance. Figure 19.2 illustrates the open-loop system and the closed-loop system (with PI control) block diagrams for a cruise-control

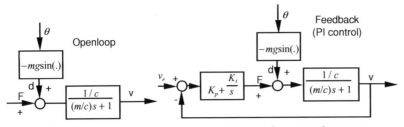

Figure 19.2. Open-loop and example closed-loop cruise-control systems.

system (see Chapter 12). As a basis for comparison, the response of each system to an uphill grade of $\theta = 10°$ is shown in Figure 19.3.

```
% Ex19_1.m
m=1000; c=85; g=9.8;
theta=pi/18;
K=1/c;     tau=m/c;
disturb=-m*g*sin(theta);
Vd=25.0; Fo=Vd/K;
Ki=100.0;    Kp=500; Ti=Kp/Ki;
vo(1)=Vd;    vc(1)=Vd;
t(1)=0.0;
integ=Vd/(K*Ki);    d=0.0;
tstep=0.1;
for i=1:999,
   t(i)=i*tstep;
   if i>300,  d=disturb; end
   vdot_o=[K*(Fo+d)-vo(i)]/tau;
   vo(i+1)=vo(i)+vdot_o*tstep;
   integ = integ + (Vd-vc(i))*tstep;
   Fc=Kp*(Vd-vc(i))+Ki*integ;
   vdot_c=[K*(Fc+d)-vc(i)]/tau;
   vc(i+1)=vc(i)+vdot_c*tstep;
end
t(1000)=100.0;
plot(t,vo,t,vc), xlabel('Time (sec)')
gtext('Openloop')
gtext('PI control')
```

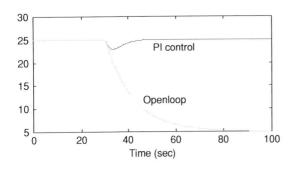

Figure 19.3. Vehicle speed under gradient disturbance.

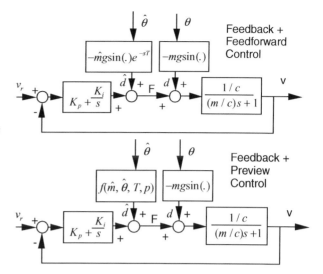

Figure 19.4. Feed-forward and preview control systems.

Figure 19.4 illustrates two distinct ways in which the site-specific grade information can be used (1) the use of feed-forward control together with the previous PI feedback controller; and (2) a similar approach that uses a feed-forward action based on preview (p) (i.e., advance knowledge of an upcoming grade). If our knowledge of the vehicle parameters (m and c), grade disturbance (Θ), and timing of the grade (T) was exact, then the effect of the grade could be canceled exactly by the feed-forward action, as shown in Figure 19.5a. However, these variables usually are not exactly known, and a more realistic performance is shown in Figure 19.5b for the feed-forward control strategy. Here, it is assumed that d is underestimated by 10 percent and that there is a 1-second delay ($T = 1$) in the application of the feed-forward compensation.

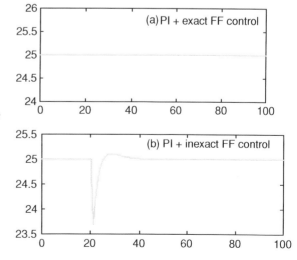

Figure 19.5. Simulation results of PI + FF cruise-control systems.

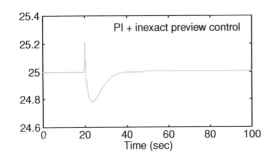

Figure 19.6. Simulation results of PI + inexact preview cruise-control system.

```
% Ex19_1a.m
m=1000.0; c=85.0; g=9.8;
theta=pi/18.0;
K=1/c;     tau=m/c;
disturb=-m*g*sin(theta);
disturb_h=disturb*0.9;
Vd=25.0;
Ki=100.0;     Kp=500.0;   Ti=Kp/Ki;
vc(1)=Vd;
t(1)=0.0;
integ=Vd/(K*Ki);    d=0.0; d_h=0.0;
tstep=0.1;

for i=1:499,
   t(i)=i*tstep;
   if i>200,  d=disturb; end
   if i>210,  d_h=disturb_h; end
   integ = integ + (Vd-vc(i))*tstep;
   Fc=Kp*(Vd-vc(i))+Ki*integ-d_h;
   vdot_c=[K*(Fc+d)-vc(i)]/tau;
   vc(i+1)=vc(i)+vdot_c*tstep;
end
t(500)=100.0;
plot(t,vc), xlabel('Time (sec)')
title('PI + inexact FF control')
```

With preview control, the driver anticipates that a grade is coming and begins to increase vehicle speed before the grade is encountered. The preview is especially important when the control signal does not affect the dynamics of the vehicle at the same point, which is true in most cases. Figure 19.6 illustrates this strategy in which a ramp preview signal is applied, starting 0.5 second before and ending 0.5 second after the disturbance occurs. Again, the same inaccuracies used previously (i.e., 10 percent underestimation) are used to obtain the results shown in Figure 19.6. It is clear from these results that the preview control provides better performance in the case of inexact knowledge. However, the improvements obtained by using site-specific information are minor in this application and probably would not justify its use.

However, these small improvements might become more attractive in applications such as ACC and platooning, in which the performance requirements are more demanding.

A vehicle cruise-control system (see Chapter 12) regulates the vehicle longitudinal velocity. As an intermediate step between platooning and cruise control, the control of the vehicle longitudinal motion (i.e., velocity and position) was considered in Chapter 12. The idea of ACC is to allow a vehicle to regulate speed when there are no automobiles in front of it. However, when a vehicle is detected, the controller regulates the relative distance between the two vehicles (i.e., "headway"). Such a system (i.e., autonomous or intelligent cruise control) could be effective, for example, in stop-and-go traffic as well as on the highway and could be used in conjunction with the Autocruise system described previously. The next section discusses the control of vehicle platoons.

```
% Exactly the same program except that
% the preview action uses the following
% smoothed signal instead of the
% step change signal in the previous
% example
  if (i>195) & (i<=205),
      d_h=disturb_h*(i-195)/10;
  end
```

19.2 Platooning

The goal in platooning is to improve throughput on congested highways by allowing groups of vehicles (e.g., 10 to 20) to travel together in tightly spaced platoons (e.g., 1-m intervals) at high speeds (e.g., 30 m/s). The spacing between platoons may be somewhat larger than current vehicle headway on the highway (e.g., 30 m). The potential advantages in terms of increased throughput are shown to be significant; however, there also are serious potential problems in terms of safety and human factors. In addition to how drivers and passengers will react to traveling at such high speeds in such close formation, there are concerns about how they will react in any type of emergency situation. In terms of safety, the vehicles in a platoon are too closely spaced to prevent collision in the case of a failure – which causes rapid deceleration of a car! However, it also can be argued that many human drivers already are practicing platooning on current highways. It is claimed that close intra-platoon spacing actually may improve safety over current highways spacings because the relative speed at impact in most cases will be small.

Platooning requires another level of control beyond individual vehicles (Figure 19.7). Two fundamentally different approaches to platooning have been suggested: (1) point-following control, in which each vehicle is assigned a particular moving slot on the highway and maintains that position; and (2) vehicle-following control, in which each vehicle in the platoon regulates its position relative to the vehicle in front of it based on information about the lead vehicle motion and locally measured variables (i.e., its own motion and headway to the vehicle in front). The two approaches are compared in Table 19.1. Here, we discuss the vehicle-following control approach,

Table 19.1. *Comparison of longitudinal control approaches*

Vehicle-follower control	Point-follower control
Advantages	**Advantages**
Extremely flexible in accommodating trains or platoons of vehicles of diverse length.	Simple vehicle control implementation – each vehicle follows a well-defined target.
Spacing between vehicles can be adjusted as speeds change.	Only one communication path needed (to roadway).
Speeds can be adjusted easily to adapt to demand shifts or incidents.	Merging, routing, and scheduling simplified by fixed, discrete spacing increments.
Communication burden of roadway is minimized.	Decouples vehicle control from system management.
Some operations can continue even when wayside equipment fails (some fault tolerance).	
Consistent with very decentralized routing and scheduling.	
Normal and emergency control modes are very similar.	
Disadvantages	**Disadvantages**
Control system on vehicle is complicated: sensing spacing and speed (and maybe acceleration) relative to predecessor; communicating information about other vehicles in platoon; must design to minimize interactions among vehicles (string stability); separate mode of operation for leading vehicle in platoon.	Slot length must be long enough for worst-case condition, limiting capacity.
Vehicles need multiple communication paths – to and from other vehicles (for control) and the roadway (for routing and scheduling).	Not easily adaptable to sudden demand shifts.
Flow instabilities are possible if something goes wrong.	Failure of a vehicle or a wayside system produces shutdown.
	Separate communication and control means needed to handle anomalies and failures.
	Not suitable for trains or platoons of variable length.

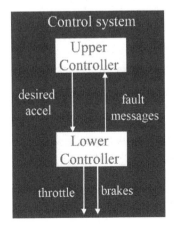

Figure 19.7. Platooning requires another upper level of control (of the platoon) beyond the control of the individual vehicle.

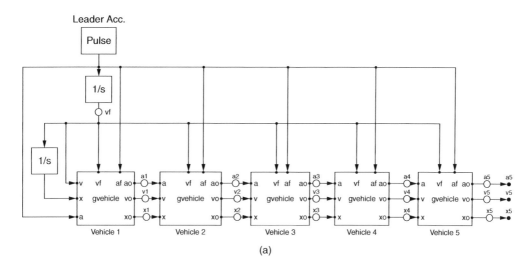

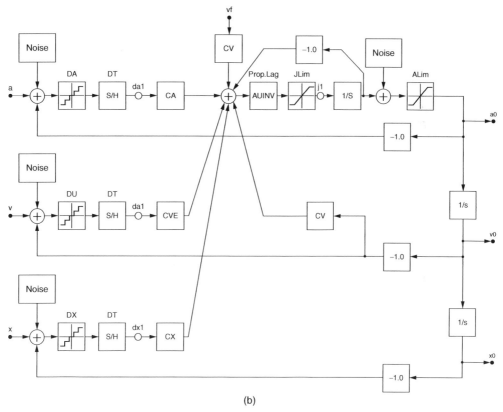

Figure 19.8. (a) Model of a platoon of five vehicles. (b) Model of an individual vehicle in the platoon (Shladover 1991a).

which is the focus of most current research and development work in the area. We note that platooning requires another level of control beyond the individual vehicle (i.e., control of the platoon).

EXAMPLE 19.2: SIMULATION OF PLATOONS. Figure 19.8a is a model for a platoon of five vehicles and a model for a single vehicle in the platoon. As shown in

Figure 19.8b, each vehicle is represented by a third-order nonlinear model. This arises from representation of the vehicle as a mass plus a propulsion system (which has a first-order lag between commanded and actual thrust). Acceleration and jerk limits are explicitly represented (i.e., Alim and Jlim) as saturation elements, and the first-order propulsion lag is represented by the block TAUINV. Thus, each vehicle is represented by equations of motion of the following form:

$$\dot{a}_0 = -(1/\tau)a_0 + (1/\tau)u$$
$$\dot{v}_0 = a_0$$
$$\dot{x}_0 = v_0$$

where we ignore nonlinearities (i.e., saturation and quantization) and noise. Studies show that to maintain asymptotic stability of the platoon, each vehicle requires direct feedback of information from the platoon leader as well as from the immediate predecessor. Figure 19.8a shows that each vehicle in the platoon is supplied with the following input data: (1) acceleration, velocity, and position (a, v, x) of the preceding vehicle; and (2) acceleration and velocity of the lead vehicle (a_f, v_f). Each vehicle then generates as output its own acceleration, velocity, and position (a_0, v_0, x_0). The controller gains are CX (for vehicle spacing errors), CVE (for velocity errors), CA (for acceleration), and CV (for difference in velocity from the commanded value for the platoon). Thus, the controller algorithm for each vehicle can be represented by:

$$u = c_x(x - x_0) + c_v(v_f - v_0) + c_{ve}(v - v_0) + c_a(a - a_0) + c_{af}(a_f - a_0)$$

Note that blocks also are included in the simulation model for quantization errors, errors due to sampling, and noise. Enhancements to and variations of this basic model are discussed in Shladover (1991). Simulation results in Figure 19.9 show various studies for a platoon velocity-change command (shown as Pulse in Figure 19.8a), which is an acceleration pulse of magnitude 1.0 m/s^2 and a duration of 2 s; this produces a commanded speed increase of 2.0 m/s. These results were obtained with position and velocity feedback but no acceleration feedback. Notice that the errors in spacing between vehicles remain less than 0.3 m for all vehicles. These results are for a propulsion lag time constant of 0.1 s, which is achievable with an electric powertrain but not with an ICE. Using the same controller with a propulsion time lag more representative of an ICE does not yield acceptable performance. Figure 19.10 shows results obtained with a time constant of 0.5 s, which is more representative of an ICE, and a controller, which uses acceleration feedback in addition to position and velocity. The control gains for the two cases (i.e., results in Figures 19.9 and 19.10) are summarized in Table 19.2.

It has been shown that maintaining the stability of a platoon of vehicles at high speeds and tight spacing requires inter-vehicle communication (Swaroop and Hedrick, 1996). A large-scale demonstration of vehicle platoons, under automatic lateral and longitudinal control, was successful, as described in Rajamani et al. (2000).

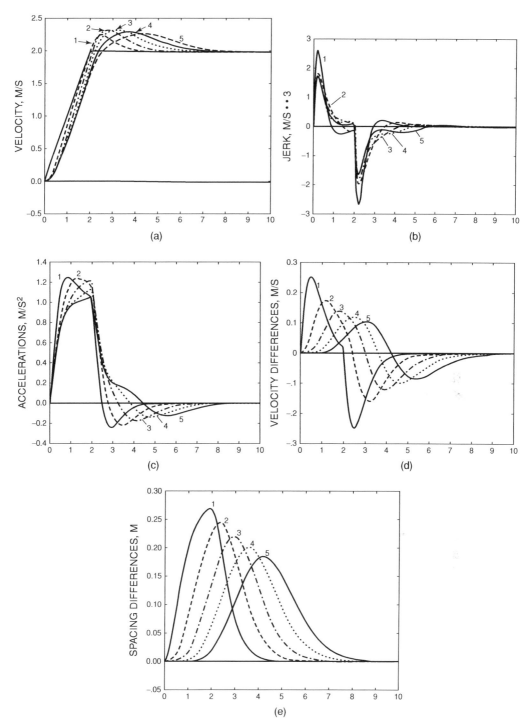

Figure 19.9. Five platoon simulation results with $\tau = 0.1$ second: (a) velocity time history, (b) jerk time history, (c) acceleration time history, (d) velocity difference time history, (e) spacing time history (Shladover 1991a).

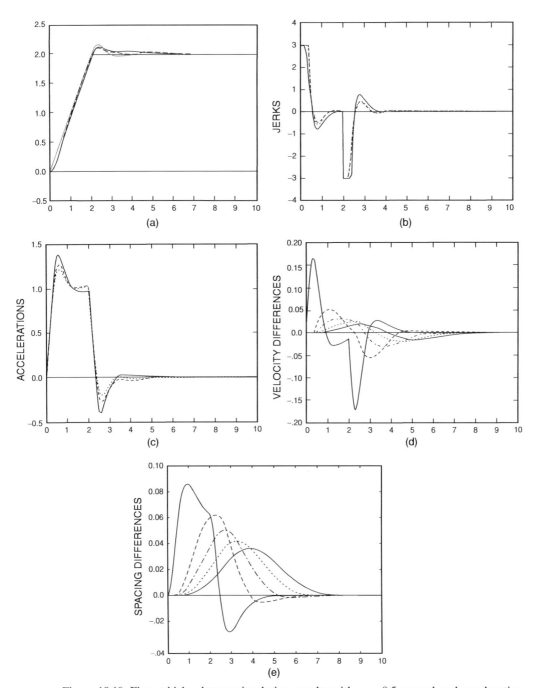

Figure 19.10. Five vehicle platoon simulation results with $\tau = 0.5$ second and acceleration feedback: (a) velocity time history, (b) jerk time history, (c) acceleration time history, (d) velocity difference time history, and (e) spacing difference time history (Shladover 1991a).

Table 19.2. *Control configurations for the two cases*

Gains on	Base vehicle model $\tau = 0.1$ (electric)	Alternative vehicle model $\tau = 0.5$ (internal combustion)
Spacing to predecessor CX (s^{-2})	3.6	18.0
Velocity difference from predecessor CVE (s^{-1})	0.9	4.5
Velocity difference from platoon leader CV (s^{-1})	2.4	12.0
Acceleration difference from predecessor CA (1)	0	2.0
Acceleration difference from platoon leader CAF (1)	0	2.0

19.3 String Stability

For a platoon of vehicles (Figure 19.11), we must consider individual vehicle stability and string stability of the platoon, which is an important new concept. If the preceding vehicle is accelerating or decelerating, then the spacing error could be nonzero; we must ensure that the spacing error attenuates as it propagates along the string of vehicles because it propagates upstream toward the tail of the string. If the position of each vehicle is denoted by x_i, as shown in Figure 19.11, and the desired spacing between vehicles is denoted by L, then the spacing error is given by:

$$\varepsilon_i = x_i - x_{i-1} + L$$

If the preceding vehicle is not accelerating or decelerating, the spacing error should converge to zero; that is:

$$\ddot{x}_i = 0 \quad \Rightarrow \quad \varepsilon_i \to 0 \quad \text{as} \quad t \to \infty$$

Thus, for a constant-spacing policy ($L = $ constant) with a proportional plus derivative controller for the vehicle level, we can ensure the stability of the individual vehicle, as desired:

$$\ddot{x}_i = -k_p \varepsilon_i - k_d \dot{\varepsilon}_i$$
$$\varepsilon_i = x_i - x_{i-1} + L \quad \Rightarrow \quad \ddot{\varepsilon}_i = \ddot{x}_i - \ddot{x}_{i-1} = \ddot{x}_i$$

Next, consider the design of the upper-level (platoon) controller, which is:

$$\Rightarrow \ddot{\varepsilon}_i + k_d \dot{\varepsilon}_i + k_p \varepsilon_i = 0$$

It is designed using the simple model:

$$\ddot{x} = \ddot{x}_{des}$$

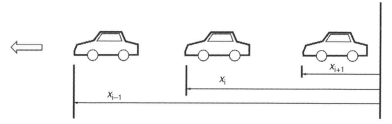

Figure 19.11. Vehicle spacing and spacing error in a platoon.

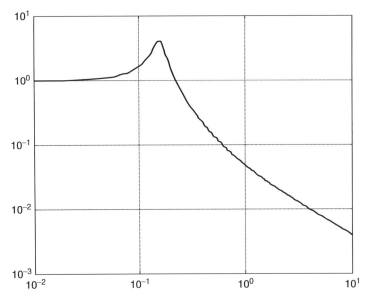

Figure 19.12. The magnitude of $G(s)$ for PD control with $L = $ constant.

whereas a more complete model can be used for robustness analysis; for example:

$$\tau \dddot{x} + \ddot{x} = \ddot{x}_{des}$$

The spacing-error transfer function can be determined as:

$$\varepsilon_i(s) = G(s)\varepsilon_{i-1}(s)$$

Conditions for the string stability of the platoon are given in Swaroop and Hendrick (1996) as follows:

$$G(s) = \frac{\varepsilon_i(s)}{\varepsilon_{i-1}(s)} \quad \text{and} \quad g(t) = L^{-1}[G(s)]$$

$$\| G(s) \|_\infty \leq 1 \quad \text{and} \quad g(t) > 0 \quad \forall t$$

If, as previously, we consider a PD controller with a constant value of L, we obtain:

$$\varepsilon_i = x_i - x_{i-1} + L$$

$$\ddot{x}_i = -k_p \varepsilon_i - k_d \dot{\varepsilon}_i$$

$$\Rightarrow \ddot{\varepsilon}_i = -k_p \varepsilon_i - k_d \dot{\varepsilon}_i + k_p \varepsilon_{i-1} + k_d \dot{\varepsilon}_{i-1}$$

The resulting transfer function is:

$$\varepsilon_i(s) = G(s)\varepsilon_{i-1}(s)$$

$$G(s) = \frac{k_d s + k_p}{s^2 + k_d s + k_p}$$

Thus, the magnitude always exceeds 1 (Figure 19.12) for all values of k_p and k_d, and we cannot have string stability for $L = $ constant and autonomous operation with PD control.

Now consider, again with $L =$ constant, a more complex control algorithm:

$$\varepsilon_i = x_i - x_{i-1} + L$$
$$\ddot{x}_{i\_des} = k_i \ddot{x}_{i-1} + k_2 \ddot{x}_\ell + k_3 \dot{\varepsilon}_i + k_4 \varepsilon_i + k_5 (\dot{x}_i - \dot{x}_\ell)$$

This controller requires inter-vehicle communication and can be designed to achieve string stability. Rajamani et al. (2000) provide an example design based on sliding-mode control.

Although not shown here, it has been demonstrated in the literature that when the constant L is replaced by a constant time-gap policy (i.e., desired spacing $= L + hv$) and a larger distance is maintained at higher speeds, then string stability can be ensured without inter-vehicle communication (i.e., a properly designed ACC should not be a problem!).

To summarize:

1. Platoons may be formed spontaneously as in ACC.
2. In AHS, platoons may be formed purposely to improve flow capacity.
3. Both ACC systems and AHS platoons must be designed to ensure string stability.
4. String stability is influenced by the headway policy. A constant-spacing policy requires lead-vehicle acceleration (through communication) to become string stable.
5. For a constant-time headway policy, inter-vehicle communication is not necessary (e.g., for a properly designed ACC).

PROBLEMS

1. Consider a platoon of vehicles, each with a PD controller (see Figure 19.11) with:

$$\varepsilon_i = x_i - x_{i-1} + L$$
$$\ddot{x}_i = -k_p \varepsilon_i - k_d \dot{\varepsilon}_i$$

Consider a constant-spacing policy and discuss the following:

    (a) stability conditions for each vehicle
    (b) string stability of a platoon of vehicles with no inter-vehicle communication
    (c) string stability of a platoon of vehicles with inter-vehicle communication

2. Consider a platoon of two vehicles (Vehicle 1: lead vehicle; Vehicle 2: following vehicle) in which each vehicle ($i = 1$ or 2) is described by the following equations:

$$\dot{a}_i = -(1/\tau_i)a_i + (1/\tau_i)u_i$$
$$\dot{v}_i = a_i$$
$$\dot{d}_i = v_i$$

where a is acceleration, v is velocity, and d is displacement.

    (a) Rewrite the open-loop equations in standard-state variable form:

$$\dot{\mathbf{x}} = \mathbf{A}\mathbf{x} + \mathbf{B}\mathbf{u}$$

where the states are defined as $x_1 = d_1, x_2 = v_1, x_3 = a_1, x_4 = d_2, x_5 = v_2$, and $x_6 = a_2$.

(b) Next, assume that the control, u_i, for each vehicle is given by:

$$u_1 = u_{10}(t)$$
$$u_2 = c_v(v_1 - v_2) + c_d(d_1 - d_2)$$

where $u_{10}(t)$ is a prescribed command input to the lead vehicle and u_2 is a feedback control law. Determine the closed-loop system equations in standard-state variable form using the same definition of the state variables as in (a).

REFERENCES

Chandler, F. E., R. Herman, and E. W. Montroll, 1958, "Traffic Dynamics: Studies in Car Following," *Operations Research*, Vol. 6, pp. 165–84.

Chang, K. S., W. Li, A. Shaikhbahai, F. Assaderaghi, and P. Varaiya, 1991, "A Preliminary Implementation for Vehicle Platoon Control System," *Proceedings of the American Control Conference*, Boston, MA, June, pp. 3078–83.

Chee, W., and M. Tomizuka, 1995, "Lane Change Maneuvers for Automated Highway Systems: Control, Estimation Algorithm and Preliminary Experimental Study," *Proceedings of the International Mechanical Engineering Congress and Exposition*, San Francisco, CA, November.

Chen, S., T. B. Sheridan, H. Kusunoki, and N. Komoda, 1995, "Car Following Measurements, Simulations, and a Proposed Procedure For Evaluating Safety," *Proceedings of the IFAC Symposium on Analysis, Design and Evaluation of Man-Machine Systems*, Pergamon Press, pp. 603–8.

Fancher, P., H. Peng, Z. Bareket, C. Assaf, and R. Ervin, 2001, "Evaluating the Influences of Adaptive Cruise Control Systems on the Longitudinal Dynamics of Strings of Highway Vehicles," *Proceedings of the IAVSD Conference*, Copenhagen, Denmark, August.

Fenton, R. E., and R. J. Mayhan, 1991, "Automated Highway Studies at The Ohio State University – An Overview," *IEEE Transactions on Vehicular Technology*, February, Vol. 40, No. 1, pp. 100–113.

Frank, A. A., S. J. Liu, and S. C. Liang, 1989, "Longitudinal Control Concepts for Automated Automobiles and Trucks Operating on a Cooperative Highway," *Proceedings of the Future Transportation Technology Conference and Exposition*, SAE SP-791, SAE Paper No. 891708.

Hedrick, J. K., D. McMahon, V. Narendran, and D. Swaroop, 1991, "Longitudinal Vehicle Controller Design for IVHS Systems," *Proceedings of the American Control Conference*, Boston, MA, June, pp. 3107–12.

Hedrick, J. K., M. Tomizuka, and P. Varaiya, 1994, "Control Issues in Automated Highway Systems," *IEEE Control Systems Magazine*, December.

Himmelspach, T., and W. Ribbens, 1989, "Radar Sensor Issues for Automotive Headway Control Applications," *Fall 1989 Summary Report for the Special Topics Course in IVHS*, Appendix G, University of Michigan.

Jurgen, R. K., 1991, "Smart Cars and Highways Go Global," *IEEE Spectrum*, May, pp. 26–36.

Kehtarnavaz, N., J. S. Lee, and N. C. Griswold, 1990, "Vision Based Convoy Following by Recursive Filtering," *Proceedings of the American Control Conference*, San Diego, CA, May, pp. 268–72.

McMahon, D. H., J. K. Hedrick, and S. E. Shladover, 1990, "Vehicle Modeling for Automated Highway Systems," *Proceedings of the American Control Conference*, San Diego, CA, May, pp. 297–303.

Mobility 2000 Workshop on Intelligent Vehicles and Highway Systems, 1990, "Final Report of the Working Group on Advanced Vehicle Control Systems (AVCS)," Dallas, TX, March.

Okuno, A., A. Kutami, and K. Fujita, 1990. "Towards Autonomous Cruising on Highways," SAE Paper No. 901484.

Rajamani, R., 2006, *Vehicle Dynamics and Control*, Springer-Verlag.

Rajamani, R., S. B. Choi, B. K. Law, J. K. Hedrick, R. Prohaska, and P. Kretz, 2000, "Design and Experimental Implementation of Longitudinal Control for a Platoon of Automated Vehicles," *ASME Journal of Dynamic Systems, Measurement and Control*, Vol. 122, No. 3, September, pp. 470–6.

Rajamani, R., H.-S. Tan, B. K. Law, and W.-B. Zhang, 2000, "Demonstration of Integrated Longitudinal and Lateral Control for the Operation of Automated Vehicles in Platoons," *IEEE Transactions on Control Systems Technology*, Vol. 8, No. 4, July, pp. 695–708.

Schwarzinger, M., T. Zielke, D. Noll, M. Brauckmann and W. von Seelen, 1992, "Vision-Based Car-Following: Detection, Tracking, and Identification," *Proceedings of the Intelligent Vehicles '92 Symposium*, Detroit, June–July.

Shladover, S., 1991a, "Longitudinal Control of Automotive Vehicles in Close-Formation Platoons," *ASME Journal of Dynamic Systems, Measurement, and Control*, Vol. 113, June, pp. 231–41.

Shladover, S., 1991b, "Research and Development Issues in Intelligent Vehicle/Highway Systems (IVHS)," in S. A. Velinsky, R. H. Fries, I. Haque, and D. Wang (eds.), *Advanced Automotive Technologies – 1991*, ASME DE-Vol. 40, New York, pp. 5–9.

Shladover, S., C. A. Desoer, J. K. Hedrick, M. Tomizuka, J. Walrand, W. B. Zhang, D. H. McMahon, H. Peng, S. Sheikholeslam, and N. McKeown, 1991, "Automatic Vehicle Control Developments in the Path Program," *IEEE Transactions on Vehicular Technology*, February, Vol. 40, No. 1, pp. 114–30.

Special Issue on IVHS, 1991, *IEEE Transactions on Vehicular Technology*, February, Vol. 40, No. 1.

Swaroop, D., and J. K. Hedrick, 1996, "String Stability of Interconnected Systems," *IEEE Transactions on Automatic Control*, Vol. 41, No. 3, March, pp. 349–57.

Swaroop, D., and J. K. Hedrick, 1999, "Constant Spacing Strategies for Platooning in Automated Highway Systems," *ASME Journal of Dynamic Systems, Measurement and Control*, Vol. 121, No. 3, September, pp. 462–70.

20 Automated Steering and Lateral Control

This chapter focuses on AVCS for ITS, which are concerned with vehicle lateral motion control. The first part of the discussion is about a potential AVCS-II or AVCS-III technology for lane sensing and lane tracking in a lane-departure warning system. It is an intermediate step to fully automated steering, which is a potential AVCS-III technology.

20.1 Lane Sensing

To automate the lateral control of vehicles, it is necessary to measure the position of a vehicle in the lane. Various schemes have been proposed and tested for this purpose, including the use of (1) an on-board vision system that detects the painted lane markings on a highway; (2) continuous magnetic wires imbedded in the center of the lane; (3) radar or ultrasonic waves to measure the distance to sidewalls; (4) a forward-looking laser to detect reflective markers; and (5) discrete magnetic markers imbedded in the roadway. Clearly, the onboard vision systems are more difficult to develop but offer the potential capability of autonomous operation. The latter four schemes require the necessary highway infrastructure to be in place. These cooperative systems, however, are likely to provide more accurate lane-position information at higher speeds with simpler and more inexpensive devices in a vehicle. Magnetic wires and markers also are more reliable under adverse weather conditions (e.g., fog, rain, and snow). Prototype systems based on the other sensing schemes (i.e., vision, laser, and radar) have not been shown to work in inclement weather. Recently, GPS-based systems have been proposed for lateral and longitudinal positioning; however, the accuracy is still not good enough, even when differential GPS is used. Furthermore, GPS may not be available in all areas (e.g., urban canyons); therefore, systems based on GPS measurement and road maps are still under development.

Because vision-based lane-sensing systems are the dominant form of sensing used in lane-detection and lateral-control projects, it certainly makes sense to describe – at least briefly – the principle of lane detection using vision systems. Of course, there are many different approaches, including edge detection, deformable template, and B-snake, which almost always use two-dimensional visible light images.

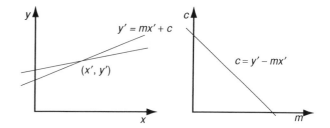

Figure 20.1. Hough transformation.

In the case of edge-detection–based methods, the operations typically involve the following steps:

1. Acquire image
2. Digitize image (e.g., $512 \times 512 \times 8$)
3. Detect edge (Sobel filter, basically a high-pass filter)
4. Thresholding (generate a binary image)
5. Noise cleaning
6. Hough transformation
7. Identify lane-marker candidates
8. Decide the lane markings

In the fourth step, dynamic thresholding can be implemented to partially solve the problem of lighting changes in the environment. Noise cleaning is the step in which human know-how and heuristics enhance the performance of the vision system. For example, because we know that lane edges should extend through to the horizon, the edges of small objects (e.g., trees and signs on the roadside) can be eliminated easily. The lane locations from a previous image also could be used to distinguish lane edges from noise. The Hough transformation in Step 6 is basically a mathematical transformation to represent curves (in particular, straight lines $y = mx + c$) in the (x,y)-coordinate in the (m,c) plane. The Hough transformation was found to be a robust method for representing straight lines and is less susceptible to noise and uncertainty.

The main goal for lane-sensing systems is to supply vehicle location and response signals to calculate appropriate steering signals (i.e., automated steering). However, intermediate systems also can be envisioned in which the driver steers the vehicle and the lane-sensing information is used for warnings, driver assist, or even intervention. A driver-monitoring system also can be imagined in which a probationary driver's steering performance is monitored, recorded, and reported. These systems could be used in driver training and evaluation or by police to monitor those with a history of alcohol or drug abuse. In one sense, these systems are less ambitious than fully automated steering. However, in terms of control-system design, they are perhaps more challenging because the vehicle–driver system dynamics must be modeled as the basis for designing a warning or intervention system; whereas in fully automated steering, the driver is not in the loop and the control design can be based on the vehicle lateral dynamics model.

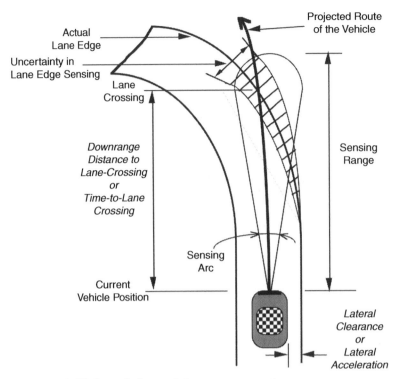

Figure 20.2. TLC metric for road-departure prevention.

Example 20.1 discusses a system for preventing run-off-road accidents (also called SVRD accidents) due to driver drowsiness and inattention, which account for nearly 40 percent of all highway fatalities in the United States each year. This system also can be categorized in ASTs that potentially could avoid the legal concerns that comprise a major obstacle for AVCS, especially AHS concepts.

EXAMPLE 20.1: PREVENTING RUN-OFF-ROAD ACCIDENTS. Here, we revisit Example 18.1 for a more in-depth discussion. In many cases, run-off-road accidents can be avoided by means of technologies that either warn the driver of an impending departure from the roadway or provide some level of control assistance to prevent an off-road trajectory. Such a system would require the means to (1) sense the immediate proximity of the vehicle to the lane edge, and (2) project the vehicle's path and the road layout for a specific distance down range from the current position. Figure 20.2 illustrates the concept of a lane-tracking margin (i.e., TLC) based on lane sensing and vehicle-path projection. It is expected that both the lane-sensing information and the vehicle-path projection will become more uncertain with range and speed. The lane sensing could be accomplished with onboard sensors, as shown in the figure, or using sensors installed in the highway, as discussed previously.

Figure 20.3 (i.e., the shaded portion) illustrates how the proposed AST for preventing run-off-road accidents could be realized. Sensed-lane information (from various possible sensing technologies) would be combined with measurements of vehicle

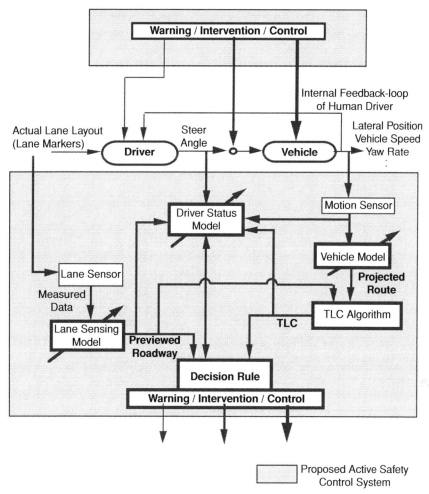

Figure 20.3. Structure of an active system to prevent SVRD accidents.

motion (e.g., vehicle velocity, acceleration, and steering-wheel displacements) to calculate the lane-tracking margin (i.e., TLC). Based on results of these calculations, the system would take one of the following actions: (1) issue a warning to the driver, (2) intervene in the vehicle steering in such a way as to complement the driver actions, or (3) assume full control of steering. The vehicle-path projection requires a model of the driver–vehicle system, not only the vehicle dynamics. The driver–vehicle model might include an adaptive driver model so that deterioration in driver response due to drowsiness and inattention can be detected. Detailed descriptions of the entire system and a prototype vehicle that uses computer vision for lane sensing are in LeBlanc et al. (1996a and 1996b).

The problem of estimating lane geometry from computer-vision images is described in detail in Lin et al. (1999). It is accomplished by using a Kalman filter to estimate the roadway geometry based on a model of the roadway and the vision measurements. Combining the measurements with a roadway model provides smooth estimates despite occasional gaps in the lane markings and outliers in the data obtained from image processing. The method also provides an estimate of lane-geometry uncertainty (e.g., uncertainty increases with range). Measurements of vehicle pitch and roll are used to compensate for camera motion and improve the estimates.

The problem of path projection, based on measurements of the vehicle dynamics, is described in Lin et al. (2000). Path projection uses a vehicle-model–based Kalman filter, together with measurements of vehicle motion (i.e., acceleration, velocity, and yaw), to project forward the vehicle path. The combination of these two (i.e., lane geometry and projected path) then can be used to obtain the TLC, as described in Lin and Ulsoy (1996). The TLC then can be used as a metric for issuing warnings to the driver, taking control actions, or emergency-steering intervention. For example, if TLC is less than an experimentally determined threshold, then a warning to the driver that road-departure is imminent is issued. These decisions (i.e., whether to warn, control, or intervene) can be enhanced if we also have information about the state of the driver (Pilutti and Ulsoy 1999, 2003). Online estimation methods can be developed based on measurement of the driver steering angle and the lateral-position information from the vision system to estimate the driver's state (i.e., alert, fatigued, or asleep). Essentially, the lane-keeping error, as seen by the vision system, is assumed to be the input to the driver block and the steering angle is the driver block output. As described in Pilutti and Ulsoy (1999), a series of driving-simulator studies was used to develop methods to identify a driver's state. Fuzzy logic then can provide a mechanism for combining the TLC and driver-state information to make improved decisions (Pilutti and Ulsoy 2003). One approach for emergency-steering intervention is differential braking, as described in Pilutti et al. (1998). As described in Chapter 5, a steering-assist control can be designed to provide for more consistent lane-keeping performance and to reduce the likelihood of SVRDs.

The overall active-safety system to prevent SVRD accidents described herein is fairly complex. However, elements of the system can be used as stand-alone systems to provide information useful in other applications. For example, a warning-only system, without including control and intervention functions, can be implemented. A simple example of this is an "electronic rumble strip" in which the vision system can be used to detect the lane edge and issue an audible warning to the driver. If an estimate of the TLC or a driver-state assessment is available, it can be used to improve the effectiveness of the simple electronic rumble strip (Pilutti and Ulsoy, 2003. A reduction in SVRDs can be achieved even with a basic system, and further reductions will be achieved as the system is enhanced. NHTSA shows that 6,335,000 accidents (with 37,081 fatalities) occurred on U.S. highways in 1998 (NHTSA 1999), of which about 30 percent were SVRDs, for a total of nearly 2 million accidents. Approximately 40 percent of the fatalities (i.e., approximately 15,000) were due to SVRDs. Even a small percentage reduction can have a significant human and economic impact.

20.2 Automated Lane-Following Control

In the fully automated steering of vehicles, it is assumed that the driver is not in the loop and that lane-sensing information is available. Referring to Figure 20.4,

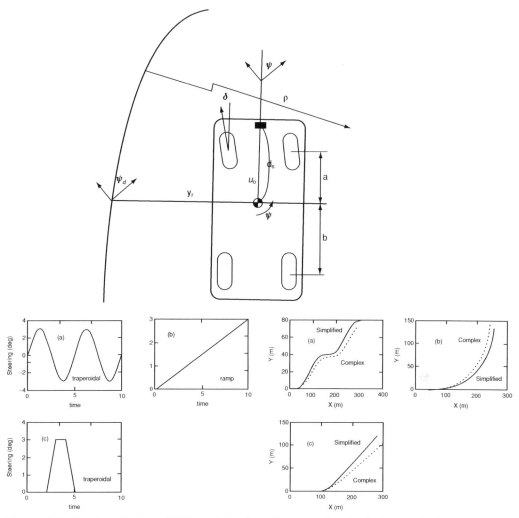

Figure 20.4. The simplified two-DOF model and results comparing simulation results for the two-DOF linear and 10-DOF nonlinear models.

which is based on the two-DOF vehicle-handling model, we can define the following set of linear-state equations for use in lateral-control design (Peng and Tomizuka 1990):

$$
\frac{d}{dt}\begin{bmatrix} y_r \\ \dot{y}_r \\ \psi - \psi_d \\ \dot{\psi} - \dot{\psi}_d \end{bmatrix} = \begin{bmatrix} 0 & 1 & 0 & 0 \\ 0 & \dfrac{A_1}{u_0} & -A_1 & \dfrac{A_2}{u_0} \\ 0 & 0 & 0 & 1 \\ 0 & \dfrac{A_3}{u_0} & -A_3 & \dfrac{A_4}{u_0} \end{bmatrix} \begin{bmatrix} y_r \\ \dot{y}_r \\ \psi - \psi_d \\ \dot{\psi} - \dot{\psi}_d \end{bmatrix} + \begin{bmatrix} 0 \\ B_1 \\ 0 \\ B_2 \end{bmatrix} \delta + \begin{bmatrix} 0 \\ A_2 - u_0^2 \\ 0 \\ A_4 \end{bmatrix} \frac{1}{R}
$$

$$(20.1)$$

Table 20.1. *Parameter values*

Parameter	Nominal value	Minimum value	Maximum value
m (kg)	1,550	0.85 (1,550)	1.15 (1,550)
I_z (kg-m^2)	3,100	0.85 (3,100)	1.15 (3,100)
C_a (N/rad)	84,000	0.2 (84,000)	2.0 (84,000)
u_0 (miles/hour)	70	25	85
a,b (m)	1.15, 1.51	–	–
d_s (m)	1.0	–	–

where y_r is the lateral deviation of the mass center of the vehicle from the reference, y is the yaw angle, y_d is the desired yaw angle obtained from the road curve (which is assumed to be known), R is the radius of curvature of the road, and:

$$A_1 = -\frac{(C_{\alpha f} + C_{\alpha r})}{m} \quad A_2 = -\frac{(C_{\alpha r}a - C_{\alpha f}b)}{m} \quad A_3 = -\frac{(C_{\alpha r}a - C_{\alpha f}b)}{I_z}$$

$$A_4 = -\frac{(C_{\alpha f}a^2 + C_{\alpha r}b^2)}{I_z} \quad B_1 = \frac{F_x + C_{\alpha f}}{m} \quad B_2 = \frac{a(F_x + C_{\alpha f})}{I_z}$$

(20.2)

The output equation is:

$$y = y_r + d_s(\psi - \psi_d) = \begin{bmatrix} 1 & 0 & d_s & 0 \end{bmatrix} \mathbf{x}$$

which represents the lateral deviation as obtained from a sensor at a distance, d_s, ahead of the mass center of the vehicle. The transfer function from the front-wheel steer angle, d, to the output y, is given by:

$$\frac{y(s)}{\delta(s)} = \frac{1}{\Delta(s)} \left[(d_s B_2 + B_1)s^2 + \frac{d_s(B_1 A_3 - B_2 A_1) + B_2 A_2 - B_1 A_4}{u_0}s + B_1 A_3 - B_2 A_1 \right]$$

where

(20.3)

$$\Delta(s) = s^2 \left[s^2 - \frac{A_1 + A_4}{u_0}s + \frac{(A_1 A_4 - A_2 A_3)}{u_0^2} + A_3 \right]$$

Figure 20.4 also shows results comparing this two-DOF model (simplified) to a nonlinear 10-DOF (i.e., more complex) model (Figure 20.5). The results show that the two-DOF linear model is suitable for controller design in lateral control studies and show the steering commands used in the simulation comparison. Listed in Table 20.1 are nominal, minimum, and maximum values of the parameters for the previous model.

EXAMPLE 20.2: CONTROLLER DESIGN FOR AUTOMATED STEERING. A block diagram of the proposed lateral control system, shown in Figure 20.6, has feed-forward and feedback parts. It also includes a parameter estimator for the tire-cornering stiffness, C_a, and a state estimator. The feed-forward controller is based on the fact that freeways are designed to have fixed curvatures; therefore, if the radius

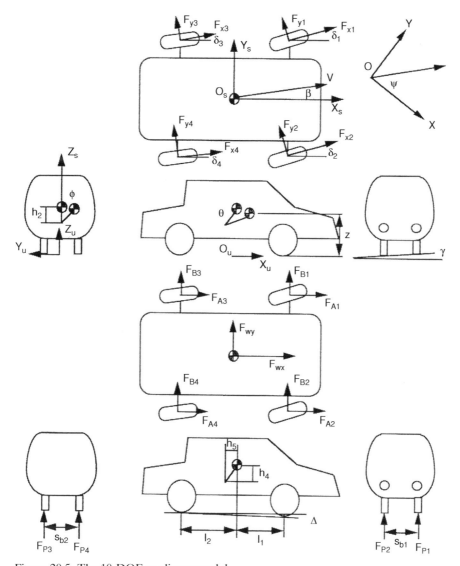

Figure 20.5. The 10-DOF nonlinear model.

of curvature R is known, then the corresponding steady-state steering angle is (from Eq. (4.44) $C_a = C_{af} = C_{ar}$, $F_x = F_{xf} = F_{xr}$, and $L = (a + b)$:

$$\delta = \frac{mu_0^2(b - a) + C_\alpha(a + b)^2}{R(C_\alpha + F_x)(a + b)} \tag{20.4}$$

An improved feed-forward controller can be designed using preview information (Peng and Tomizuka 1991). The feedback controller uses state-feedback estimates and is designed using the frequency-shaped linear quadratic control design approach (Peng and Tomizuka 1990). The details of the design procedure are omitted here. A least-squares algorithm is used to estimate the tire-cornering stiffness, C_a.

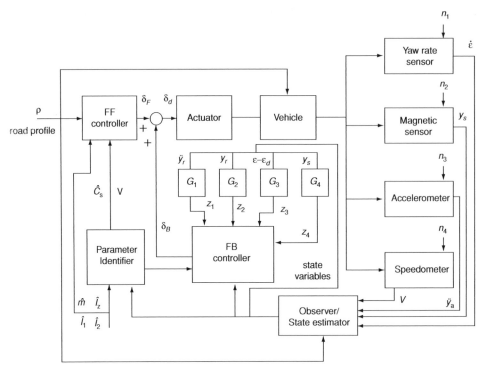

Figure 20.6. Block diagram of lateral control system in Example 20.2.

Figure 20.7 shows simulation results for a wind gust of -500 N and a moment of -200 N-m in the lateral direction for 3 seconds. Figure 20.8 shows simulations using a discrete marker scheme, in which the markers are placed at 4-meter intervals and all other measurements are made at 100 Hz. Figure 20.9 shows simulation results in which a portion of the road is icy so that the value of C_a decreases to the minimum value for 1 second in the curved section of the road. Experimental results are in Peng et al. (1993).

20.3 Automated Lane-Change Control

If the reference/sensing system can supply continuous vehicle lateral displacement and yaw measurement between lanes, the control algorithm developed for lane following also can be used for lane change. Because there is no fixed path to be followed for a lane-change maneuver, a trajectory-planning task must be performed first to define a smooth path. Chee and Tomizuka (1995) found that a virtual desired trajectory (VDT) defined by a trapezoidal acceleration profile easily can satisfy ride-quality constraints (described by jerk and acceleration limits) and time optimality. When the effective range of the reference/sensing systems is less than the full-lane width, the tracking error and other vehicle states must be estimated using the state observer. Other than these two extra components (i.e., VDT and state observer), the system architecture and hard requirement are almost identical to a lane-following system.

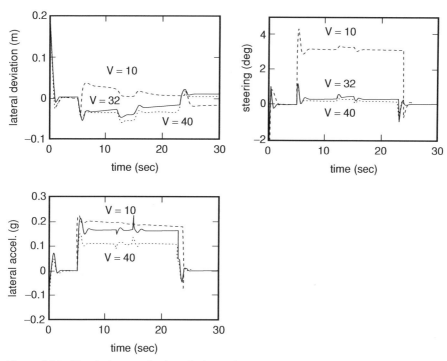

Figure 20.7. Simulation results for wind-gust input.

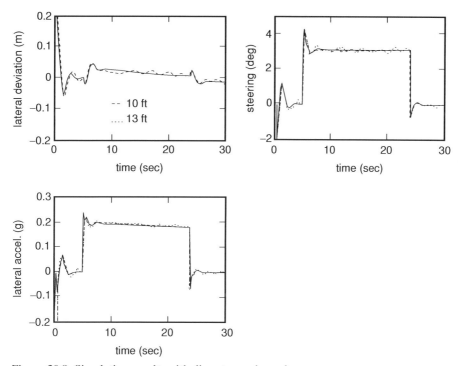

Figure 20.8. Simulation results with discrete marker scheme.

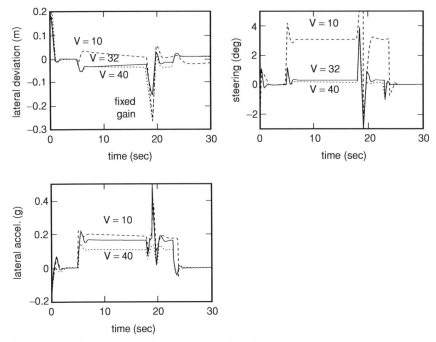

Figure 20.9. Simulation results with icy road section.

PROBLEMS

1. A simple method for calculating the TLC was given in Problem 18.3 for a straight roadway. Determine the error in TLC that results from using this method on a segment of roadway with constant curvature, R.

2. Derive the feed-forward control-steering input given in Eq. (20.4) and its numerical value for the nominal parameters given in Table 20.1. How does this value change with the road curvature, R, with the vehicle characteristics (i.e., a, b, C_a, and m) and with the driving characteristics u_0 and F_x?

3. Show that using the nominal parameter values in Table 20.1, the transfer function in Eq. (20.3) is:

$$\frac{Y(s)}{\delta(s)} = \frac{8.5355s^2 + 3.1331s + 39.061}{s^2(s^2 + 0.65831s + 1.0816)}$$

and design a PID feedback controller based on this transfer function.

4. Combine the feed-forward controller in Problem 2 with the feedback controller in Problem 3. Compare the closed-loop performance of your system via simulation to the results presented in Figure 20.7.

REFERENCES

Ackermann, J., and W. Sienel, 1990, "Robust Control for Automatic Steering," *Proceedings of the American Control Conference*, San Diego, CA, May, pp. 795–800.

Chee, W., and M. Tomizuka, 1995, "Lane Change Maneuvers for Automated Highway Systems: Control, Estimation Algorithm and Preliminary Experimental Study," *Proceedings of the International Mechanical Engineering Congress and Exposition*, San Francisco, CA, November.

Fenton, R. E., and R. J. Mayhan, 1991, "Automated Highway Studies at the Ohio State University," *IEEE Transactions on Vehicular Technology*, Vol. 40, No. 1, pp. 100–113.

Kimbrough, S., 1991, "Coordinated Braking and Steering Control for Emergency Stops and Accelerations," in S. A. Velinsky, R. H. Fries, I. Haque, and D. Wang (eds.), *Advanced Automotive Technologies-1991*, ASME DE-Vol. 40, New York, pp. 229–44.

LeBlanc, D. J., R. D. Ervin, G. E. Johnson, P. J. Th. Venhovens, G. Gerber, R. DeSonia, C.-F. Lin, T. Pilutti, and A. G. Ulsoy, 1996a, "CAPC: An Implementation of a Road-Departure Warning System," *IEEE Control Systems Magazine*, Vol. 16, No. 6, December, pp. 61–71.

LeBlanc, D. J., P. J. Th. Venhovens, C.-F. Lin, T. Pilutti, R. D. Ervin, A. G. Ulsoy, C. MacAdam, and G. E. Johnson, 1996a, "Warning and Intervention System to Prevent Road-Departure Accidents," *Vehicle System Dynamics*, Vol. 25, Suppl., pp. 383–96.

Lin, C. F., and A. G. Ulsoy, 1996, "Time to Lane Crossing Calculation and Characterization of Its Associated Uncertainty," *ITS Journal*, Vol. 3, No. 2, pp. 85–98.

Lin, C. F., A. G. Ulsoy, and D. J. LeBlanc, 1999, "Lane Geometry Perception and the Characterization of Its Associated Uncertainty," *ASME Journal of Dynamic Systems, Measurement and Control*, Vol. 121, No. 1, March, pp. 1–9.

Lin, C. F., A. G. Ulsoy, and D. J. LeBlanc, 2000, "Vehicle Dynamics and External Disturbance Estimation for Vehicle Path Prediction," *IEEE Transactions on Control System Technology*, Vol. 8, No. 3, May, pp. 508–18.

Margolis, D. L., M. Tran, and M. Ando, 1991, "Integrated Torque and Steering Control for Improved Vehicle Handling," in S. A. Velinsky, R. H. Fries, I. Haque, and D. Wang (eds.), *Advanced Automotive Technologies-1991*, ASME DE-Vol. 40, New York, pp. 267–90.

Matsumoto, N., and M. Tomizuka, 1990, "Vehicle Lateral Velocity and Yaw Rate Control with Two Independent Control Inputs," *Proceedings of the American Control Conference*, San Diego, CA, May, pp. 1868–75.

NHTSA, 1999, National Highway Traffic Safety Administration, *Traffic safety facts 1998*. Report DOT HS-808-983. US Department of Transportation, Washington, DC.

Peng, H., W. Zhang, S. Shladover, M. Tomizuka and A. Aral, 1993, "Magnetic-Marker Based Lane Keeping: A Robustness Experimental Study," *Proceedings of the SAE International Congress and Exposition*, Detroit, MI, SAE Technical Paper 930556.

Peng, H., and M. Tomizuka, 1990, "Vehicle Lateral Control for Highway Automation," *Proceedings of the American Control Conference*, San Diego, CA, May, pp. 788–94.

Peng, H., and M. Tomizuka, 1991, "Preview Control for Vehicle Lateral Guidance in Highway Automation," *Proceedings of the American Control Conference*, Boston, MA, June, pp. 3090–5.

Pilutti, T., and A. G. Ulsoy, 1999, "Identification of Driver State for Lane Keeping Tasks," *IEEE Transactions on Systems, Man and Cybernetics*, Vol. 29, No. 5, September, pp. 486–502.

Pilutti, T., and A. G. Ulsoy, 2003, "Fuzzy Logic Based Virtual Rumble Strip for Road Departure Warning Systems," *IEEE Transactions on Intelligent Transportation Systems*, Vol. 4, No. 1, March, pp. 1–12.

Pilutti, T., A. G. Ulsoy, and D. Hrovat, 1998, "Vehicle Steering Intervention Through Differential Braking," *ASME Journal of Dynamic Systems, Measurement and Control*, Vol. 120, No. 3, September, pp. 314–21.

Taheri, S., and H. Law, 1990, "Investigation of a Combined Slip Control Braking and Closed-Loop Four-Wheel Steering System for an Automobile During Combined Hard Braking and Severe Steering," *Proceedings of the American Control Conference*, San Diego, CA, May, pp. 1862–7.

APPENDICES

Review of Control-Theory Fundamentals

A.1 Review of Feedback Control

This appendix, a brief review of important concepts and methods from control theory, is intended to help readers review that material through self-study. A background in dynamic systems and control is essential for an understanding of the material presented in this textbook on automotive control systems. Furthermore, it is assumed that readers are familiar with the computational tools available for simulation and control in the MATLAB/Simulink environment. An excellent online tutorial is available at the Web site www.engin.umich.edu/class/ctms/.

Definitions and Motivation

System. A group of objects that are combined to function as an integrated part for a specific objective (e.g., an engine, a car, or a group of vehicles).

Inputs of a system are the means by which the state of the system can be changed.

Outputs of a system are the means by which the state of the system is manifested (e.g., in the speed control of a car, throttle [input] and speed [output] and in the directional control of a car, steering [input] and yaw rate [output]).

Control. This directs a system's inputs so that the outputs behave in the manner desired.

As an example, consider the idealized vehicle-speed control problem shown in Figure A.1, in which the goal is to control the speed of an automobile by adjustment of the throttle only.

Block-Diagram Representation of Control Systems

Block-diagram form is a simple way to present the interconnection of subsystems in a control system. In block-diagram form, components of the system are connected by signal flows.

The block diagram in Figure A.2, which includes physical elements and signals commonly seen on control systems, can be simplified to the form shown in Figure A.3 for controller-design purposes.

Control the speed of an automobile (throttle only)

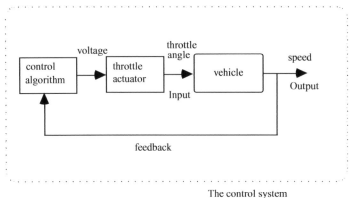

Figure A.1. An idealized control system.

Benefits and Drawbacks of Feedback

In this section, we perform several straightforward analyses to illustrate the benefits and drawbacks of using feedback. Mathematically, if we have perfect knowledge of the plant to be controlled and there are no external disturbances, we could "invert" the plant and obtain the open-loop controller for perfect tracking (i.e., $y(t) = r(t)$ for all t). In the following discussion, this simple idea is illustrated by an algebraic example.

> **EX. A.1.1: FEED-FORWARD CONTROL DESIGN.** Suppose we are designing a cruise-control system considering only steady-state response (i.e., disregard transient response). Assuming that after extensive experiments on a flat test track we found that for each 1 degree of throttle-pedal displacement, the steady-state vehicle speed increases by 4 miles/hour. The open-loop control design is then:

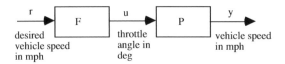

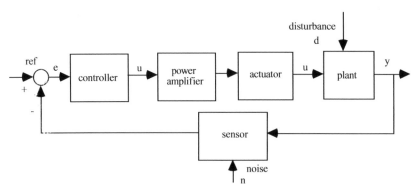

Figure A.2. Block diagram of typical control systems.

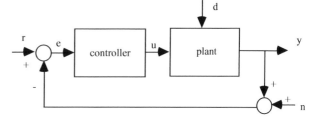

Figure A.3. Simplified block diagram of a control system.

Now, because $P = 4$, we should have $F = 0.25 (= 1/P)$; indeed, if we invert the plant and obtain the controller F, the vehicle speed y will be equal to the desired vehicle speed at steady-state. This is why some people refer to the open-loop-control design problem as the "plant-inversion" problem. This also can be understood from the following general equation:

$$y = Pu = P(Fr) = PFr$$

Because the control-design purpose is to make $y = r$, it is obvious that we should let $PF = 1$ or $F = \frac{1}{P}$. However, the open-loop, or feed-forward, control-design method however is vulnerable to external disturbances and plant uncertainties.

EX. A.1.2: OPEN-LOOP, DISTURBANCE. For the same problem illustrated in Ex. A.1.1, suppose there exist external disturbances (e.g., road gradient and wind gust) that were not considered in the control-design process and the road-gradient angle was found to reduce the vehicle speed at 1.5 mile/hour/degree. The vehicle response then should be represented by the following diagram.

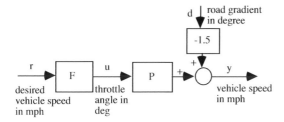

Suppose $P = 4$ and $F = 0.25$; then we have $y = r - 1.5d$. In other words, the controller does not achieve the design objective when a slope is present, and the effect of slope is as large as the uncontrolled case. This is because there is no feedback. There are at least two ways to solve the external disturbance problem, as follows:

1. Measure the disturbance d: This usually is not practical or too expensive. Sometimes "disturbance observers" are constructed to estimate the magnitude of disturbance from its effect on plant output.
2. Feedback: Because the control-design objective is to make y follow r, we intuitively should adjust the control signal u according to the observed error signal $(r-y)$. When $(r-y)$ becomes larger, the control u also should be increased (assuming that we know the larger control signal u will increase the output signal y).

EX. A.1.3: PURE FEEDBACK, DISTURBANCE. The following figure shows that increased feedback action (larger C) reduces the effect of disturbances, which is evident from the following examples:

$$C = 1 \quad \Rightarrow \quad y = \frac{4}{5}r - \frac{1.5}{5}d$$

$$C = 100 \quad \Rightarrow \quad y = \frac{400}{401}r - \frac{1.5}{401}d$$

$$C \rightarrow \infty \quad \Rightarrow \quad y \rightarrow r$$

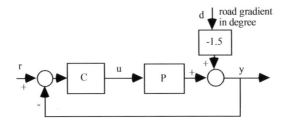

$$y = 4C(r - y) - 1.5d \quad \Rightarrow \quad y = \frac{4C}{4C + 1}r - \frac{1.5}{4C + 1}d$$

There is a limit, however, on the size of feedback gain that can be used due to both practical (e.g., actuator or other hardware limits) and stability considerations. The stability limit is explained further in Ex. A.1.9.

EX. A.1.4: FEEDBACK + FEED-FORWARD, DISTURBANCE.

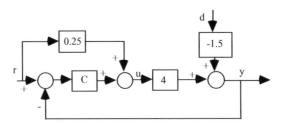

$$y = 4[0.25r + C(r - y)] - 1.5d \quad \Rightarrow \quad y = r - \frac{1.5}{4C + 1}d$$

It is shown in this equation that we can make $y = r$ in the case when there is no external disturbance by adding the feed-forward term. Comparing the results of Ex. A.1.3 and Ex. A.1.4, it can be seen that the feed-forward control term cannot reduce the deviation caused by the disturbance. The reason is that the feed-forward control signal is not aware of the existence and magnitude of the disturbance. In general, if we invert the plant to obtain the feed-forward controller, we can obtain a perfect match between the reference and output signals in the case in which there is no external disturbance. This is obvious from the following example.

EX. A.1.5: FEEDBACK + FEED-FORWARD, DISTURBANCE, GENERAL CASE.

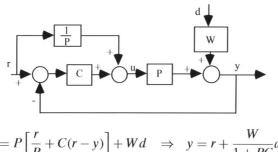

$$y = P\left[\frac{r}{P} + C(r - y)\right] + Wd \quad \Rightarrow \quad y = r + \frac{W}{1 + PC}d$$

EX. A.1.6: OPEN-LOOP, PLANT UNCERTAINTY. For the same problem illustrated in Ex. A.1.1, suppose there is a mismatch between the perceived plant $\hat{P}$ and the true plant $P = \hat{P}(1 + \Delta)$. How does this affect the control-system output?

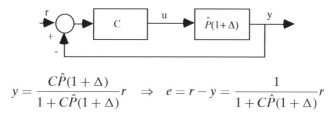

Because the open-loop design simply inverts the perceived plant, we have $F = \frac{1}{\hat{P}}$. Therefore:

$$y = \hat{P}(1 + \Delta) \cdot \frac{1}{\hat{P}} \cdot r = (1 + \Delta)r$$

The difference between the desired output and true output is: $e = r - y = \Delta r$.

EX. A.1.7: CLOSED LOOP, PLANT UNCERTAINTY.

$$y = \frac{C\hat{P}(1 + \Delta)}{1 + C\hat{P}(1 + \Delta)}r \quad \Rightarrow \quad e = r - y = \frac{1}{1 + C\hat{P}(1 + \Delta)}r$$

The error signal can be reduced if we increase C.

The results of Ex. A.1.1 through Ex. A.1.7 are summarized in the following conclusions. A large loop gain CP, also known as the loop-transfer function, in a feedback system ensures that:

1. The output closely follows the reference signal.
2. Desensitization of the output y to the disturbance, d.
3. Desensitization of the I/O map from r to y due to plant variations, Δ.

Pitfalls of Feedback

Analysis results in the previous two sections seem to suggest that we should use a feedback-control configuration with as large a control gain as possible. Large feedback gains, however, can cause problems, which are illustrated in the following examples.

EX. A.1.8: MEASUREMENT NOISE. When the output signal is contaminated by measurement noise (which is true for almost all practical cases), the error signal can be obtained as in the following equation:

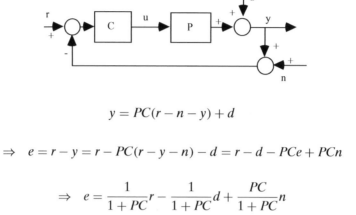

$$y = PC(r - n - y) + d$$

$$\Rightarrow \quad e = r - y = r - PC(r - y - n) - d = r - d - PCe + PCn$$

$$\Rightarrow \quad e = \frac{1}{1 + PC}r - \frac{1}{1 + PC}d + \frac{PC}{1 + PC}n$$

A major control-design objective is to reduce the error signal as much as possible. We are not too concerned about the first term because it can be eliminated easily by a feed-forward controller (if there is one). The second and third terms clearly show the dilemma posed by nature: To reduce the error signal, we would like to make the following:

1. $\frac{1}{1+PC} \equiv S$ (sensitivity function) small, to reduce the effect of the disturbance
2. $\frac{PC}{1+PC} \equiv T$ (complementary sensitivity function) small, to reduce the effect of measurement noise

However, we have:

$$S + T = 1$$

In other words, there is a tradeoff between disturbance-rejection and noise-rejection capabilities in the design of feedback controllers: We cannot make both S and T small (at all frequencies).

EX. A.1.9: INSTABILITY. Examples A.1.1 through A.1.8 illustrate the nature of control-design problems based on static (i.e., algebraic) plant and control systems; however, the real physical systems that we are dealing with are mostly dynamic. Dynamic control systems pose another pitfall that is not present for static systems: the instability problem. In the following discussion, we use a MATLAB example to illustrate this phenomenon. The plant to be controlled in this example is assumed to be governed by the following dynamic equation:

$$\dddot{y} + 3\ddot{y} + 3\dot{y} + y = u$$

Suppose the controller designed is of simple proportional form; then, the block diagram is:

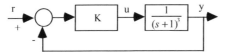

where the design task is to find the appropriate K that gives satisfactory results. The following MATLAB program simulates the step response of the closed-loop system:

```
>> K=1;
>> den = [1 3 3 1+K];
>> num = K;
>> t=0:0.1:30;
>> y=step(num,den,t);
>> plot(t,y);
>> title(['K=' num2str(K)]);
>> xlabel('Time (sec)');
>> ylabel('y');
```

The following figures show that as K increases, the response becomes increasingly more oscillatory. The system finally becomes unstable when $K > 8$.

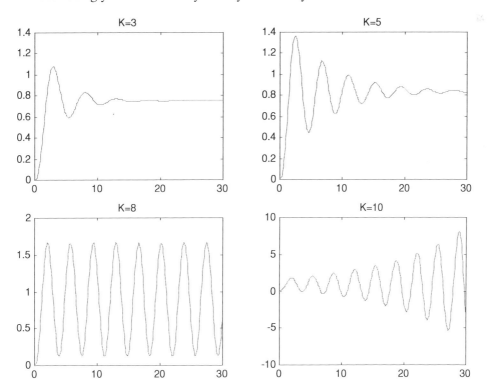

State-Space Realization

Dynamic equations of the plant in ordinary differential equation form can be put into a special form – "state-equation" form – which is suitable for control-design purposes. We first introduce the following two definitions.

State variables: A minimum set of variables, $x_1(t), \ldots, x_n(t)$, such that knowledge of these variables at any time t_o, plus information about the input signals $u(\tau)$, $\tau \geq t_o$, and the system dynamic equations is sufficient to determine the state of the system at any time $t > t_o$.

State equations: A collection of first-order ordinary differential equations that represents the relationship between input and state variables. In general, these equations can be nonlinear and/or time varying, as shown here:

$$\frac{dx_1(t)}{dt} = f_1(x_1, \cdots x_n, u_1, \cdots, u_m)$$

$$\vdots$$

$$\frac{dx_n(t)}{dt} = f_n(x_1, \cdots x_n, u_1, \cdots, u_m)$$

For linear systems, we can group the state equations into matrix form:

$$\frac{d\mathbf{x}(t)}{dt} = \mathbf{A}\mathbf{x}(t) + \mathbf{B}\mathbf{u}(t)$$

where matrices $\mathbf{A}$ and $\mathbf{B}$ are constant or time-varying. However, in any case, they should not depend on $\mathbf{x}$ and $\mathbf{u}$. Assuming that the number of state variables is n, then the size of matrix $\mathbf{A}$ is $n \times n$ and $\mathbf{B}$ is $n \times m$. Again, nonlinear systems cannot be represented in the matrix form shown previously.

A.2 Mathematical Background and Design Techniques

Laplace Transforms

The major reasons to use the Laplace Transform and examine the dynamic system response in the s-domain (s: the Laplace operator) are as follows:

1. Calculus operations (i.e., differentiation and integration) can be replaced by simpler algebraic operations $(+, -, *, /)$.
2. Free response and forced response of the dynamic system can be solved simultaneously.

Definition: Given a function $f(t)$ that satisfies:

$$\int_{0^-}^{\infty} \left| f(t)e^{-\sigma t} \right| dt < \infty$$

for some positive real σ. The Laplace Transform of $f(t)$ is defined as:

$$F(s) \equiv \int_{0^-}^{\infty} f(t)e^{-st} dt = L[f(t)]$$

As long as the time function $f(t)$ does not grow faster than exponentially, its Laplace Transform always will exist. This condition is satisfied by all impulse response signals of physical systems.

EX. A.2.1: Find the Laplace Transform of the unit-step function:

$$F(s) = \int_{0^-}^{\infty} 1 \cdot e^{-st} dt = \frac{-1}{s} e^{-st} \Big|_0^{\infty} = \frac{1}{s}$$

EX. A.2.2: Find the Laplace Transform of the function $f(t) = e^{-at}$:

$$F(s) = \int_{0^-}^{\infty} e^{-at} \cdot e^{-st} dt = \frac{-1}{s+a} e^{-(s+a)t} \Big|_0^{\infty} = \frac{1}{s+a}$$

EX. A.2.3: Find the Laplace Transform of the function $f(t) = \sin \omega t$:

$$F(s) = \int_{0^-}^{\infty} \sin \omega t \cdot e^{-st} dt = \int_{0^-}^{\infty} \frac{1}{2j} (e^{j\omega t} - e^{-j\omega t}) \cdot e^{-st} dt$$

$$= \frac{1}{2j} \left[\frac{1}{s - j\omega} - \frac{1}{s + j\omega} \right] = \frac{\omega}{s^2 + \omega^2}$$

Important Properties of the Laplace Transform

Time domain operators: $\dfrac{d^n}{dt^n} \leftrightarrow s^n$, $\underbrace{\displaystyle\iint \cdots \int}_{n} \leftrightarrow \dfrac{1}{s^n}$

$$L\left\{ \frac{df(t)}{dt} \right\} = \int_0^{\infty} \frac{df}{dt} e^{-st} dt = f(t) e^{-st} \Big|_0^{\infty} + s \int_0^{\infty} f(t) e^{-st} dt = -f(0) + sF(s) \quad \text{(A.1)}$$

$$L\left\{ \frac{d^2 f(t)}{dt^2} \right\} = sL\left\{ \frac{df}{dt} \right\} - f'(0) = s^2 F(s) - sf(0) - f'(0)$$

$$L\left\{ \int_0^t f(\tau) d\tau \right\} = \int_0^{\infty} \int_0^t f(\tau) d\tau e^{-st} dt$$

$$= \frac{-1}{s} \int_0^t f(\tau) d\tau \; e^{-st} \Big|_0^{\infty} + \frac{1}{s} \int_0^{\infty} f(t) e^{-st} dt = \frac{F(s)}{s}$$

Time shift: Time delay τ_d in time domain $\Rightarrow e^{-s\tau_d}$ in s-domain

$$L\{f(t - \tau_d)\} = \int_0^{\infty} f(t - \tau_d) e^{-st} dt = \int_{\tau_d}^{\infty} f(t - \tau_d) e^{-st} dt$$

$$= \int_0^{\infty} f(\sigma) e^{-s(\sigma + \tau_d)} d\sigma = e^{-s\tau_d} F(s)$$

Initial Value Theorem (use Eq. (A.1)):

$$\lim_{s \to \infty} \int_0^{\infty} \frac{df}{dt} e^{-st} dt = \lim_{s \to \infty} [-f(0) + sF(s)]$$

$$\Rightarrow \quad 0 = \lim_{s \to \infty} [-f(0) + sF(s)] \quad \Rightarrow \quad \lim_{t \to 0} f(t) = \lim_{s \to \infty} [sF(s)]$$

Final Value Theorem (use Eq. (A.1)):

$$\lim_{s \to 0} \int_0^{\infty} \frac{df}{dt} e^{-st} dt = \lim_{s \to 0} [-f(0) + sF(s)]$$

$$\Rightarrow \quad f(\infty) - f(0) = \lim_{s \to 0} [-f(0) + sF(s)] \quad \Rightarrow \quad \lim_{t \to \infty} f(t) = \lim_{s \to 0} [sF(s)]$$

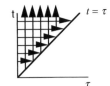

Figure A.4. Double integration in different orders.

Convolution Integral

Given the I/O relationship of a linear time invariant system:

$$y(t) = \int_0^t u(t - \tau)g(\tau)d\tau$$

$$L[y(t)] = \int_0^\infty \int_0^t u(t - \tau)g(\tau)d\tau e^{-st}dt = \int_0^\infty \int_0^t u(t - \tau)g(\tau)e^{-st}d\tau dt$$

$$\sigma = t - \tau = \int_0^\infty \int_\tau^\infty u(t - \tau)g(\tau)e^{-st}dt d\tau = \int_0^\infty g(\tau) \int_\tau^\infty u(t - \tau)e^{-st}dt d\tau$$

$$= \int_0^\infty g(\tau) \int_0^\infty u(\sigma)e^{-s\sigma}d\sigma e^{-s\tau}d\tau = G(s)U(s) \tag{A.2}$$

where the change of integration limits can be understood from Figure A.4:

Eq. (A.2) is very important. The time response of a linear time invariant system is governed by the complex convolution operation between input and output signals. In the s-domain, however, the I/O relationship becomes a simple multiplication. In other words, it is challenging to do any serious analysis and design work in the time domain but it is straightforward in the s-domain.

We usually refer to $G(s)$ as the *transfer function* because it operates on the input signal to get the output signal (Figure A.5). For zero initial conditions, the transfer function represents the ratio of output to input in the Laplace domain (i.e., s-domain).

EX. A.2.4: I/O RELATIONSHIP OF CLOSED-LOOP SYSTEM. This example is intended to show why it is important to use s-domain representation rather than time domain.

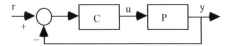

To determine the output $y(t)$ under a certain input signal $r(t)$, we could:

1. Integrate over time; this can be achieved using MATLAB.
2. Construct the time-domain I/O relationship using convolution.

$$y(t) = \int_0^t u(t - \tau_1)p(\tau_1)d\tau_1$$

$$u(t) = \int_0^t [r(t - \tau_2) - y(t - \tau_2)]c(\tau_2)d\tau_2$$

$$\Rightarrow \quad y(t) = \int_0^t \int_0^{t-\tau_1} [r(t - \tau_1 - \tau_2) - y(t - \tau_1 - \tau_2)]c(\tau_2)d\tau_2 p(\tau_1)d\tau_1$$

Figure A.5. I/O relationship in s-domain.

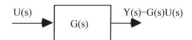

Because $y(t)$ is involved on both sides of the equation, analytical solutions usually cannot be obtained.

3. Obtain the s-domain relationship.

$$Y(s) = P(s)U(s)$$

$$U(s) = C(s)(R(s) - Y(s)) \qquad (A.3)$$

$$\Rightarrow \ Y(s) = \frac{P(s)C(s)}{1 + P(s)C(s)} R(s)$$

The time domain function $y(t)$ then can be easily obtained by taking the inverse Laplace Transform of the right-hand side of Eq. (A.3).

Transfer Function of Linear Time Invariant Systems

By using the Laplace Transform, differential equations can be written as algebraic equations. This is achieved by replacing the time-domain differentiation operator $\frac{d}{dt}$ by the Laplace Transform operator s. In general, if the dynamic equation of a linear time invariant system is:

$$\frac{d^n}{dt^n}y(t) + a_1\frac{d^{n-1}}{dt^{n-1}}y(t) + \cdots + a_n y(t) = b_0\frac{d^m}{dt^m}u(t) + \cdots + b_m u(t)$$

then we have in the s-domain ($\frac{d}{dt} \to s$)

$$s^n Y(s) + a_1 s^{n-1} Y(s) + \cdots + a_n Y(s) = b_0 s^m U(s) + \cdots + b_m U(s)$$

$$15pt] \ \Rightarrow \ Y(s) = \underbrace{\frac{b_0 s^m + b_1 s^{m-1} + \cdots + b_m}{s^n + a_1 s^{n-1} + \cdots + a_{n-1}s + a_n}}_{\text{transfer function}} U(s) \qquad (A.4)$$

Eq. (A.4) showsthe relationship between the system dynamic equation and the transfer function for one given high-order ordinary differential equation (ODE). In many cases, we have several lower-order ODEs. It may be easier to obtain the state-equation form of the system first. The transfer function then can be obtained from the state-space form $\dot{x}(t) = Ax(t) + Bu(t)$ and $y(t) = Cx(t) + Du(t)$.

An important characteristic of the system to be obtained from the transfer function is the DC gain. For the system shown in Eq. (A.4), if the input $u(t)$ is a unit-step input, then the output $y(t)$ at steady-state has the value b_m/a_n.

Figure A.6. I/O relationship of the transfer function.

FACT A.2.5: A PHYSICAL INTERPRETATION OF THE TRANSFER FUNCTION. When a physical system is described by the transfer function $G(s)$ (i.e., the impulse response $\equiv g(t)$, $G(s) = L[g(t)]$) and when the input signal (i.e., assumed to be slow compared to the dynamics of $G(s)$) is in the form $u(t) = e^{s_1 t}$, then the output signal (under zero initial conditions and assuming that the system modes settle quickly) also is an exponential signal, amplified in magnitude by the system-transfer function (i.e., $y(t) = G(s_1)e^{s_1 t}$).

Proof:

$$y(t) = \int_0^\infty g(\tau)u(t-\tau)d\tau = \int_0^\infty g(\tau)e^{s_1(t-\tau)}d\tau$$

$$= \int_0^\infty g(\tau)e^{-s_1\tau}d\tau e^{s_1 t} = G(s_1)e^{s_1 t} \tag{A.5}$$

This relationship is illustrated in Figure A.6.

FACT A.2.6: FREQUENCY RESPONSE. For a stable system $G(s)$, if we apply a sinusoidal input $\sin(\omega t)$, then the steady-state output $y(t)$ (after the transient settles) is $G(j\omega)\sin(\omega t) = |G(j\omega)|\sin(\omega t + \angle G(j\omega))$.

Proof: Because $\sin(\omega t) = \frac{1}{2j}(e^{j\omega t} - e^{-j\omega t})$, applying Eq. (A.5),

$$y(t) = \frac{1}{2j}[G(j\omega)e^{j\omega t} - G(-j\omega)e^{-j\omega t}] = \frac{1}{2j}[|G(j\omega)| \, e^{j(\omega t + \angle G(j\omega))}$$

$$- |G(-j\omega)| \, e^{j(-\omega t + \angle G(-j\omega))}] = |G(j\omega)|\sin(\omega t + \angle G(j\omega))$$

In other words, the frequency response is a special case of the relationship described in Eq. (A.5) by letting the real part of the complex number s be zero. The interpretation of the system frequency response is the following: When the input signal to a linear system is sinusoidal, the output also is a sinusoidal signal, amplified by $|G(j\omega)|$ (i.e., *gain* of the system at this frequency ω), with a phase shift of $\angle G(j\omega)$ (i.e., *phase* of the system at this frequency ω). To obtain the frequency response, we simply substitute $s = j\omega$ into the transfer function $G(s)$. The magnitude (i.e., gain) and phase of the resulting function $G(j\omega)$ then determine the frequency response of the system.

EX. A.2.7: MODELING A MASS-SPRING-DAMPER SYSTEM. Consider the modeling of a mass-spring-damper system in terms of both state equations and a transfer function.

$$\dot{\mathbf{x}} = \mathbf{Ax} + \mathbf{Bu}$$

$$\mathbf{y} = \mathbf{Cx} + \mathbf{Du} \qquad y = G(s)u$$

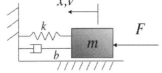

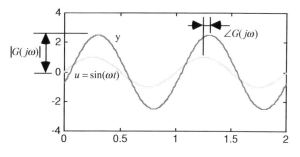

Figure A.7. Graphical interpretation of system frequency response.

The state equations for $x_1 = x$, $x_2 = v$ can be written as:

$$\frac{d}{dt}\left\{ \begin{array}{c} x \\ v \end{array} \right\} = \left[\begin{array}{cc} 0 & 1 \\ -\dfrac{k}{m} & -\dfrac{b}{m} \end{array} \right] \left\{ \begin{array}{c} x \\ v \end{array} \right\} + \left\{ \begin{array}{c} 0 \\ \dfrac{1}{m} \end{array} \right\} F$$

$$y = [1 \quad 0] \left\{ \begin{array}{c} x \\ v \end{array} \right\} + 0F$$

The transfer function is given by:

$$Y(s)/U(s) = G(s) = \frac{1}{ms^2 + bs + k}$$

Block Diagram

The block diagram of dynamic systems (in the s-domain) has the same simple algebraic relationship as the static systems introduced in Section A.1. Therefore, simplification of the block diagram in s-domain becomes straightforward.

EX. A.2.8: THREE BASIC OPERATIONS: SERIES, PARALLEL, AND FEEDBACK FORM

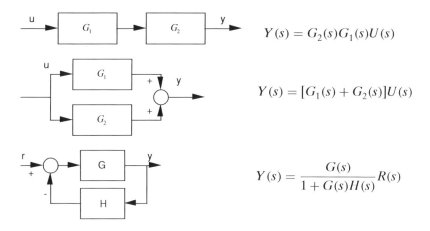

$$Y(s) = G_2(s)G_1(s)U(s)$$

$$Y(s) = [G_1(s) + G_2(s)]U(s)$$

$$Y(s) = \frac{G(s)}{1 + G(s)H(s)}R(s)$$

EX. A.2.9: BLOCK-DIAGRAM ALGEBRA. We can combine signal loops and shift connection points using the three basic operations shown in Ex. A.2.7. Examples include the following:

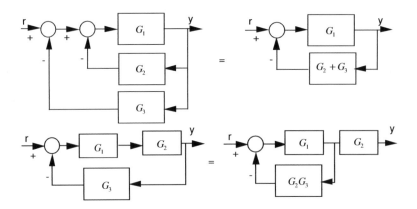

Using the basic relationship presented in Ex. A.2.8 and Ex. A.2.9, we easily can simplify complicated systems and obtain the transfer function of a closed-loop system.

Poles and Zeros

Given a transfer function:

$$\frac{Y(s)}{U(s)} = \frac{b_0 s^m + b_1 s^{m-1} + \cdots + b_m}{s^n + a_1 s^{n-1} + \cdots + a_{n-1}s + a_n} = \frac{b(s)}{a(s)}$$

If $b(s)$ and $a(s)$ are co-prime (which usually is true for most physical systems), then the poles and zeros are defined as:

Poles $\forall s_i, a(s_i) = 0$
Zeros $\forall s_i, b(s_i) = 0$

For example, the finite poles and zeros of the transfer function $\frac{2s+1}{s^2+3s+2}$ are $(-1,-2)$ and (-0.5), respectively. The equation $a(s) = s^2 + 3s + 2 = 0$ is referred to as the *characteristic equation* of the system, the roots of which are the poles of the system. The poles represent the "modes" of the system under impulse input (i.e., impulse response, also known as natural response). There is a close relationship between the location of poles and time-domain response.

EX. A.2.10: TIME RESPONSE, REAL POLES. The impulse response of the system $G(s) = \frac{2s+1}{s^2+3s+2}$ is:

$$Y(s) = G(s)U(s) = \frac{2s+1}{s^2+3s+2} \cdot 1$$

$$\Rightarrow \quad y(t) = L^{-1}\left[\frac{2s+1}{s^2+3s+2}\right] = L^{-1}\left[\frac{-1}{s+1} + \frac{3}{s+2}\right] = -e^{-t} + 3e^{-2t}$$

where the inverse Laplace Transform uses tables that are available in many automatic-control textbooks. It is clear that the poles correspond to the "modes" of system response. A real pole located at $s = -a$ corresponds to the time-domain mode of e^{-at}. What about complex poles? Complex poles always exist in complex conjugate pairs. The reason is that differential equations of physical systems have only real parameters.

EX. A.2.11: TIME RESPONSE, COMPLEX POLES. Find the inverse Laplace Transform of the function $G(s) = \frac{5s+3}{s^2+2s+10}$

$$G(s) = \frac{5s+3}{s^2+2s+10} = \frac{5(s+1)}{(s+1)^2+3^2} + \frac{-2}{(s+1)^2+3^2}$$

$$\Rightarrow g(t) = L^{-1}\left[\frac{5(s+1)}{(s+1)^2+3^2} + \frac{-\frac{2}{3}\cdot 3}{(s+1)^2+3^2}\right] = 5e^{-t}\cos 3t - \frac{2}{3}e^{-t}\sin 3t$$

Ziegler–Nichols Tuning Rules

Two empirical rules were proposed by Ziegler and Nichols (Franklin et al. 2010): (1) the *step-response method*, which requires that the open-loop step response of the system is roughly S-shaped; and (2) the *ultimate-sensitivity method*, which examines the closed-loop response of the plant. These two methods are summarized in the following discussion.

Step-Response Method

Examine the step response of the open-loop system. Draw the steepest slope line and define two quantities, R and L.

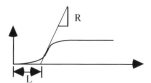

Intuitively, L represents a "delay" of the system; when L is large, the control should be conservative (i.e., smaller PI gains, larger D gains). The steepest slope R represents the "bandwidth" of a system. When R is large, the control gain again should be conservative. Based on extensive experiments, Ziegler and Nichols recommend the following gains:

P control $\quad K = \dfrac{1}{RL}$

PI control $\quad K = \dfrac{0.9}{RL} \quad T_I = 3.3L$

PID control $\quad K = \dfrac{1.2}{RL} \quad T_I = 2L \quad T_D = 0.5L$

Ultimate-Sensitivity Method

Examine the closed-loop response of the P-controlled system (i.e., controller C consists of only proportional gain). Increase control gain C until the plant is marginally stable, with constant amplitude oscillatory motion. If we denote the value of this gain (i.e., ultimate P-gain) K_u and the oscillation period P_u, then the recommended PID gains are as follows:

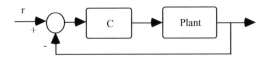

P control	$K = 0.5K_u$		
PI control	$K = 0.45K_u$	$T_I = 0.83P_u$	
PID control	$K = 0.6K_u$	$T_I = 0.5P_u$	$T_D = 0.125P_u$

These recommended values should be viewed as the "initial starting point." They usually have smaller damping than generally is desired, and fine-tuning these values to suit specific needs typically is necessary.

EX. A.2.12: Design a PID controller for the plant:

$$P(s) = \frac{1000}{(s+1)(s+4)(s+10)}$$

Obtain the open-loop step response of the plant from MATLAB. From the plot, we have $R = 16$ and $L = 0.23$. Thus, the PID controller is obtained from the Ziegler–Nichols rules (i.e., the step-response method)

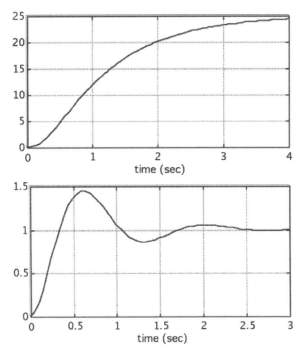

where the control parameters are $K_p = 0.326$, $K_i = 0.709$, and $K_d = 0.0375$. This design has too much overshoot, a long rise time, and a long settling time; fine-tuning may be needed.

```
% Example program to plot closed-
% loop response of PID controlled % system
R=16;
L=0.23;
K=1.2/R/L;
```

```
Ti=2*L;
Td=0.5*L;
Ki=K/Ti;
Kd=K*Td;

num_c=[Kd K Ki];
den_c=[1e-10 1 0]; % to avoid causality problem

num_p=1000;
den_p=[1 15 54 40];
[num_op, den_op] = series(num_c, den_c, num_p, den_p);
[num_cl, den_cl] = cloop(num_op, den_op, -1);
t=0:0.05:4;
y_p=step(num_p, den_p, t);
y_cl=step(num_cl, den_cl, t);
plot(t,y_p)
xlabel('Time (sec)')
grid
pause
plot(t(1:61),y_cl(1:61)) % plot up to 3 sec
xlabel('Time (sec)')
grid
```

Root-Locus Methods

Root locus is a technique to study the locations of roots of algebraic equations when one parameter is varying. One application of the root-locus method to control design examines the effect of controller gains on the locations of closed-loop poles. The *canonical form* for the root-locus method looks like this:

$$a(s) + Kb(s) = 0, \quad \text{or equivalently } 1 + K\frac{b(s)}{a(s)} = 0$$

where K is the parameter that is varying and $a(s)$ and $b(s)$ are monic polynomials of s. When we study the closed-loop pole locations, the first step is to rewrite the characteristic equation of the closed-loop transfer function into canonical form; this process usually is straightforward.

EX. A.2.13: CANONICAL FORM FOR ROOT-LOCUS ANALYSIS. For the mass-spring-damper system:

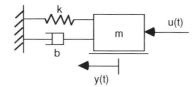

If a PID control law is used, the unity-feedback closed-loop transfer function is:

$$G_c(s) = \frac{Y(s)}{Y_d(s)} = \frac{K_D s^2 + K_p s + K_I}{ms^3 + (b + K_D)s^2 + (k + K_p)s + K_I}$$

The characteristic equation is then:

$$ms^3 + (b + K_D)s^2 + (k + K_p)s + K_I = 0 \qquad (A.6)$$

If we want to check the locations of the closed-loop poles as K_p varies (whereas K_I and K_D are fixed), Eq. (A.6) should be rewritten first in the following canonical form:

$$a(s) + Kb(s) = 0$$

where for this example, we have:

$$a(s) = s^3 + \frac{(b + K_D)}{m}s^2 + \frac{k}{m}s + \frac{K_I}{m} \quad b(s) = s \quad \text{and} \quad K = \frac{K_p}{m}$$

EX. A.2.14: Design a PID controller for the plant:

$$P(s) = \frac{1000}{(s + 1)(s + 4)(s + 10)}$$

PID design was presented previously using the Ziegler–Nichols rules. Because these rules are empirical, fine-tuning usually is necessary. Trial and error can be used, but a more systematic way can be based on the root-locus method. We could fix the Ki and Kd gains and tune Kp for better response. Then, we fix Kp, Ki, and adjust Kd, etc. The following MATLAB program illustrates this process. Notice that the characteristic equation of the closed-loop system is:

$$(s + 1)(s + 4)(s + 10) + \left(K_p + \frac{K_I}{s} + K_d s\right)1000 = 0$$

Therefore, the corresponding K, $a(s)$, and $b(s)$ of the canonical form is:

	a(s)	b(s)	K
vary Kp	$s(s + 1)(s + 4)(s + 10) + 1000(K_I + K_d s^2)$	s	$1000K_p$
vary Ki	$s(s + 1)(s + 4)(s + 10) + 1000(K_p s + K_d s^2)$	1	$1000K_I$
vary Kd	$s(s + 1)(s + 4)(s + 10) + 1000(K_p s + K_I)$	s^2	$1000K_D$

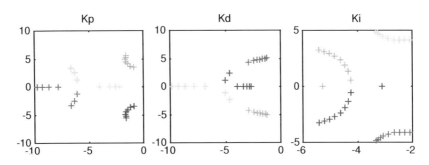

```
% Example to plot closed-loop
% response of PID controlled system
R=16;
L=0.23;
K=1.2/R/L;
Ti=2*L;
Td=0.5*L;
```

```
Kp=K; % 0.3261
Ki=K/Ti; % 0.7089
Kd=K*Td; % 0.0375

% tuning Kp
num_kp=[1 0];
den_kp=[1 15 54 40 0]+[0 0 1000*Kd 0 1000*Ki];
K=700*Kp:100*Kp:1300*Kp;
R=rlocus(num_kp, den_kp, K);
plot(R,'+'), title('Kp')
Kp=Kp*1.1; %Select new Kp
pause;
% tuning Kd
num_kd=[1 0 0];
den_kd=[1 15 54 40 0]+[0 0 0 1000*Kp 1000*Ki];
K=800*Kd:100*Kd:1400*Kd;
R=rlocus(num_kd, den_kd, K);
plot(R,'+'), title('Kd')
Kd=Kd*1.4; %Select New Kd
pause;

% tuning Ki
num_ki=1;
den_ki=[1 15 54 40 0]+[0 0 1000*Kd 1000*Kp 0];
K=800*Ki:50*Ki:1200*Ki;
R=rlocus(num_ki, den_ki, K);
plot(R,'+'), title('Ki')
Ki=Ki*0.95; %Select New Ki
```

After one round of root locus, a new set of control gains that should perform better than the original design is obtained, as shown on the following plot.

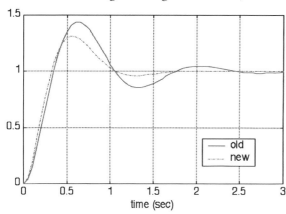

```
% Example program to plot closed-loop
% response of PID controlled system
R=16;
L=0.23;
K=1.2/R/L;
```

```
Ti=2*L;
Td=0.5*L;
Kp=K; % 0.3261
Ki=K/Ti; % 0.7089
Kd=K*Td; % 0.0375

num_c1=[Kd Kp Ki];
num_c2=[Kd*1.4 Kp*1.1 Ki*0.95];
den_c=[1e-5 1 0]; % to avoid casuality problem
num_p=1000;
den_p=[1 15 54 40];
[num_op1, den_op1] = series(num_c1, den_c, num_p, den_p);
[num_op2, den_op2] = series(num_c2, den_c, num_p, den_p);
[num_cl1, den_cl1] = cloop(num_op1, den_op1, -1);
[num_cl2, den_cl2] = cloop(num_op2, den_op2, -1);
t=0:0.05:4;
y_cl1=step(num_cl1, den_cl1, t);
y_cl2=step(num_cl2, den_cl2, t);
plot(t(1:61),y_cl1(1:61), t(1:61),y_cl2(1:61),'-.')
xlabel('time (sec)')
legend('old','new')
grid
```

State-Feedback Control

Consider a plant modeled in terms of linear-state equations, as follows:

$$\dot{\mathbf{x}} = \mathbf{A}\mathbf{x} + \mathbf{B}\mathbf{u} \qquad (A.7)$$

with a state-feedback control law of the form:

$$\mathbf{u} = -\mathbf{K}\mathbf{x} \qquad (A.8)$$

Substituting Eq. (A.8) into Eq. (A.7) yields the closed-loop system equations:

$$\dot{\mathbf{x}} = (\mathbf{A} - \mathbf{B}\mathbf{K})\mathbf{x} = \mathbf{A}_c\mathbf{x} \qquad (A.9)$$

Thus, the closed-loop system dynamics is governed by the closed-loop coefficient matrix $\mathbf{A}_c$. The control gains, $\mathbf{K}$, can be selected to achieve the desired closed-loop pole locations. This is facilitated by the use of commands such as *acker* (or *place*) in MATLAB, which are illustrated in the following example.

EX. A.2.15: Consider the following state-equation representation of an undamped second-order system with a natural frequency of 1 rad/s and a state-feedback controller:

$$\begin{Bmatrix} \dot{x}_1 \\ \dot{x}_2 \end{Bmatrix} = \begin{bmatrix} 0 & 1 \\ -1 & 0 \end{bmatrix} \begin{Bmatrix} x_1 \\ x_2 \end{Bmatrix} + \begin{bmatrix} 0 \\ 1 \end{bmatrix} u$$

$$u = -\mathbf{K}\mathbf{x} = -[K_1 \quad K_2] \begin{Bmatrix} x_1 \\ x_2 \end{Bmatrix} = -K_1 x_1 - K_2 x_2$$

Find the state-feedback control law that places both of the closed-loop poles at -2. In other words, by selecting these poles, we double the natural frequency and increase the damping ratio from 0 to 1. The desired poles are at $(s + 2)(s + 2) = s^2 + 4s + 4$ and the closed-loop characteristic equation is obtained from:

$$\det[s\mathbf{I} - (\mathbf{A} - \mathbf{BK})] = s^2 + K_2 s + (1 + K_1) = 0$$

Consequently, we obtain $K_1 = 3$ and $K_2 = 4$. The same result is obtained using the following series of MATLAB commands and Simulink model:

```
>>A=[0,1;-1,0]; B=[0;1];
>> p=[-2;-2];
>>K=acker(A,B,p)
```

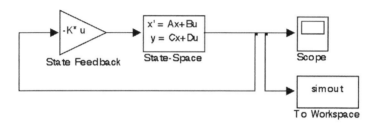

The response $y = [x_1, x_2]$ of the closed-loop system to initial conditions $\mathbf{x}(0) = [1\ 0]$ is shown here.

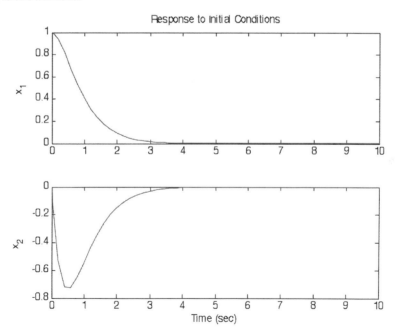

Limitations of the state-feedback control approach, as presented previously, are as follows:

1. The system must be controllable.
2. All of the states must be measureable.
3. No tracking of a reference input is possible.
4. No integral action to increase system type is included.

All of these limitations, except the requirement that the plant be controllable, can be addressed by extensions of the basic state-feedback method presented here (Franklin et al. 2010). These extensions include the use of an observer (or state estimator), the inclusion of a reference input in the control u, and augmentation of the state equations of the plant to include integral action. Also, optimal control methods can be used to determine the controller gains, **K**, rather than using pole-placement methods. The concepts of controllability and observability are fundamental to the state-feedback controller-design approach and are described in Franklin et al. (2010).

REFERENCE

Franklin, G. F., J. D. Powell, and A. Emami-Naeini, 2010, *Feedback Control of Dynamic Systems*, Pearson Prentice-Hall.

APPENDIX B

Two-Mass Three-Degree-of-Freedom Vehicle Lateral/Yaw/Roll Model

One factor that greatly affects the handling behavior of vehicles is the vehicle roll motion (i.e., rotation ϕ about the x-axis). In the development of the bicycle model (see Chapter 4), it was assumed that a vehicle consists of one rigid body and that the roll DOF was neglected. This assumption is valid only when the vehicle body does not roll excessively. In reality, suspension systems separate vehicles into rolling and nonrolling masses. The rolling (i.e., sprung) mass rolls relative to the nonrolling (i.e., unsprung) mass, and considerable weight shift may occur. The change in tire normal forces on the two sides of the vehicle can significantly change the tire lateral force generated. Furthermore, the existence of the extra DOF (i.e., roll) significantly changes the dynamic equations. Due to these reasons, it becomes necessary to include the roll DOF to accurately predict vehicle handling response under high lateral-acceleration level. In this appendix, a three-DOF model is developed that includes lateral, yaw, and roll DOF.

A side view of the vehicle is shown in Figure B.1. The vehicle is divided into rolling (m_R) and nonrolling (m_{NR}) masses. Point O is assumed to be the overall CG of the vehicle (combining m_R and m_{NR}). A coordinate system xyz is assumed to be fixed on the vehicle, with the following features:

- The x-axis is horizontal and passes through the CG of m_{NR}.
- The z-axis is vertical and passes through the CG of vehicle (Point O).
- The roll-axis connects the roll center at the rear axle and front axle and is pointing somewhat downward with an angle θ_R.

From the definition of vehicle CG, we have $m_{NR} \cdot e = m_R \cdot c$.

The equations of motion can be derived using Lagrange's approach (i.e., from the energy viewpoint). For each DOF, we obtain a second-order ODE from Lagrange's equation:

$$\frac{d}{dt}\left(\frac{\partial T}{\partial \dot{q}_r}\right) - \left(\frac{\partial T}{\partial q_r}\right) = Q_r \tag{B.1}$$

where T is the system kinetic energy and Q_r is the generalized force that includes both conservative and nonconservative forces. In other words, the conservative forcing terms associated with potential energy are absorbed in the Q_r term. Assume that the generalized inertial coordinate XY (with origin fixed at the Point O) and the

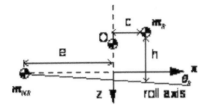

Figure B.1. Side view of the three-DOF vehicle model.

vehicle-fixed local coordinate xy defined previously are related by the yaw angle; their relationship is described by the following equations (Figure B.2):

$$u = \dot{X} \cos \psi + \dot{Y} \sin \psi \tag{B.2}$$

$$v = -\dot{X} \sin \psi + \dot{Y} \cos \psi \tag{B.3}$$

In this discussion, assume that there is a series of inertial coordinates defined on the moving Point O and that the angle ψ is always small.

The kinetic energy of the vehicle consists of two parts: the energy stored in the rolling and the nonrolling masses. In equation form, we have $T = T_R + T_{NR}$ or, more precisely:

$$T_{NR} = \frac{1}{2} m_{NR} \vec{V}_{NR} \cdot \vec{V}_{NR} + \frac{1}{2} (I_{zz})_{NR} r^2 \tag{B.4}$$

because there are no pitch and roll motions for the nonrolling unit. From the fact that $\vec{V}_{NR} = u\vec{i} + (v - er)\vec{j}$, we have:

$$T_{NR} = \frac{1}{2} m_{NR} [u^2 + (v - er)^2] + \frac{1}{2} (I_{zz})_{NR} r^2 \tag{B.5}$$

Similarly, for the rolling mass, we have $T_R = \frac{1}{2} m_R \vec{V}_R \cdot \vec{V}_R + \frac{1}{2} \vec{\omega}_R \cdot \vec{H}_R$. The velocity and inertia terms, respectively, are:

$$\vec{V}_R = (u - h\phi r)\vec{i} + v + hp + cr)\vec{j} \tag{B.6}$$

$$\vec{\omega}_R = p\vec{i} + (r + \theta_R p)\vec{k} \tag{B.7}$$

$$\begin{bmatrix} H_{Rx} \\ H_{Ry} \\ H_{Rz} \end{bmatrix} = \begin{bmatrix} I_{xx} & 0 & -I_{xz} \\ 0 & I_{yy} & 0 \\ -I_{xz} & 0 & I_{zz} \end{bmatrix}_R \begin{bmatrix} p \\ 0 \\ r + \theta_R p \end{bmatrix} = \begin{bmatrix} (I_{xx})_R p - (I_{xz})_R (r + \theta_R p) \\ 0 \\ -(I_{xz})_R p + (I_{zz})_R (r + \theta_R p) \end{bmatrix} \tag{B.8}$$

Therefore:

$$T_R = \frac{1}{2} m_R [(u - h\phi r)^2 + (v + hp + cr)^2]$$
$$+ \frac{1}{2} (I_{xx})_R p^2 - (I_{xz})_R (r + \theta_R p) p + \frac{1}{2} (I_{zz})_R (r + \theta_R p)^2 \tag{B.9}$$

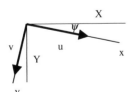

Figure B.2. Relationship of the two coordinate systems.

The total energy of the vehicle is the sum of the energy of the two masses – that is, Eqs. (B.5) and (B.9). Now that the total energy is precisely defined, the dynamic equations can be obtained by applying Lagrange's method to each of the three DOF. For the *lateral* DOF, we obtain:

$$\frac{\partial T}{\partial \dot{Y}} = \frac{\partial T}{\partial u}\frac{\partial u}{\partial \dot{Y}} + \frac{\partial T}{\partial v}\frac{\partial v}{\partial \dot{Y}} \tag{B.10}$$

$$= m_{NR}u\sin\psi + m_{NR}(v - er)\cos\psi + m_R(u - h\phi r)\sin\psi$$
$$+ m_R(v + hp + cr)\cos\psi$$
$$= (m_{NR} + m_R)[u\sin\psi + v\cos\psi] + (cm_R - em_{NR})r\cos\psi$$
$$+ m_R[hp\cos\psi + h\phi r\sin\psi]$$

and

$$\frac{\partial T}{\partial Y} = 0 \tag{B.11}$$

By neglecting second- and higher-order angular terms and using the fact that $m_{NR} \cdot e = m_R \cdot c$, we obtain:

$$\sum F_y = \frac{d}{dt}\left(\frac{\partial T}{\partial \dot{Y}}\right) = m(\dot{v} + ur) + m_R h\dot{p} \tag{B.12}$$

For the vehicle *yaw* direction (note that $r = \dot{\psi}$), we obtain:

$$\frac{\partial T}{\partial r} = -em_{NR}(v - er) + (I_{zz})_{NR}r - h\phi m_R(u - h\phi r) + cm_R(v + hp + cr)$$
$$- (I_{xz})_R p + (I_{zz})_R(r + \theta_R p)$$
$$= (cm_R - em_{NR})v + [(I_{zz})_{NR} + (I_{zz})_R + m_{NR}e^2 + m_R c^2 + h^2\phi^2 m_R]r$$
$$+ [-(I_{xz})_R + (I_{zz})_R\theta_R + cm_R h]\,p - h\phi m_R u \equiv I_z r + I_{xz}p - h\phi m_R u \tag{B.13}$$

$$\frac{\partial T}{\partial \psi} = \frac{\partial T}{\partial u}\frac{\partial u}{\partial \psi} + \frac{\partial T}{\partial v}\frac{\partial v}{\partial \psi}$$
$$= \frac{\partial T}{\partial u}(-\dot{X}\sin\psi + \dot{Y}\cos\psi) + \frac{\partial T}{\partial v}(-\dot{X}\cos\psi - \dot{Y}\sin\psi)$$
$$= \frac{\partial T}{\partial u}v - \frac{\partial T}{\partial v}u = m_{NR}[uv - (v - er)u]$$
$$+ m_R[(u - h\phi r)v - (v + hp + cr)u]$$
$$= (em_{NR} - cm_R)ur - m_R[hpu + h\phi rv] \tag{B.14}$$

Therefore:

$$\sum M_z = I_z\dot{r} + I_{xz}\,\dot{p} \tag{B.15}$$

Last but not least, the vehicle *roll equation* is shown here. Because the nonrolling mass is not rolling (i.e., has no contribution to the dynamics), the subscript R is added to the kinetic energy T to clearly show this point. The partial derivatives are as follows:

$$\frac{\partial T_R}{\partial p} = m_R h(v + hp + cr) + (I_{xx})_R p - (I_{xx})_R (r + 2\theta_R p) + (I_{zz})_R (r + \theta_R p)\theta_R$$

$$= \left[h^2 m_R + (I_{xx})_R - 2\theta_R (I_{xz})_R + \theta_R^2 (I_{zz})_R \right] p$$
$$+ \left[-(I_{xz})_R + \theta_R (I_{zz})_R + m_R hc \right] r + m_R hv$$
$$\equiv I_x p + I_{xz} r + m_R hv \tag{B.16}$$

$$\frac{\partial T_R}{\partial \phi} = -hrm_R(u - h\phi r) \tag{B.17}$$

The roll-dynamic equation is then:

$$\sum M_x|_{rollaxis} = I_x \dot{p} + I_{xz} \dot{r} + hm_R \dot{v} + hm_R ru = m_R h(\dot{v} + ur) + I_x \dot{p} + I_{xz} \dot{r} \tag{B.18}$$

In summary, the dynamic equations for the three-DOF (i.e., lateral, yaw, and roll) are:

$$\boxed{\begin{aligned} \sum F_y &= m(\dot{v} + ur) + m_R h \dot{p} \\ \sum M_z &= I_z \dot{r} + I_{xz} \dot{p} \\ \sum M_x|_{rollais} &= I_x \dot{p} + m_R h(\dot{v} + ur) + I_{xz} \dot{r} \end{aligned}}$$

Comparing this to the two-DOF model in Eqs. (4.45) and (4.46), the extra terms (i.e., $m_R h \dot{p}$ in the lateral dynamics and $I_{xz} \dot{p}$ in the yaw dynamics) clearly show the effect of the roll DOF on the vehicle dynamic response. The more important influence, however, lies on the excitation side (i.e., the "F" side of $F = ma$) of the equation. The external forces acting on the vehicle (or the rolling mass for the roll equation) are as follows:

$$\sum F_y = C_{\alpha f}\alpha_f + C_{\alpha r}\alpha_r + C_{\gamma f}\frac{\partial \gamma_f}{\partial}\phi$$
$$= C_{\alpha f}\left(\delta_f - \beta - \frac{ar}{u}\right) + C_{\alpha r}\left(\frac{\partial \delta_r}{\partial \phi}\phi - \beta + \frac{br}{u}\right) + C_{\gamma f}\frac{\partial \gamma_f}{\partial \phi}\phi \tag{B.19}$$

$$\sum M_z = a\left[C_{\alpha f}\alpha_f + C_{\gamma f}\frac{\partial \gamma_f}{\partial \phi}\phi \right] - bC_{\alpha r}\alpha_r + \frac{\partial M_z}{\partial \alpha_f}\alpha_f + \frac{\partial M_z}{\partial \alpha_r}\alpha_r$$
$$\approx aC_{\alpha f}\left(\delta_f - \beta - \frac{ar}{u}\right) + aC_{\gamma f}\frac{\partial \gamma_f}{\partial \phi}\phi - bC_{\alpha r}\left(\frac{\partial \delta_r}{\partial \phi}\phi - \beta + \frac{br}{u}\right) \tag{B.20}$$

$$\sum M_x = (m_R gh - K_R)\phi - c_R\dot{\phi} \tag{B.21}$$

where in Eq. (B.21), K_R is the overall suspension stiffness and C_R is the overall suspension damping. The approximation of Eq. (B.20) is justified because the aligning moment generated from a tire is approximately two orders of magnitude smaller than the yaw moment generated by its corresponding lateral force.

Comparing Eqs. (B.19) and (B.20) with the corresponding equations for the two-DOF vehicle model, it can be seen that two extra types of variables were introduced:

(1) force and moment variables due to roll-induced camber thrust changes; and (2) rear-wheel steering angle generated by the roll motion of the sprung mass. Commonly, stability derivatives are defined based on these force equations. The major benefits are (1) sources of these forces are clear, and (2) the notation is unified. Essentially, the stability derivatives are nothing but partial derivatives of the external forcing terms with respect to the state and input variables. The definitions of the stability derivatives are highlighted by the following two equations:

$$\sum F_y = \frac{\partial \sum F_y}{\partial \beta}\beta + \frac{\partial \sum F_y}{\partial r}r + \frac{\partial \sum F_y}{\partial \phi}\phi + \frac{\partial \sum F_y}{\partial \delta_f}\delta_f$$
$$\equiv Y_\beta \beta + Y_r r + Y_\phi \phi + Y_\delta \delta_f \tag{B.22}$$

$$\sum M_z = \frac{\partial \sum M_z}{\partial \beta}\beta + \frac{\partial \sum M_z}{\partial r}r + \frac{\partial \sum M_z}{\partial \phi}\phi + \frac{\partial \sum M_z}{\partial \delta_f}\delta_f$$
$$\equiv N_\beta \beta + N_r r + N_\phi \phi + N_\delta \delta_f \tag{B.23}$$

From Eqs. (B.19) and (B.20), the stability derivatives are then:

$$
\begin{aligned}
&Y_\beta = -(C_{\alpha f} + C_{\alpha r}) & &Y_r = \frac{-aC_{\alpha f} + bC_{\alpha r}}{u_0} \\[2mm]
&Y_\phi = C_{\alpha r}\frac{\partial \delta_r}{\partial \phi} + C_{\gamma f}\frac{\partial \gamma_f}{\partial \phi} & &Y_\delta = C_{\alpha f} \\[2mm]
&N_\beta = -aC_{\alpha f} + bC_{\alpha r} & &N_r = -\left(\frac{a^2 C_{\alpha f} + b^2 C_{\alpha r}}{u_0}\right) \\[2mm]
&N_\phi = aC_{\gamma f}\frac{\partial \gamma_f}{\partial \phi} - bC_{\alpha r}\frac{\partial \delta_r}{\partial \phi} & &N_\delta = aC_{\alpha f}
\end{aligned}
$$

where we now have denoted the constant longitudinal velocity $u = u_0$. For the roll dynamics, the equation can be rearranged as:

$$\sum M_x = (m_R g h - K_R)\phi - c_R \dot{\phi} \equiv L_\phi \phi + L_p p \tag{B.24}$$

Based on the newly defined variables, the dynamic equations are then:

$$Y_\beta \beta + Y_r r + Y_\phi \phi + Y_\delta \delta = m(u\dot{\beta} + u_0 r) + m_R h \dot{p} \tag{B.25}$$

$$N_\beta \beta + N_r r + N_\phi \phi + N_\delta \delta = I_z \dot{r} + I_{xz} \dot{p} \tag{B.26}$$

$$L_\phi \phi + L_p p = m_R h(\dot{v} + u_0 r) + I_x \dot{p} + I_{xz} \dot{r} \tag{B.27}$$

In matrix form, we obtain:

$$
\begin{bmatrix} mu_o & 0 & m_R h & 0 \\ 0 & I_z & I_{xz} & 0 \\ m_R h u_o & I_{xz} & I_x & 0 \\ 0 & 0 & 0 & 1 \end{bmatrix}
\begin{bmatrix} \dot{\beta} \\ \dot{r} \\ \dot{p} \\ \dot{\phi} \end{bmatrix}
+
\begin{bmatrix} -Y_\beta & mu_0 - Y_r & 0 & -Y_\phi \\ -N_\beta & -N_r & 0 & -N_\phi \\ 0 & m_R h u_0 & -L_p & -L_\phi \\ 0 & 0 & -1 & 0 \end{bmatrix}
\begin{bmatrix} \beta \\ r \\ p \\ \phi \end{bmatrix}
=
\begin{bmatrix} Y_\delta \\ N_\delta \\ 0 \\ 0 \end{bmatrix}\delta
$$
$$\tag{B.28}$$

If the roll-related states (i.e., p and ϕ) are dropped, with $v = \beta u_0$, Eq. (B.28) reduces to the two-DOF bicycle model shown in Eq. (4.58).

Equation (B.28) is not in standard-state-equation form but rather the form $\mathbf{E}\dot{\mathbf{x}} + \mathbf{F}\mathbf{x} = \mathbf{G}\delta$, where $\mathbf{x} = [\beta \; r \; p\phi]^{\mathrm{T}}$. In fact, one of the combined inertia terms (I_z) is state-dependent. In other words, Eq. (B.28) appears to be linear but actually is nonlinear. Nevertheless, if we ignore the influence of roll angle on I_z (which introduces only minimal error), Eq. (B.28) can be transformed into a state-space equation $\dot{\mathbf{x}} = \mathbf{A}\mathbf{x} + \mathbf{B}\delta$, where $\mathbf{A} = -\mathbf{E}^{-1}\mathbf{F}$ and $\mathbf{B} = \mathbf{E}^{-1}\mathbf{G}$. The state and input matrices $\mathbf{A}$ and $\mathbf{B}$ can be obtained symbolically but they also can be calculated numerically.

Index

Made in the USA
Lexington, KY
01 October 2016